T3-BNY-851

Energy Methods in Finite Element Analysis

WILEY SERIES IN NUMERICAL METHODS IN ENGINEERING

Consulting Editors
R. H. Gallagher, *College of Engineering, University of Arizona*
and
O. C. Zienkiewicz, *Department of Civil Engineering, University College Swansea*

Rock Mechanics in Engineering Practice
Edited by K. G. Stagg and O. C. Zienkiewicz both of University College Swansea

Optimum Structural Design: Theory and Applications
Edited by R. H. Gallagher, Cornell University and O. C. Zienkiewicz, University College Swansea

Finite Elements in Fluids
Vol. 1 Viscous Flow and Hydrodynamics
Vol. 2 Mathematical Foundations, Aerodynamics and Lubrication
Edited by R. H. Gallagher, Cornell University; J. T. Oden, University of Texas; C. Taylor and O. C. Zienkiewicz, University College Swansea

Finite Elements for Thin Shells and Curved Members
Edited by D. G. Ashwell, University College Cardiff and R. H. Gallagher, Cornell University

Finite Elements in Geomechanics
Edited by G. Gudehus, Universität Karlsruhe

Numerical Methods in Offshore Engineering
Edited by O. C. Zienkiewicz, R. W. Lewis, and K. G. Stagg, all of University College Swansea

Finite Elements in Fluids
Vol. 3
Edited by R. H. Gallagher, O. C. Zienkiewicz, J. T. Oden, M. Morandi Cecchi and C. Taylor

Energy Methods in Finite Element Analysis
Edited by R. Glowinski, E. Rodin and O. C. Zienkiewicz

Energy Methods in Finite Element Analysis

Edited by

R Glowinski
University of Paris

E. Y. Rodin
Washington University, St. Louis

O. C. Zienkiewicz
University College Swansea

A Wiley–Interscience Publication

JOHN WILEY & SONS
Chichester · New York · Brisbane · Toronto

***Library of Congress Cataloging in Publication Data*:**

Main entry under title:

Energy methods in finite element analysis.

(Wiley series in numerical methods in engineering)
'A Wiley–Interscience publication.'
'Dedicated to the memory of Professor B. Fraeijs de Veubeke.'
Includes index.
1. Finite element method. I. Glowinski, R.
II. Rodin, Ervin Y. III. Zienkiewicz, O. C.
IV. Fraeijs de Veubeke, B.
TA347.F5E53 624'.171'01515 78–13642

ISBN 0 471 99723 4

Typeset in Northern Ireland at The Universities Press (Belfast) Ltd.
and printed by The Pitman Press, Bath.

Contributing Authors

D. J. ALLMAN — *Royal Aircraft Establishment, Farnborough, U.K.*

J. H. ARGYRIS — *University of Stuttgart, Stuttgart, F.R.G.*

P. G. BERGAN — *The Norwegian Institute of Technology, Trondheim, Norway.*

P. BETTESS — *University of Wales, Swansea, Wales, U.K.*

F. BREZZI — *University of Pavia, Italy.*

P. BROWN — *Leicester University, Leicester, U.K.*

S. CESCOTTO — *Laboratoire de Mécanique des Matériaux et Statique des Constructions, Université de Liège, Liège, Belgium.*

E. CHRISTIANSEN — *Odense University, Denmark.*

P. G. CIARLET — *Université Pierre et Marie Curie, Laboratoire d'Analyse Numérique, Paris, France.*

P. DESTUYNDER — *Centre de Mathématiques Appliquées, Ecole Polytechnique, Palaiseau, France.*

P. C. DUNNE — *University of Stuttgart, Stuttgart, F.R.G.*

G. FONDER — *Laboratoire de Mécanique des Matériaux et Statique des Constructions, Université de Liège, Liège, Belgium.*

F. FREY — *Laboratoire de Mécanique des Matériaux et Statique des Constructions, Université de Liège, Liège, Belgium.*

M. GERADIN — *LTAS, University of Liège, Liège, Belgium.*

R. GLOWINSKI — *Université Pierre et Marie Curie, Laboratoire d'Analyse Numérique, Paris, and IRIA/LABORIA, Rocquencourt, Le Chesnay, France.*

I. HOLAND — *The Norwegian Institute of Technology, Trondheim, Norway.*

B. M. IRONS — *University of Calgary, Calgary, Canada.*

C. JOHNSON — *Chalmers University, Göteborg, Sweden.*

D. W. KELLY — *University of Wales, Swansea, Wales, U.K.*

H. MATTHIES — *Massachusetts Institute of Technology, Cambridge, U.S.A.*

B. MERCIER	*Centre de Mathématiques Appliquées, Ecole Polytechnique, Palaiseau, France.*
L. S. D. MORLEY	*Royal Aircraft Establishment, Farnborough, U.K.*
J. T. ODEN	*University of Texas, Austin, Texas, U.S.A.*
O. PIRONNEAU	*IRIA/LABORIA, Rocquencourt, Le Chesnay, France*
A. R. S. PONTER	*Brown University, Rhode Island, U.S.A.*
P. A. RAVIART	*Université Pierre et Marie Curie, Laboratoire d'Analyse Numérique, Paris, France.*
A. SAMUELSSON	*Chalmers University, Göteborg, Sweden.*
T. H. SÖREIDE	*The Norwegian Institute of Technology, Trondheim, Norway.*
G. STRANG	*Massachusetts Institute of Technology, Cambridge U.S.A.*
R. L. TAYLOR	*University of California, Berkeley, California, U.S.A.*
J. M. THOMAS	*Université Pierre et Marie Curie, Laboratoire d'Analyse Numérique, Paris, France.*
G. B. WARBURTON	*University of Nottingham, Nottingham, U.K.*
O. C. ZIENKIEWICZ	*University of Wales, Swansea, Wales, U.K.*

Preface

This volume is dedicated to the memory of Professor B. Fraeijs de Veubeke, whose contributions to the theory of variational and energy methods and their use in the finite element method form a cornerstone for its development.

After a biography of Professor Fraeijs de Veubeke giving information about his life, his scientific career and his main contributions to science, this book contains nineteen chapters presenting various aspects of energy and variational methods in finite element analysis.

Chapter 1 by J. T. Oden gives the mathematical foundations of variational mechanics. It describes the various formulations existing for a given problem with a detailed discussion of the classical variational principles, the dual principles and their applications to elastostatics.

In Chapter 2, by P. G. Ciarlet and P. Destuynder, it is proved *without a priori assumptions*, that the classical two-dimensional linear models in elastic plate theory are indeed limits of the standard three-dimensional models of linear elasticity. This result is proved using variational formulations of the elastic problems and singular perturbation methods.

In Chapter 3, A. Samuelsson introduces the concept of 'global constant strain condition' to study non-conforming finite elements and shows its relationship to the well known 'patch test'.

In Chapter 4, G. B. Warburton gives a survey of the recent developments in structural dynamics computational methods via finite elements. Modal methods and numerical integration methods are described with their main properties, and their use is discussed with many details.

In Chapter 5, by O. C. Zienkiewicz, D. W. Kelly, and P. Bettess, one studies how standard finite element methods and boundary integral methods can be coupled in order to solve, for example, boundary value problems on unbounded domains; various examples from fluid mechanics, electrotechnics, etc., illustrate the possiblities of this new class of methods.

The chapter from 6 to 11 are all related to complementary energy methods and dual variational principles applied to finite element approximations.

In Chapter 6, D. J. Allman discusses the use of compatible and equilibrium models and finite elements, applied to the stretchings of elastic plates. A new triangular equilibrium element is introduced and the properties of the associated matrix is studied in details.

In Chapter 7, L. S. D. Morley, starting from Koiter's first approximation shell theory, develops an approximation of elastic shell problems, based on a new finite element stiffness formulation.

In Chapter 8, R. L. Taylor and O. C. Zienkiewicz show that the computational difficulties associated with complementary energy methods can be overcome by an appropriate penalty technique. Numerical examples illustrate the feasibility of the method. The two following chapters are more mathematical in nature.

In Chapter 9, P. A. Raviart and J. M. Thomas give the theoretical foundations of the dual finite element models for second order linear elliptic problems.

In Chapter 10, F. Brezzi does the same for fourth order linear elliptic problems taking the biharmonic plate bending problem to illustrate the various possible approaches.

Chapter 11, by C. Johnson and B. Mercier, is dedicated to a class of mixed equilibrium finite element methods. Applications to problems in linear elasticity, plasticity, and fluid dynamics (Navier–Stokes equations) are given. The next two chapter deal with incompressible media.

Chapter 12, by J. H. Argyris and P. C. Dunne describes a new finite element approximation for incompressible or near-incompressible materials. The method is based on displacement models and allows the use of low order elements.

Chapter 13, by R. Glowinski and O. Pironneau, discusses the approximation of the Stokes problem for incompressible fluids, by means of low order conforming elements to approximate velocity and pressure. This method is based on a new variational formulation of the Stokes problem and leads to approximate problems which can be easily solved by standard finite element Poisson solvers. Chapters 14 to 18 are related to nonlinear problems.

In Chapter 14, P. G. Bergan, I. Holand, and T. H. Soreide present a new incremental method for solving problems in nonlinear finite element analysis. The method is based on the use of a 'current stiffness parameter' which is a normalized measure of the incremental stiffness in the direction of motion. Several numerical tests illustrate the feasibility of the method.

In Chapter 15, S. Cescotto, F. Frey, and G. Fonder give a unified approach for Lagrangian description in nonlinear, large displacement, and large strain structural problems. Total and updated Lagrangian descriptions appear as particular cases of this more general theory.

In Chapter 16, B. M. Irons analyses some of the difficulties arising from curve fitting and shows how nonlinear effects can make difficult the numerical solution of problems apparently easy to solve.

In Chapter 17, H. Matthies, G. Strang, and E. Christiansen analyse from a mathematical point of view a fairly difficult infinite dimensional saddle-point problem, which describes the duality between the static and kinematic theories of limit analysis. The theoretical difficulty lies in the fact that the natural functional spaces to study this class of problems, of great interest in perfect plasticity, are L_1 and spaces of functions of bounded variation. They introduce the space of functions of bounded deformation, required because Korn's inequality fails in L_1.

In Chapter 18, A. R. S. Ponter and P. Brown discuss a new finite element method for computing the deformation of creeping structures. The finite element formulation is discussed in detail for a strain-hardening creep relationship and computed solutions are presented for a thermally loaded plate.

The last chapter by M. Geradin concerns modal analysis. The biorthogonal Lanczos algorithm is proposed as a very efficient tool to compute the upper eigenvalue spectrum of an arbitrary square matrix.

We hope that this volume which contains important contributions relevant to engineering, mechanics, and numerical analysis will provide a valuable tool and a source of inspiration to engineers, research scientists and other interested by the various aspects of the finite element methodology and its applications to science, a field to which Professor Fraeijs de Veubeke has so greatly contributed.

A list of the main publications of Professor Fraeijs de Veubeke is given as an Appendix at the end of this book.

May 1978

R. GLOWINSKI
E. Y. RODIN
O. C. ZIENKIEWICZ

Bandouin Fraeijs de Veubeke
1917–1976

Biography of Professor B. Fraeijs de Veubeke

Professor B. Fraeijs de Veubeke was born in Reigate, England, in 1917 of Belgian parents. He graduated from the University of Louvain as a civil and mechanical engineer in 1940, and in 1944 from the University of Liège as an aeronautical engineer. It is to this latter field that he retained allegiance throughout his career—and his association with aeronautics became stronger when in 1944, he joined the Royal Air Force as a volunteer, retiring from this service in 1946.

After six years of practical work on aircraft design and testing, he joined the University of Liège where he later was appointed to the Chair of Aerospace Engineering and Continuum Mechanics. He remained in this post to the last, establishing a notable centre of research, leading active industrial collaboration, and inspiring many students.

His research work ranged widely, and his contributions to aerodynamics, theory of vibrations, and optimal control are of great importance. To each he added original thought and the stamp of his personality. However, his work in the field of variational methods and their application to finite element analysis of elastic structures are his most notable contributions. His now classical paper presented in 1964 established firmly the use of upper and lower bound energy procedures, and on this work much of his own research and that of others is based. Indeed, he and his colleagues carried out the full development of finite elements founded on complementary energy principles, as well as the extension of these principles to dynamics and large deformation analysis.

Although many honours have been given to him in recognition of his pioneering work, he remained a modest man and a true friend to many. His untimely death in 1976 was a cause of grief to this scientific colleagues and friends in all parts of the world; they will cherish his memory.

It is to him therefore that the contributors dedicate their chapters in this volume.

Contents

Chapter 1

The Classical Variational Principles of Mechanics

J. T. Oden

1.1 INTRODUCTION

The last twenty years have been marked by some of the most significant advances in variational mechanics of this century. These advances have been made in two independent camps. First and foremost, the entire theory of partial differential equations has been recast in a 'variational' framework that has made it possible to significantly expand the theory of existence, uniqueness, and regularity of solutions of both linear and nonlinear boundary-value problems. In this regard, the treatise of Lions and Magenes[1] on linear partial differential equations, the works of Lions,[2] Brezis,[3] and Browder[4] on nonlinear equations, and of Duvaut and Lions[5] and Lions and Stampacchia[6] on variational inequalities should be mentioned. Secondly, the underlying structure of classical variational principles of mechanics are better understood. The work of Vainberg[7] led to important generalizations of classical variational notions related to the minimization of functionals defined on Banach spaces, and an excellent memoir on variational theory and differentiation of operators on Banach spaces was contributed by Nashed.[8] Generalizations of Hamiltonian theory were described by Noble[9] and Noble and Sewell[10] and this led to generalization of the concept of complementary variational principles by Arthurs[11] and others. Tonti[12,13] showed that many of the linear equations of mathematical physics share a common structure that leads naturally to several dual and complementary variational principles, and these notions were further developed by Oden and Reddy.[14] A quite general theory of complementary and dual variational principles for linear problems in mathematical physics was given in the monograph of Oden and Reddy,[15] and a more complete historical account of the subject, together with additional references, can be found in that work.

This chapter deals with a general theory of variational methods for linear problems in mechanics and mathematical physics. This work is partly expository in nature, since one of its principal missions is to develop, in a rather tutorial way, complete with examples, a rather general mathematical framework for both the modern theory of variational boundary-value problems and the 'classical' primal, dual, and complementary variational principles of mechanics. However, many of the ideas appear to be new, and generalized Green's formulas are given which make it possible to further generalize the recent theories of Oden and Reddy.[15]

1.2 MATHEMATICAL PRELIMINARIES

1.2.1 Transposes and adjoints of linear operators

It is frequently important to *distinguish* between the *transpose* of a linear operator on Hilbert spaces, and its *adjoint*. We set the stage for this discussion by introducing some notation.

Let $\mathscr{U}, \mathscr{V}$ be Hilbert spaces with inner products (u_1, u_2) and (v_1, v_2), respectively.

Let $\mathscr{U}', \mathscr{V}'$ be the topological duals of $\mathscr{U}$ and $\mathscr{V}$, i.e. the spaces of continuous linear functionals on $\mathscr{U}$ and $\mathscr{V}$, respectively.

Let $\langle \cdot, \cdot \rangle_{\mathscr{U}}$, $\langle \cdot, \cdot \rangle_{\mathscr{V}}$ denote duality pairings on $\mathscr{U}' \times \mathscr{U}$ and $\mathscr{V}' \times \mathscr{V}$, respectively, i.e. if $l \in \mathscr{U}'$ and $q \in \mathscr{V}'$ we write

$$l(u) = \langle l, u \rangle_{\mathscr{U}} \quad \text{and} \quad q(v) = \langle q, v \rangle_{\mathscr{V}}$$

We shall denote by A a continuous linear operator from $\mathscr{U}$ into $\mathscr{V}$.

Now, following the standard arguments, we note that since $Au \in \mathscr{V}$, the linear functional $q(Au) = \langle q, Au \rangle_{\mathscr{V}}$ also identifies a continuous linear functional on $\mathscr{U}$, i.e. a correspondence exists between $q \in \mathscr{V}'$ and the elements of the dual space $\mathscr{U}'$. We describe this correspondence by introducing an operator $A' : \mathscr{V}' \to \mathscr{U}'$ such that

$$(A'q)(u) = q(Au) \quad \text{or} \quad \langle A'q, u \rangle_{\mathscr{U}} = \langle q, Au \rangle_{\mathscr{V}} \tag{1.1}$$

The operator A' is called the *transpose* of A; it is clearly linear and continuous from $\mathscr{V}'$ into $\mathscr{U}'$.

Now the fact that $\mathscr{U}$ and $\mathscr{V}$ are Hilbert spaces allows us to enter into a considerable amount of additional detail in describing A' and other relationships between $\mathscr{U}$ and $\mathscr{V}$. Since $\mathscr{U}$ is a Hilbert space, we know from the Riesz representation theorem that for every linear functional $l \in \mathscr{U}'$ there exists a unique element $u_l \in \mathscr{U}$ such that

$$\langle l, u \rangle_{\mathscr{U}} = (u_l, u)_{\mathscr{U}} \qquad \forall u \in \mathscr{U}$$

Indeed, this relationship defines an isometric isomorphism $K_{\mathscr{U}} : \mathscr{U} \to \mathscr{U}'$ such that

$$\langle K_{\mathscr{U}} u_0, u \rangle_{\mathscr{U}} = (u_0, u)_{\mathscr{U}} \qquad \forall u_0, u \in \mathscr{U} \tag{1.2}$$

and $K_{\mathcal{U}}$ is called the *Riesz map* corresponding to the space $\mathcal{U}$. Similarly, if $K_{\mathcal{V}}$ is the Riesz map corresponding to the space $\mathcal{V}$, we have $\langle K_{\mathcal{V}}v_0, v\rangle_{\mathcal{V}} = (v_0, v)_{\mathcal{V}}\ \forall v_0,\ v \in \mathcal{V}$. In view of the definitions of the Riesz maps, we see that

$$\langle K_{\mathcal{V}}v, Au\rangle_{\mathcal{V}} = (v, Au)_{\mathcal{V}} \qquad \forall v \in \mathcal{V}, u \in \mathcal{U}$$

and

$$\langle A'K_{\mathcal{V}}v, u\rangle_{\mathcal{U}} = (K_{\mathcal{U}}^{-1}A'K_{\mathcal{V}}v, u)_{\mathcal{U}} \qquad \forall v \in \mathcal{V}, u \in \mathcal{U}$$

Thus, we discover in a very natural way that for each continuous linear operator $A : \mathcal{U} \to \mathcal{V}$ satisfying the above relations there corresponds an operator $A^* : \mathcal{V} \to \mathcal{U}$ given by the composition

$$A^* = K_{\mathcal{U}}^{-1}A'K_{\mathcal{V}} \tag{1.3}$$

such that

$$(A^*v, u)_{\mathcal{U}} = (v, Au)_{\mathcal{V}} \tag{1.4}$$

The operator A^* is linear and continuous and is called the adjoint of A.

The following theorem establishes some important properties of the transpose.

Theorem 1.2.1 *Let $A' \in \mathcal{L}(\mathcal{V}', \mathcal{U}')$ (here $\mathcal{L}(\mathcal{V}', \mathcal{U}')$ is the space of continuous linear operators mapping $\mathcal{V}'$ into $\mathcal{U}'$) denote the transpose of a continuous linear operator A mapping Hilbert spaces $\mathcal{U}$ into $\mathcal{V}$. Then the following hold:*

(i)

$$\mathcal{N}(A') = \mathcal{R}(A)^{\perp}$$

where $\mathcal{N}(A')$ is the null space of A' and $\mathcal{R}(A)^{\perp}$ is the orthogonal complement of the range $\mathcal{R}(A)$ of A in $\mathcal{V}'$.

(ii) *A' is injective if and only if $\mathcal{R}(A)$ is dense in $\mathcal{V}$.*

Proof: This is a well-known theorem; see, for example, Dunford and Schwartz.[16] ■

1.2.2 Pivot Spaces

The Riesz map $K_{\mathcal{U}} : \mathcal{U} \to \mathcal{U}'$ is an isometric isomorphism from $\mathcal{U}$ onto its dual $\mathcal{U}'$. Consequently, it is possible to identify $\mathcal{U}$ with its dual. In many instances, particularly in the theory of linear boundary-value problems, we encounter collections of Hilbert spaces satisfying $\mathcal{U}_m \subset \mathcal{U}_{m-1} \subset \ldots \subset \mathcal{U}_2 \subset \mathcal{U}_1 \subset \mathcal{U}_0$, the inclusions being dense and continuous. When one member of this set is identified with its dual, say $\mathcal{U}_0$, we call it the *pivot space*, and write

$$\mathcal{U}_0 = \mathcal{U}_0'$$

The term 'pivot' arises from the fact that

$$\ldots \subset \mathcal{U}_2 \subset \mathcal{U}_1 \subset \mathcal{U}_0 = \mathcal{U}_0' \subset \mathcal{U}_1' \subset \mathcal{U}_2' \subset \ldots$$

i.e. $\mathcal{U}_0$ provides a 'pivot' between the spaces $\mathcal{U}_i$, $i \geqslant 0$, and their duals.

1.2.3 Sobolev spaces

The notion of a Sobolev space is fundamental to the modern theory of boundary-value problems, and most of the applications of our theory can be developed within the framework of Sobolev spaces.

Let Ω be a smooth,† open, bounded domain in $\mathbb{R}^n$. We shall denote by $H^m(\Omega)$ the Sobolev space of order m, which is a linear space of functions (or equivalence classes of functions) defined by

$$H^m(\Omega) = \{u : u \text{ and all of its distributional derivatives } D^\alpha u \text{ of order } \leqslant m \text{ are in } L_2(\Omega),\ m \geqslant 0\} \tag{1.5}$$

Here we employ standard multi-index notations (see, Oden and Reddy[19]). Clearly, $H^0(\Omega) = L^2(\Omega)$.

Now the Sobolev spaces (1.5) are Hilbert spaces. Indeed, $H^m(\Omega)$ is complete with respect to the norm associated with the inner product

$$(u, v)_{H^m(\Omega)} = \int_\Omega \sum_{|\alpha| \leqslant m} D^\alpha u \overline{D^\alpha v}\, dx \tag{1.6}$$

and the norm on $H^m(\Omega)$ is

$$\|u\|_{H^m(\Omega)} = \sqrt{(u, u)_{H^m(\Omega)}} = \left\{ \sum_{|\alpha| \leqslant m} \|D^\alpha u\|^2_{L_2(\Omega)} \right\}^{\frac{1}{2}} \tag{1.7}$$

The subspace of $H^m(\Omega)$ consisting of those functions in $H^m(\Omega)$ whose derivatives of order $\leqslant m-1$ vanish on the boundary $\partial\Omega$ of Ω is denoted $H_0^m(\Omega)$:

$$H_0^m(\Omega) = \{u \in H^m(\Omega) : D^\alpha u|_{\partial\Omega} = 0,\ |\alpha| \leqslant m-1\} \tag{1.8}$$

Equivalently, $H_0^m(\Omega)$ can be defined as the completion of the space $C_0^\infty(\Omega)$ of infinitely differentiable functions with compact support in Ω with respect to the H^m-Sobolev norm defined in (1.8)

The so-called *negative Sobolev spaces* are defined as the duals of the $H_0^m(\Omega)$ spaces:

$$H^{-m}(\Omega) = (H_0^m(\Omega))', \qquad m \geqslant 0 \tag{1.9}$$

† We shall assume throughout that Ω is simply-connected with a C^∞-boundary $\partial\Omega$. However, most of the results and examples we cite subsequently hold if $\partial\Omega$ is only Lipschitzian. See, for example, Nečas[17] or Adams.[18]

The Sobolev spaces are important in making precise the 'degree of smoothness' of functions. The following list summarizes some of their most important properties:

(i)

$$H^m(\Omega) \subset H^k(\Omega), \qquad m \geq k \geq 0 \tag{1.10}$$

Indeed,

$$H^m(\Omega) \subset H^{m-1}(\Omega) \subset \ldots \subset H^2(\Omega) \subset H^1(\Omega) \subset H^0(\Omega) \tag{1.11}$$

The remarkable fact is that these inclusions are continuous and dense. Indeed, $H^m(\Omega)$ is dense in $H^k(\Omega)$, $m \geq k \geq 0$, and $\|u\|_{H^k(\Omega)} \leq \|u\|_{H^m(\Omega)}$

(ii) Obviously, if m is sufficiently large, the elements of $H^m(\Omega)$ can be identified with continuous functions. Just how large m must be in order that each $u \in H^m(\Omega)$ be continuous is determined by one of the so-called Sobolev embedding theorems: If Ω is smooth (e.g. if $\Omega \subset \mathbb{R}^n$ satisfies the cone condition) and if $m > n/2$, then $H^m(\Omega)$ is continuously and compactly embedded in the space $C^0(\bar{\Omega})$ of continuous functions.

(iii) Sobolev spaces $H^s(\partial\Omega)$ can be defined for classes of functions whose domain is the boundary $\partial\Omega$. There is an important relation between these boundary spaces and those containing functions defined on Ω. We define the *trace operators* γ_j as

$$\gamma_j u \equiv \left.\frac{\partial^j u}{\partial n^j}\right|_{\partial\Omega}, \qquad 0 \leq j \leq m-1 \tag{1.12}$$

i.e. they are the normal derivatives at $\partial\Omega$ of order $\leq m-1$. Let $\varphi \in L_2(\partial\Omega)$ and define the norm†

$$\|\varphi\|_{H^{m-j-1/2}(\partial\Omega)} \equiv \inf_{u \in H^m(\Omega)} \{\|u\|_{H^m(\Omega)};\ \varphi = \gamma_j u\} \tag{1.13}$$

The completion of $L_2(\partial\Omega)$ with respect to this norm is a Hilbert space of boundary functions denoted

$$H^{m-j-1/2}(\partial\Omega), \qquad 0 \leq j \leq m-1 \tag{1.14}$$

Two important properties of these spaces are called the *trace properties* of Sobolev spaces:

(iii.1) The operators γ_j can be extended to continuous linear operators mapping $H^m(\Omega)$ onto $H^{m-j-1/2}(\partial\Omega)$, i.e. there exist constants $C_j > 0$ such that

$$\|\gamma_j u\|_{H^{m-j-1/2}} \leq C_j \|u\|_{H^m(\Omega)}, \qquad 0 \leq j \leq m-1 \tag{1.15}$$

† There are other more constructive ways of defining these spaces. See, for example, Oden and Reddy.[19]

(iii.2) The kernel of the collection of trace operators $\boldsymbol{\gamma}=(\gamma_0, \gamma_1, \ldots, \gamma_{m-1})$ is $H_0^m(\Omega)$ and $H_0^m(\Omega)$ is dense in $H^m(\Omega)$;

$$\gamma_j(H_0^m(\Omega))=0, \qquad j=0,1,\ldots,m-1 \tag{1.16}$$

1.3 GREEN'S FORMULAE FOR OPERATORS ON HILBERT SPACES

1.3.1 A general comment

One of the most important applications of the notion of adjoints of linear operators involves cases in which it makes sense to distinguish between spaces of functions defined on the interior of some domain and spaces of functions defined on the boundary of a domain. The introduction of boundary values is obviously essential in the study of boundary value problems in Hilbert spaces, and it leads us to the idea of an abstract Green's formula for linear operators.

In this section, we develop a general and abstract Green's formula which extends those given previously in the literature. Our format resembles that of Aubin,[20] who developed Green's formulae for elliptic operators of even order. Our results involve formal operators associated with bilinear forms $B:\mathcal{H}\times\mathcal{G}$, $\mathcal{H}$ and $\mathcal{G}$ being Hilbert spaces (see Section 1.3.3) and reduce to those of Aubin when collapsed to the very special case, $\mathcal{G}=\mathcal{H}$.

1.3.2 Abstract trace property

An abstraction of the idea of boundary values of elements in Hilbert spaces is embodied in the concept of spaces with a trace property. A Hilbert space $\mathcal{H}$ is said to have the trace property if and only if the following conditions hold:

(i) $\mathcal{H}$ is contained in a larger Hilbert space $\mathcal{U}$ which has a weaker topology than $\mathcal{H}$.
(ii) $\mathcal{H}$ is dense in $\mathcal{U}$ and $\mathcal{U}$ is a pivot space, i.e.

$$\mathcal{H}\subset\mathcal{U}=\mathcal{U}'\subset\mathcal{H}' \tag{1.17}$$

(iii) There exists a linear operator γ that maps $\mathcal{H}$ onto another Hilbert space $\partial\mathcal{H}$ such that the kernel $\mathcal{H}_0$ of γ is dense in $\mathcal{U}$, i.e.

$$\ker\gamma=\mathcal{H}_0\subset\mathcal{H};\,\mathcal{H}_0\subset\mathcal{U}=\mathcal{U}'\subset H_0' \tag{1.18}$$

The space $\partial\mathcal{H}$ corresponds to a space of boundary values, and the operator γ extends the elements of $\mathcal{H}$, which can be thought of as functions defined on the interior of some domain, onto the space of boundary values $\partial\mathcal{H}$. The operator γ is sometimes called the trace operator.

The spaces $H^m(\Omega)$ and the operators γ_j in (1.12)–(1.16) are examples: $H^m(\Omega)$ is dense in $L_2(\Omega)$, $m \leqslant 0$, $L_2(\Omega)$ can be identified with its dual, extensions of the trace operators γ_j of (1.12) map $H^m(\Omega)$ onto the boundary spaces $H^{m-j-1/2}(\partial\Omega)$, $0 \leqslant j \leqslant m-1$, and $\ker \gamma = \ker(\gamma_0, y_1, \ldots, \gamma_{m-1}) = H_0^m(\Omega)$ is dense in $L_2(\Omega)$.

1.3.3 Bilinear forms and associated operators

Let $\mathcal{H}$ and $\mathcal{G}$ denote two real Hilbert spaces (the extension of our results to complex spaces is trivial), and let both $\mathcal{H}$ and $\mathcal{G}$ have the trace property, i.e.

$$\left.\begin{array}{ll} \mathcal{H} \subset \mathcal{U} = \mathcal{U}' \subset \mathcal{H}', & \mathcal{G} \subset \mathcal{V} = \mathcal{V}' \subset \mathcal{G}' \\ \gamma : \mathcal{H} \to \partial\mathcal{H}, & \gamma^* : \mathcal{G} \to \partial\mathcal{G} \\ \ker \gamma = \mathcal{H}_0, & \ker \gamma^* = \mathcal{G}_0 \\ \mathcal{H}_0 \subset \mathcal{U} = \mathcal{U}' \subset \mathcal{H}_0', & \mathcal{G}_0 \subset \mathcal{V} = \mathcal{V}' \subset \mathcal{G}_0' \end{array}\right\} \tag{1.19}$$

The inclusions $\mathcal{H} \subset \mathcal{U}$, $\mathcal{G} \subset \mathcal{V}$, $\mathcal{H}_0 \subset \mathcal{U}$, $\mathcal{G}_0 \subset \mathcal{V}$, are dense and continuous.

Next, we introduce an operator B which maps pairs $(u, \mathbf{v})$, $u \in \mathcal{H}$ and $\mathbf{v} \in \mathcal{G}$, linearly and continuously into real numbers:

$$B : \mathcal{H} \times \mathcal{G} \to \mathbb{R} \tag{1.20}$$

We denote the values of B in $\mathbb{R}$ by $B(u, \mathbf{v})$, and we refer to B as a *continuous bilinear form* on $\mathcal{H} \times \mathcal{G}$. That B is *bilinear* mean that

$$\left.\begin{array}{l} B(\alpha u_1 + \beta u_2, v) = \alpha B(u_1, v) + \beta B(u_2, v) \\ B(u, \alpha v_1 + \beta v_2) = \alpha B(u, v_1) + \beta B(u, v_2) \end{array}\right\} \tag{1.21}$$

$\forall u, u_1, u_2 \in \mathcal{H}$, $v, v_1, v_2 \in \mathcal{G}$, $\alpha, \beta \in \mathbb{R}$. That B is *continuous* means that it is bounded, i.e. there exists a positive constant M such that

$$B(u, v) \leqslant M \|u\|_{\mathcal{H}} \|v\|_{\mathcal{G}} \qquad \forall u \in \mathcal{H}, \qquad \forall v \in \mathcal{G} \tag{1.22}$$

Now let u be fixed element of $\mathcal{H}$ and let $v \in \mathcal{G}_0$. Then $B(u, v)$ describes a continuous linear functional l_u on the space $\mathcal{G}_0$ for each choice of $u \in \mathcal{H} : B(u, v) = l_u(v)$, $v \in \mathcal{G}_0$. The linear functional l_u depends linearly and continuously on u, and we describe this dependence in terms of a linear operator $Au \equiv l_u$. Thus

$$B(u, v) = \langle Au, v \rangle_{\mathcal{G}}, \qquad v \in \mathcal{G}_0 \tag{1.23}$$

The operator A is called the *formal operator associated with the bilinear form* B. Clearly

$$A \in \mathcal{L}(\mathcal{H}, \mathcal{G}_0') \tag{1.24}$$

In a similar manner, if we fix $v \in \mathscr{G}$, $B(u, v)$ defines a continuous linear functional on $\mathscr{H}_0$, and we define a continuous linear operator A^* by

$$B(u, \mathbf{v}) = \langle A^* v, u \rangle_{\mathscr{H}}, \qquad u \in \mathscr{H}_0 \tag{1.25}$$

The operator A^* is known as the formal *adjoint* of A, and

$$A^* \in \mathscr{L}(\mathscr{G}, \mathscr{H}_0') \tag{1.26}$$

The fact that $\mathscr{H}_0 = \ker \gamma$ and $\mathscr{G}_0 = \ker \gamma^*$ enables us to establish the following fundamental lemma.

Theorem 1.3.1 Let $\mathscr{H}$ and $\mathscr{G}$ denote the Hilbert spaces with the trace properties described above, and let B denote a continuous bilinear form from $\mathscr{H} \times \mathscr{G} \to \mathbb{R}$ with formal association operators $A \in \mathscr{L}(\mathscr{H}, \mathscr{G}_0')$ and $A^* \in \mathscr{L}(\mathscr{G}, \mathscr{H}_0')$. Moreover, let $\mathscr{H}_A$ and $\mathscr{G}_{A^*}$ denote subspaces

$$\left.\begin{aligned} \mathscr{H}_A &= \{u \in \mathscr{H} : Au \in \mathscr{V}\} \qquad (\mathscr{V} = \mathscr{V}' \subset \mathscr{G}_0') \\ \mathscr{G}_{A^*} &= \{v \in \mathscr{G} : A^* v \in \mathscr{U}\} \qquad (\mathscr{U} = \mathscr{U}' \subset \mathscr{H}_0') \end{aligned}\right\} \tag{1.27}$$

Then there exists uniquely defined operators $\delta \in \mathscr{L}(\mathscr{H}_A, \partial\mathscr{G}')$ and $\delta^* \in \mathscr{L}(\mathscr{G}_{A^*}, \partial\mathscr{H}')$ such that the following formulae hold:

$$\left.\begin{aligned} B(u, v) &= (v, Au)_{\mathscr{V}} + \langle \delta u, \gamma^* v \rangle_{\partial\mathscr{G}}, \qquad u \in \mathscr{H}_A, \qquad v \in \mathscr{G} \\ B(u, v) &= (A^* v, u)_{\mathscr{U}} + \langle \delta^* v, \gamma u \rangle_{\partial\mathscr{H}}, \qquad u \in \mathscr{H}, \qquad v \in \mathscr{G}_{A^*} \end{aligned}\right\} \tag{1.28}$$

Here $\langle \cdot, \cdot \rangle_{\partial\mathscr{H}}$ and $\langle \cdot, \cdot \rangle_{\partial\mathscr{G}}$ denote duality pairings on $\partial\mathscr{H}' \times \partial\mathscr{H}$ and $\partial\mathscr{G}' \times \partial\mathscr{G}$ respectively.

For a proof of the theorem, see Oden.[22]

Let $B : \mathscr{H} \times \mathscr{G} \to \mathbb{R}$ be a continuous bilinear form on Hilbert spaces $\mathscr{H}$ and $\mathscr{G}$ having the trace property. Let in addition, A be the formal operator associated with $B(\cdot, \cdot)$ and let A^* be its formal adjoint. The operators $\gamma \in \mathscr{L}(\mathscr{H}, \partial\mathscr{H})$ and $\delta \in \mathscr{L}(\mathscr{H}_A, \partial\mathscr{G}')$ described above are called the *Dirichlet operator* and the *Neumann operator*, respectively, *corresponding to the operator* A. Likewise, the operators $\gamma^* \in \mathscr{L}(\mathscr{G}, \partial\mathscr{G})$ and $\delta^* \in \mathscr{L}(\mathscr{G}_{A^*}, \partial\mathscr{H}')$ described above are called the Dirichlet and Neumann operators, respectively, corresponding to the operator A^*.

1.3.4 Green's formulae

The relationships derived in Theorem 1.3.1. are called Green's formulae for the bilinear form $B(\cdot, \cdot)$:

$$\left.\begin{aligned} B(u, \mathbf{v}) &= (\mathbf{v}, Au)_{\mathscr{V}} + \langle \delta u, \gamma^* \mathbf{v} \rangle_{\partial\mathscr{G}}, \qquad u \in \mathscr{H}_A, \qquad \mathbf{v} \in \mathscr{G} \\ B(u, \mathbf{v}) &= (A^* \mathbf{v}, u)_{\mathscr{U}} + \langle \delta^* \mathbf{v}, \gamma u \rangle_{\partial\mathscr{H}}, \qquad u \in \mathscr{H}, \qquad \mathbf{v} \in \mathscr{G}_{A^*} \end{aligned}\right\} \tag{1.29}$$

If we take $u \in \mathscr{H}_A$ and $\mathbf{v} \in \mathscr{G}_{A^*}$, we obtain the abstract Green's formula for the operator $A \in \mathscr{L}(\mathscr{H}, \mathscr{G}_0') \cap \mathscr{L}(\mathscr{H}_A, \mathscr{V})$:

$$(A^*\mathbf{v}, u)_{\mathscr{U}} = (\mathbf{v}, Au)_{\mathscr{V}} + \langle \delta u, \gamma^* \mathbf{v} \rangle_{\partial\mathscr{G}} - \langle \delta^* \mathbf{v}, \gamma u \rangle_{\partial\mathscr{H}}, \qquad u \in \mathscr{H}_A, \qquad \mathbf{v} \in \mathscr{G}_{A^*} \tag{1.30}$$

The collection of boundary terms,

$$\Gamma(u, \mathbf{v}) = \langle \delta u, \gamma^* \mathbf{v} \rangle_{\partial\mathscr{G}} - \langle \delta^* \mathbf{v}, \gamma u \rangle_{\partial\mathscr{H}} \tag{1.31}$$

is called the bilinear concomitant of A; $\Gamma : \mathscr{H}_A \times \mathscr{G}_{A^*} \to \mathbb{R}$.

Example 1.3.1 Consider the case in which Ω is a smooth open bounded subset of $\mathbb{R}^2$ with a smooth boundary $\partial\Omega$ and

$$\mathscr{H} = \mathscr{G} = H^1(\Omega) = \{u : u, u_x, u_y \in L_2(\Omega)\}; \qquad \mathscr{U} = \mathscr{V} = L_2(\Omega)$$

Let $a = a(x, y)$, $b = b(x, y)$, and $c = c(x, y)$ be sufficiently smooth functions of x and y (e.g. a, b, $c \in C^1(\bar{\Omega})$), and define the bilinear form $B : H^1(\Omega) \times H^1(\Omega) \to \mathbb{R}$ by

$$B(u, v) = \int_\Omega (a \nabla u \cdot \nabla v + bvu_x + cvu_y)\, dx\, dy$$

where $\nabla u = \text{grad } u = (u_x, u_y)$. Thus,

$$\partial\mathscr{H} = \partial\mathscr{G} = H^{1/2}(\partial\Omega); \qquad \ker \gamma = H_0^1(\Omega) = \{u \in H^1(\Omega); u = 0 \text{ on } \partial\Omega\}$$

In this case,

$$B(u, v) = \int_\Omega v(-\nabla \cdot (a\nabla u) + bu_x + cu_y)\, dx\, dy + \oint_{\partial\Omega} a \frac{\partial u}{\partial n} v\, ds$$

Thus, the formal operator corresponding to $B(\cdot, \cdot)$ is

$$Au = -\nabla \cdot (a \nabla u) + b \frac{\partial u}{\partial x} + c \frac{\partial u}{\partial y}$$

if we take

$$u \in \mathscr{H}_A = \{u \in H^1(\Omega) : Au \in L_2(\Omega)\}$$

Also, we now have

$$\delta u = a \frac{\partial u}{\partial n}\bigg|_{\partial\Omega}, \qquad \gamma^* v = v|_{\partial\Omega}$$

Similarly, if $v \in \mathscr{G}_{A^*} = \mathscr{H}_A$, we have

$$B(u, v) = \int_\Omega u\left(-\nabla \cdot (a\nabla v) - \frac{\partial bv}{\partial x} - \frac{\partial cv}{\partial y}\right) dx\, dy + \\ + \oint_{\partial\Omega} \left(au \frac{\partial v}{\partial n} + bvun_x + cvun_y\right) ds$$

where n_x, n_y are the components of the unit outward normal $\mathbf{n}$ to $\partial\Omega$. Thus,

$$A^*v = -\boldsymbol{\nabla}\cdot(a\boldsymbol{\nabla}v) - \frac{\partial bv}{\partial x} - \frac{\partial cv}{\partial y}$$

$$\gamma u = u|_{\partial\Omega}, \qquad \delta^* v = \left[a\frac{\partial v}{\partial n} + (bn_x + cn_y)v \right]\Big|_{\partial\Omega}$$

The bilinear concomitant is then

$$\Gamma(u, v) = \oint \left[av\frac{\partial u}{\partial n} - au\frac{\partial v}{\partial n} - (cn_y + bn_x)uv \right] ds \quad \blacksquare$$

Example 1.3.2 Let Ω be as in Example 1.3.1 and define

$$B(u, \mathbf{v}) = \int_\Omega \text{grad}\, u \cdot \mathbf{v}\, dx\, dy$$

as a bilinear form on $\mathcal{H}\times\mathcal{G}$, where

$$\mathcal{H} = H^1(\Omega) = \{u\colon u, u_x, u_y, \in L_2(\Omega)\}$$
$$\mathcal{G} = \mathbf{H}^1(\Omega) = \{\mathbf{v} = (v_1, v_2)\colon v_1, v_{1x}, v_{1y}, v_2, v_{2x}, v_{2y} \in L_2(\Omega)\}$$

$H^1(\Omega)$ is dense in $\mathcal{U} = L_2(\Omega)$ and $\mathbf{H}^1(\Omega)$ is dense in $\mathcal{V} = \mathbf{L}_2(\Omega) = L_2(\Omega)\times L_2(\Omega)$. In this case, we may take $\gamma^*\mathbf{v} = \mathbf{v}\cdot\mathbf{n}|_{\partial\Omega}$, but $\delta = 0$. Indeed, if the formal operator associated with the given bilinear form is

$$Au = \text{grad}\, u$$

then $\mathcal{H}_A = \mathcal{H} = H^1(\Omega)$ and $C_A : \mathcal{H}_A \to \mathcal{G}_0^\perp$ is identically zero. We next note that

$$B(u, \mathbf{v}) = \int_\Omega u(-\text{div}\, \mathbf{v})\, dx\, dy + \oint_{\partial\Omega} \mathbf{v}\cdot\mathbf{n}u\, ds$$

Thus, $\mathcal{G}_{A^*} = \mathbf{H}^1(\Omega)$ and

$$A^*\mathbf{v} = -\text{div}\, \mathbf{v}; \qquad \gamma u = u|_{\partial\Omega}; \qquad \delta^*\mathbf{v} = \mathbf{v}\cdot\mathbf{n}|_{\partial\Omega}$$

The Green's formula is the classical relation,

$$\oint_\Omega (\text{grad}\, u\cdot\mathbf{v} + u\, \text{div}\, \mathbf{v})\, dx\, dy = \oint_{\partial\Omega} u\mathbf{v}\cdot\mathbf{n}\, ds \quad \blacksquare$$

1.3.5 Mixed boundary condition

An abstract Green's formula appropriate for operators with mixed boundary conditions can be obtained by introducing some additional operators and spaces.

Let

$$\left.\begin{aligned}
&\pi_1^* = \text{a continuous linear projection defined on } \partial\mathcal{G} \text{ into itself}\\
&\pi_2^* = I - \pi_1^*,\ I \text{ being the identity map from } \partial\mathcal{G} \text{ onto itself}\\
&\gamma_1^* = \pi_1^*\gamma^*, \qquad \gamma_2^* = \pi_2^*\gamma^* \quad \text{and} \quad \gamma^* = \gamma_1^* + \gamma_2^*\\
&\mathcal{J} = \ker \gamma_1^* = \{\mathbf{v} \in \mathcal{G} : \pi_1^*\gamma^*\mathbf{v} = 0\}
\end{aligned}\right\} \tag{1.32}$$

The space $\mathcal{J}$ is a closed linear subspace of $\mathcal{G}$ with the property

$$\mathcal{G}_0 \subset \mathcal{J} \subset \mathcal{G} \tag{1.33}$$

The operators γ_1^* and γ_2^* effectively decompose $\partial\mathcal{G}$ into two subspaces

$$\left.\begin{aligned}
\partial\mathcal{G}_1 = \gamma_1^*(\mathcal{G}), \qquad \partial\mathcal{G}_2 = \gamma_2^*(\mathcal{G})\\
\partial\mathcal{G} = \partial\mathcal{G}_1 + \partial\mathcal{G}_2
\end{aligned}\right\} \tag{1.34}$$

A similar collection of operators can be introduced on $\partial\mathcal{H}$:

$$\left.\begin{aligned}
&\pi_1 = \text{a continuous linear projection of } \partial\mathcal{H} \text{ into } \partial\mathcal{H}\\
&\pi_2 = I' - \pi_1 (I' : \partial\mathcal{H} \to \partial\mathcal{H})\\
&\gamma_1 = \pi_1\gamma, \qquad \gamma_2 = \pi_2\gamma, \qquad \gamma = \gamma_1 + \gamma_2\\
&\mathcal{F} = \ker \gamma_2 = \{u \in \mathcal{H} : \pi_2\gamma u = 0\}\\
&\mathcal{H}_0 \subset \mathcal{F} \subset \mathcal{H}\\
&\partial\mathcal{H}_1 = \gamma_1(\mathcal{H}), \qquad \partial\mathcal{H}_2 = \gamma_2(\mathcal{H}), \qquad \partial\mathcal{H} = \partial\mathcal{H}_1 + \partial\mathcal{H}_2
\end{aligned}\right\} \tag{1.35}$$

The Green formulae (1.28) now yield

$$\left.\begin{aligned}
B(u, \mathbf{v}) &= (\mathbf{v}, Au)_{\mathcal{V}} + \langle \delta u, \gamma_2^*\mathbf{v}\rangle_{\partial\mathcal{G}_2}, \qquad u \in \mathcal{H}_A, \qquad \mathbf{v} \in \mathcal{J}\\
B(u, \mathbf{v}) &= (A^*\mathbf{v}, u)_{\mathcal{U}} + \langle \delta^*\mathbf{v}, \gamma_1 u\rangle_{\partial\mathcal{H}_1}, \qquad \mathbf{v} \in \mathcal{G}_{A^*}, \qquad u \in \mathcal{F}
\end{aligned}\right\} \tag{1.36}$$

We observe for $\mathbf{v} \in \mathcal{J}$,

$$\begin{aligned}
\langle \delta u, \gamma^*\mathbf{v}\rangle_{\partial\mathcal{G}} &= \langle \delta u, \pi_1^*\gamma^*\mathbf{v}\rangle_{\partial\mathcal{G}_1} + \langle \delta u, \pi_2^*\gamma^*\mathbf{v}\rangle_{\partial\mathcal{G}_2}\\
&= \langle \delta u, \pi_2^*\gamma^*\mathbf{v}\rangle_{\partial\mathcal{G}_2}\\
&= \langle \pi_2^{*\prime}\delta u, \gamma^*\mathbf{v}\rangle_{\partial\mathcal{G}_2} = \langle \delta_2 u, \gamma^*\mathbf{v}\rangle_{\partial\mathcal{G}_2}
\end{aligned} \tag{1.37}$$

where

$$\delta_2 = \pi_2^{*\prime}\delta, \qquad \delta_2 \in \mathcal{L}(\mathcal{H}_A, \partial\mathcal{G}_2') \tag{1.38}$$

Finally, collecting (1.36) and (1.37), we arrive at the abstract Green's formula,

$$(A^*\mathbf{v}, u)_{\mathcal{U}} = (\mathbf{v}, Au)_{\mathcal{V}} - \langle \delta^*\mathbf{v}, \gamma_1 u\rangle_{\partial\mathcal{H}_1} + \langle \delta_2 u, \gamma^*\mathbf{v}\rangle_{\partial\mathcal{H}_2} \qquad \forall u \in \mathcal{H}_A \cap \mathcal{F}, \qquad \mathbf{v} \in \mathcal{G}_{A^*} \cap \mathcal{J} \tag{1.39}$$

1.4 ABSTRACT VARIATIONAL BOUNDARY-VALUE PROBLEMS

1.4.1 Some linear boundary-value problems

The Green's formulae and various properties of the bilinear forms $B(\cdot, \cdot)$ described in the previous section provide the basis for a theory of boundary-value problems involving linear operators on Hilbert spaces. We shall continue to use the notations and conventions of the previous section: $\mathscr{H}$ and $\mathscr{G}$ are Hilbert spaces with the trace property, densely embedded in pivot spaces $\mathscr{U}$ and $\mathscr{V}$, respectively, and $\gamma^*: \mathscr{G} \to \partial\mathscr{G}$, $\ker \gamma^* = \mathscr{G}_0$, $\gamma: \mathscr{H} \to \partial\mathscr{H}$, $\ker \gamma = \mathscr{H}_0$, etc.

Let $B: \mathscr{H} \times \mathscr{G} \to \mathbb{R}$ be a continuous bilinear form and let A be the formal operator associated with $B(u, v)$. We consider three types of boundary-value problems associated with A.

(1) *The Dirichlet problem for A.* Given data $f \in \mathscr{V}$ and $g \in \partial\mathscr{H}$, the problem of finding $u \in \mathscr{H}_A$ such that

$$\left.\begin{aligned} Au &= f \\ \gamma u &= g \end{aligned}\right\} \tag{1.40}$$

is called the *Dirichlet problem* for the operator A.

(2) *The Neuman problem for A.* Given data $f \in \mathscr{V}$ and $s \in \partial\mathscr{G}'$, the problem of finding $u \in \mathscr{H}_A$ such that

$$\left.\begin{aligned} Au &= f \\ \delta u &= s \end{aligned}\right\} \tag{1.41}$$

is called the *Neumann problem* for the operator A.

(3) *The mixed problem for A.* Given data $f \in \mathscr{V}$, $g \in \partial\mathscr{H}_1$, and $s \in \partial\mathscr{G}_2'$, the problem of finding $u \in \mathscr{H}_A$ such that

$$\left.\begin{aligned} Au &= f \\ \gamma_1 u &= g \\ \delta_2 u &= s \end{aligned}\right\} \tag{1.42}$$

is called the *mixed problem* for the operator A.

Now the bilinear form $B: \mathscr{H} \times \mathscr{G} \to \mathbb{R}$ described in the previous section can be used to construct variational boundary-value problems analogous to those for A. We shall consider the following variational problems:

(1) *The variational Dirichlet problem for A.* Given data $f \in \mathscr{V}$ and $g \in \partial\mathscr{H}$, find $w \in \mathscr{H}_0 = \ker \gamma$ such that

$$B(w, \mathbf{v}) = (f, \mathbf{v})_{\mathscr{V}} - B(\gamma^{-1}g, \mathbf{v}) \ \forall \mathbf{v} \in \mathscr{G}_0 \tag{1.43}$$

where γ^{-1} is a right inverse of γ.

(2) *The variational Neumann problem for A.* Given data $f \in \mathcal{V}$ and $s \in \partial\mathcal{G}'$, find $u \in \mathcal{H}$ such that

$$B(u, \mathbf{v}) = (f, \mathbf{v})_{\mathcal{V}} + \langle s, \gamma^* \mathbf{v} \rangle_{\partial\mathcal{G}} \; \forall \mathbf{v} \in \mathcal{G} \tag{1.44}$$

(3) *The variational mixed problem for A.* Given data $f \in \mathcal{V}$, $g \in \partial\mathcal{H}_1$, and $s \in \partial\mathcal{G}_2'$, find $w \in \ker \gamma_1 = \ker \pi_1 \gamma$ such that

$$B(w, \mathbf{v}) = (f, \mathbf{v})_{\mathcal{V}} - B(\gamma_1^{-1} g, \mathbf{v}) + \langle s, \gamma^* \mathbf{v} \rangle_{\partial\mathcal{G}_2} \; \forall \mathbf{v} \in \mathcal{J} \tag{1.45}$$

where γ_1^{-1} is a right inverse of γ_1 and $\mathcal{J} = \ker \gamma_1^* = \ker \pi_1 \gamma^*$.

Remark. There are, of course, several other abstract boundary-value problems for the operator A that could be mentioned. For example, a second bilinear form $b : \partial\mathcal{H} \times \partial\mathcal{G} \to \mathbb{R}$ could be introduced at this point which would permit the construction of oblique boundary conditions and which would lead to a formulation more general than (1.42). However, such technical generalizations obscure the simple structure of the ideas we wish to clarify here, and so they are omitted. ■

Theorem 1.4.1 The Dirichlet problem (1.40) *for the operator A and the variational Dirichlet problem* (1.43) *are equivalent in the following sense. Let* γ^{-1} *be the inverse map of* γ; *if u is a solution of* (1.40) *then* $w = u - \gamma^{-1} g$ *is a solution of* (1.43). *Conversely, if w is a solution of* (1.43), *then* $u = w + \gamma^{-1} g$ *is a solution of* (1.40). *Moreover, the Neumann problem* (1.41) *for the operator A is equivalent to the variational Neumann problem* (1.44) *in the sense that any solution of* (1.41) *is also a solution of* (1.44) *and, conversely, any solution of* (1.44) *is a solution of* (1.41). *Likewise, problems* (1.42) *and* (1.45) *are equivalent in a similar sense.* ■

For a proof see Reference 22.

Example 1.4.1 Let $\mathcal{H} = \mathcal{G} = H^1(\Omega)$, $\mathcal{U} = \mathcal{V} = L_2(\Omega)$, $\mathcal{H}_0 = \mathcal{G}_0 = H_0^1(\Omega)$, and $\partial\mathcal{H} = \partial\mathcal{G} = H^{1/2}(\partial\Omega)$, where Ω is a smooth open bounded domain in $\mathbb{R}^2$. The bilinear form

$$B(u, v) = \int_\Omega (\nabla u \cdot \nabla v + \alpha uv) \, dx \, dy$$

where α is a non-negative constant, is a continuous bilinear form from $H^1(\Omega) \times H^1(\Omega)$ into $\mathbb{R}$. The formal operator associated with $B(\cdot, \cdot)$ is $A = -\Delta + \alpha$, where $\Delta = \nabla \cdot \nabla = \nabla^2 = \partial^2/\partial x^2 + \partial^2/\partial y^2$ is the Laplacian operator. We denote

$$\mathcal{H}_A = \mathcal{G}_{A^*} = H^1(A, \Omega) = \{u \in H^1(\Omega) : -\Delta u + \alpha u \in L_2(\Omega)\}$$

The Dirichlet problem for A is to find $u \in H^1(A, \Omega)$ such that

$$\begin{aligned} -\Delta u + \alpha u = f \quad \text{in} \quad \Omega, \qquad & f \in L_2(\Omega) \\ u = g \quad \text{on} \quad \partial\Omega, \qquad & g \in H^{1/2}(\partial\Omega) \end{aligned}$$

In view of Theorem 1.4.1, this problem is equivalent to the problem of finding $w \in H_0^1(\Omega)$ such that

$$\int_\Omega (\nabla w \cdot \nabla v + \alpha w v) \, dx \, dy = \int_\Omega f v \, dx \, dy - \int_\Omega (\nabla w_0 \cdot \nabla v + \alpha w_0 v) \, dx \, dy \qquad \forall v \in H_0^1(\Omega)$$

where w_0 is any function in $H^1(\Omega)$ such that $w_0 = g$ on $\partial\Omega$. Then $u = w + w_0$ is the solution of the Dirichlet problem for A.

The Neumann problem for A is to find $u \in H^1(A, \Omega)$ such that

$$\begin{aligned} -\Delta u + \alpha u = f \quad \text{in} \quad \Omega, \qquad & f \in L_2(\Omega) \\ \frac{\partial u}{\partial n} = s \quad \text{on} \quad \partial\Omega, \qquad & s \in (H^{1/2}(\partial\Omega))' \end{aligned}$$

and it is equivalent to seeking $u \in H^1(\Omega)$ such that

$$\int_\Omega (\nabla u \cdot \nabla v + \alpha u v) \, dx \, dy = \int_\Omega f v \, dx \, dy + \oint_{\partial\Omega} s v \, ds \quad \forall v \in H^1(\Omega)$$

where the contour integral $\oint_{\partial\Omega}$ denotes duality pairing on $(H^{1/2}(\partial\Omega))' \times H^{1/2}(\partial\Omega)$. ■

We observe that boundary conditions enter the statement of a variational boundary-value problem in two distinct ways. The *essential* or *stable* boundary conditions enter by simply defining the spaces $\mathcal{H}_0$ and $\mathcal{G}_0$ on which the problem is posed. The *natural* or *unstable* boundary conditions are introduced in the definition of the bilinear form $B(u, \mathbf{v})$ and are defined on the spaces $\mathcal{H}_A$ and $\mathcal{G}_{A^*}$.

1.4.2 Compatibility of the data

We shall discuss briefly here the issue of the compatibility of the data f, g, s with various boundary-value problems. The ideas are derived from the classical theorem (recall Theorem 1.2.1).

Theorem 1.4.2 Let A be a bounded linear operator from a Hilbert space $\mathcal{H}$ into a Hilbert space $\mathcal{G}$. Let A^* be its adjoint. Let $\mathcal{N}(A)$, $\mathcal{N}(A^*)$ denote the null *spaces of A and A^* respectively and $\mathscr{R}(A)$ and $\mathscr{R}(A^*)$ denote the ranges of A and A^* respectively. Then*

$$\left.\begin{aligned} \mathscr{R}(A)^\perp &= \mathscr{N}(A^*), \qquad \overline{\mathscr{R}(A)} = \mathscr{N}(A^*)^\perp \\ \mathscr{R}(A^*)^\perp &= \mathscr{N}(A), \qquad \overline{\mathscr{R}(A^*)} = \mathscr{N}(A)^\perp \end{aligned}\right\} \tag{1.46}$$

where $\perp$ denotes the orthogonal complement of the spaces indicated and an overbar indicates the closure.

Proof: For a proof see, for example, Taylor.[21] ■

We shall first address the compatibility question in connection with the Dirichlet problem for the operator A: find $u \in \mathscr{H}_A$ such that

$$Au = f, \qquad f \in \mathscr{V}$$
$$\gamma u = g, \qquad g \in \partial\mathscr{H}$$

Let $A^* \in \mathscr{L}(\mathscr{G}_{A^*}, \mathscr{H}'_0)$ denote the formal adjoint of A and let $\gamma^* \in \mathscr{L}(\mathscr{G}, \partial\mathscr{G})$, $\ker \gamma^* = \mathscr{G}_0$. Then the adjoint Dirichlet problem corresponding to A is the problem of finding $v \in \mathscr{G}_{A^*}$ such that

$$A^*\mathbf{v} = f^*, \qquad f^* \in \mathscr{U}$$
$$\gamma^*\mathbf{v} = g^*, \qquad g^* \in \partial\mathscr{G}$$

We also introduce the null spaces,

$$\mathscr{N}(A, \gamma) = \{u \in \mathscr{H}_A : Au = 0, \gamma u = 0\}$$
$$\mathscr{N}(A^*, \gamma^*) = \{\mathbf{v} \in \mathscr{G}_{A^*} : A^*\mathbf{v} = 0, \gamma^*\mathbf{v} = 0\}$$

We shall assume that these spaces are finite-dimensional.

Now if $\mathscr{N}(A, \gamma)$ is finite-dimensional, it is closed in $\mathscr{U}$ and we have

$$\mathscr{U} = \mathscr{N}(A, \gamma) \oplus \mathscr{N}(A, \gamma)^\perp$$

where

$$\mathscr{N}(A, \gamma)^\perp = \{u \in \mathscr{U} : (u, v)_{\mathscr{U}} = 0 \ \forall v \in \mathscr{N}(A, \gamma)\}$$

Then A can be regarded as continuous linear operator from $\mathscr{N}(A, \gamma)^\perp$ onto its range $\mathscr{R}(A) \subset \mathscr{V}$. By the Banach theorem, a continuous inverse A^{-1} exists from $\mathscr{R}(A)$ onto $\mathscr{N}(A, \gamma)^\perp$, and $\mathscr{R}(A)$ is closed in $\mathscr{V}$. Thus $\mathscr{V} = \mathscr{R}(A) \oplus \mathscr{R}(A)^\perp$ where $\mathscr{R}(A)^\perp = \{\mathbf{v} \in \mathscr{V} : (\mathbf{v}, \mathbf{w})_{\mathscr{V}} = 0 \ \forall \mathbf{w} \in \mathscr{R}(A)\}$. The Dirichlet problem for A clearly has at least one solution whenever the data $f \in \mathscr{R}(A)$ and $g \in \mathscr{R}(\gamma)$. Data satisfying these requirements are said to be compatible with the operators (A, γ).

A convenient test for compatibility of the data is given in the following theorem.

Theorem 1.4.3 *Let $f \in \mathcal{V}$ and $g \in \partial\mathcal{H}$ be data in a Dirichlet problem for the operator A. Then a necessary and sufficient condition that there exists a solution of the Dirichlet problem is that*

$$(f, \mathbf{v})_{\mathcal{V}} - \langle \delta^* \mathbf{v}, g \rangle_{\partial\mathcal{H}} = 0 \qquad \forall \mathbf{v} \in \mathcal{N}(A^*, \gamma^*) \tag{1.47}$$ ■

For the Neumann problems

$$Au = f, \qquad A^*\mathbf{v} = f^*$$
$$\delta u = s, \qquad \delta^*\mathbf{v} = s^*$$

we define

$$\mathcal{N}(A, \delta) = \{u \in \mathcal{H}_A : Au = 0, \delta u = 0\}$$
$$\mathcal{N}(A^*, \delta^*) = \{\mathbf{v} \in \mathcal{G}_A{}^* : A^*\mathbf{v} = 0, \delta^*\mathbf{v} = 0\}$$

and, analogously, have

Theorem 1.4.4 *A necessary and sufficient condition for the existence of at least one solution of the Neumann problem for the operator A is that the data (f, s) satisfy*

$$(f, \mathbf{v})_{\mathcal{V}} + \langle s, \gamma^* \mathbf{v} \rangle_{\partial\mathcal{G}} = 0 \qquad \forall \mathbf{v} \in \mathcal{N}(A^*, \delta^*) \tag{1.48}$$ ■

A similar compatibility condition can be developed for mixed boundary-value problems:

Theorem 1.4.5 *A necessary and sufficient condition for the existence of a solution to the mixed boundary-value problem for A is that the data (f, g, s) satisfy*

$$(f, \mathbf{v})_{\mathcal{V}} - \langle \delta^* \mathbf{v}, g \rangle_{\partial\mathcal{H}_1} + \langle s, \gamma^* \mathbf{v} \rangle_{\partial\mathcal{G}_2} = 0 \qquad \forall \mathbf{v} \in \mathcal{N}(A^*, \gamma_1^*, \delta_2^*) \tag{1.49}$$

where

$$\mathcal{N}(A^*, \gamma_1^*, \delta_2^*) = \{\mathbf{v} \in \mathcal{G}_{A^*} : A^*\mathbf{v} = 0, \gamma_1^*\mathbf{v} = 0, \delta_2^*\mathbf{v} = 0\} \tag{1.50}$$ ■

Whenever the compatibility conditions hold, a solution to the Dirichlet, Neumann, or mixed problems may exist, but it will not necessarily be unique. The solution u is unique, of course, whenever $\mathcal{N}(A, \gamma) = \{0\}$, for the Dirichlet problem, whenever $\mathcal{N}(A, \delta) = \{0\}$ for the Neumann problem, and when $\mathcal{N}(A, \gamma, \delta) = \{0\}$ for the mixed problem. However, these conditions often do not hold. We can, however, force the solutions to either class of problems to be unique by imposing an additional condition.

Theorem 1.4.6 *Let $u \in \mathcal{H}_A$ be a solution to the Dirichlet problem for the operator A. Then u is the only solution to this problem if*

$$(u, w)_{\mathcal{U}} = 0 \qquad \forall w \in \mathcal{N}(A, \gamma) \tag{1.51}$$

Likewise, a solution u of the Neumann (mixed) problem for A is unique if $(u, w)_{\mathcal{U}} = 0\ \forall w \in \mathcal{N}(A, \delta)(\forall w \in \mathcal{N}(A, \gamma, \delta))$, *respectively.* ■

1.4.3 Existence theory

The theory presented thus far suggests the following general setting for linear variational boundary-value problems.

—Let $\mathcal{H}$ and $\mathcal{G}$ be arbitrary Hilbert spaces (now $\mathcal{H}$ and $\mathcal{G}$ are not necessarily the spaces appearing earlier in this section—they do not necessarily have the trace property).
—$B : \mathcal{H} \times \mathcal{G} \to \mathbb{R}$ is a bilinear form. Then find an element $u \in \mathcal{H}$ such that

$$B(u, v) = f(v) \qquad \forall v \in \mathcal{G}$$

where $f \in \mathcal{G}'$.

The essential question here is: What conditions can be imposed so that we are guaranteed that a unique solution exists which depends continuously on the choice of data f? This question was originally resolved for certain choices of B by Lax and Milgram. A more general form of their classic theorem made popular by Babuška[23] (see also Nečas[17] and Oden and Reddy[19]) is given as follows.

Theorem 1.4.7 *Let* $B : \mathcal{H} \times \mathcal{G} \to \mathbb{R}$ *be a bilinear functional on* $\mathcal{H} \times \mathcal{G}$, $\mathcal{H}$ *and* $\mathcal{G}$ *being Hilbert spaces, which has the following three properties.*
(i) *There exists a constant* $M > 0$ *such that*

$$B(u, v) \leq M \|u\|_{\mathcal{H}} \|v\|_{\mathcal{G}} \qquad \forall u \in \mathcal{H}, \qquad v \in \mathcal{G}$$

where $\|\cdot\|_{\mathcal{H}}$ *and* $\|\cdot\|_{\mathcal{G}}$ *denote the norms on* $\mathcal{H}$ *and* $\mathcal{G}$, *respectively,*
(ii) *There exists a constant* $\gamma > 0$ *such that*

$$\inf_{\substack{u \in \mathcal{H} \\ \|u\|_{\mathcal{H}} = 1}} \sup_{\substack{v \in \mathcal{G} \\ \|v\|_{\mathcal{G}} \leq 1}} |B(u, v)| \geq \gamma > 0$$

(iii) *We have*

$$\sup_{u \in \mathcal{H}} |B(u, v)| > 0, \qquad v \neq 0$$

Then there exists a unique solution to the problem of finding $u \in \mathcal{H}$ *such that*

$$B(u, v) = f(v) \qquad \forall v \in \mathcal{G}, \qquad f \in \mathcal{G}'$$

Moreover, the solution u_0 *depends continuously on the data; in fact,*

$$\|u_0\|_{\mathcal{H}} \leq \frac{1}{\gamma} \|f\|_{\mathcal{G}'}$$ ■

Property (i) $B(\cdot, \cdot)$ is, of course, a continuity requirement; $B(\cdot, \cdot)$ is assumed to be a bounded linear functional on $\mathcal{H}$ and on $\mathcal{G}$. Properties (ii) and (iii) serve to establish the existence of a continuous inverse of the associated operator A.

Corollary 1.4.8 *Let $B: \mathcal{H} \times \mathcal{H} \to \mathbb{R}$ be a bilinear form on a real Hilbert space $\mathcal{H}$. Let there exist positive constants M and γ such that*

$$B(u, v) \leqslant M\|u\|_{\mathcal{H}}\|v\|_{\mathcal{H}} \qquad \forall u, v \in \mathcal{H}$$

$$B(u, u) \geqslant \gamma\|u\|^2_{\mathcal{H}}$$

Then there exists a unique $u \in \mathcal{H}$ such that

$$B(u, v) = f(v) \qquad \forall v \in \mathcal{H}, \qquad f \in \mathcal{H}'$$

and

$$\|u\|_{\mathcal{H}} \leqslant \frac{1}{\gamma}\|f\|_{\mathcal{H}'} \qquad \blacksquare$$

1.5 CONSTRUCTION OF VARIATIONAL PRINCIPLES

Let W be a real Banach space and W' its topological dual. If $\mathcal{P}$ is an operator from W into W', *not necessarily linear*, we may consider the abstract problem of finding $u \in W$ such that

$$\mathcal{P}(u) = 0, \qquad 0 \in W' \tag{1.52}$$

Now it is well known that in many cases an alternative problem can be formulated, equivalent to (1.52), which involves seeking a $u \in W$ such that

$$K'(u) = 0$$

where K is an appropriate functional defined on W and $K'(u)$ is the Gâteaux derivative of K at u, i.e.

$$\lim_{\epsilon \to 0} \frac{1}{\epsilon}(K(u + \epsilon\eta) - K(u)) = \langle K'(u), \eta\rangle$$

where $\langle \cdot, \cdot \rangle$ denotes duality pairing on $W' \times W$. Thus, if there exists a Gâteaux differentiable functional $K: W \to \mathbb{R}$ such that

$$\mathcal{P} = K' \tag{1.53}$$

then (1.52) is equivalent to the *classical variational problem* of finding elements $u \in W$ which are critical points of the functional K. We say that $\mathcal{P}$ is the gradient of K and we sometimes write $\mathcal{P} = \text{grad } K$.

Any operator $\mathcal{P}: W \to W'$ for which there exists a functional $K: W \to \mathbb{R}$ such that $\mathcal{P} = \text{grad } K$ is called a *potential operator*. It is well known (see, for example, Vainberg[7] or Nashed[8]) that if a continuous Gâteaux differential

$\delta\mathcal{P}(u, \eta)$ of $\mathcal{P}$ exists, then a necessary and sufficient condition that $\mathcal{P}$ be potential is that it be symmetric in the sense that

$$\langle \delta\mathcal{P}(u, \eta), \zeta\rangle = \langle \delta\mathcal{P}(u, \zeta), \eta\rangle \qquad \forall u, \eta, \xi \in W$$

Given a potential operator $\mathcal{P}: W \to W'$, the problem of determining a functional K such that (1.53) holds is called the *inverse problem of the calculus of variations.* Its solution is provided by the following theorem, the proof of which can be found in the monograph of Vainberg[7].

Theorem 1.5.1 Let $\mathcal{P}: W \to W'$ be a potential operator on the Banach space W. Then $\mathcal{P}$ is the gradient of the functional $K: W \to \mathbb{R}$ given by

$$K(u) = \int_0^1 \langle \mathcal{P}(u_0 + s(u - u_0)), u_0\rangle \, ds + K_0 \tag{1.54}$$

where u_0 is a fixed point in W, $K_0 = K(u_0)$, and $s \in [0, 1]$. ■

By an appropriate identification of the space W and the duality $\langle\cdot, \cdot\rangle$, all of the classical variational principles of mathematical physics can be constructed using (1.54). A lengthy list of applications of (1.54) to this end can be found in Chapter 5 of Oden and Reddy[17].

1.6 THE CLASSICAL VARIATIONAL PRINCIPLES

1.6.1 A general class of boundary-value problems

We shall now describe a general class of abstract boundary-value problems that is encountered with remarkable frequency in linear problems of mathematical physics and mechanics. Continuing to use the notation of the previous section, we introduce a linear, symmetric, operator

$$E: \mathcal{V}' \to \mathcal{V} \tag{1.55}$$

which effects a continuous, isometric isomorphism of the dual of the pivot space $\mathcal{V}'$ onto $\mathcal{V}$, which has a continuous inverse, $E^{-1}: \mathcal{V} \to \mathcal{V}'$. If $\mathbf{v} \in \mathcal{V}'$, we shall denote the elements in $E(\mathcal{V}')$ by

$$\boldsymbol{\sigma} = E\mathbf{v} \tag{1.56}$$

The Green's formula (1.30) can now be written

$$(A^*\boldsymbol{\sigma}, u)_{\mathcal{U}} = (Au, \boldsymbol{\sigma})_{\mathcal{V}} + \langle \delta^*\boldsymbol{\sigma}, \gamma_1 u\rangle_{\partial\mathcal{H}_1} - \langle \delta u, \gamma_2^*(\boldsymbol{\sigma})\rangle_{\partial\mathcal{G}_2} \qquad \forall u \in \mathcal{H}_A, \qquad \forall \boldsymbol{\sigma} \in \mathcal{G}_{A^*} \tag{1.57}$$

Now let us consider the following abstract problem. Given $f \in \mathscr{U}'$, $g \in \partial\mathscr{H}_1$, and $s \in \partial\mathscr{G}_2$, find $u \in \mathscr{H}_A$ such that

$$\left.\begin{aligned} A^*EAu &= f \\ \gamma_1 u &= g \\ \gamma_2^* EAu &= s \end{aligned}\right\} \tag{1.58}$$

where we have used the notations of (1.32) and (1.35). We shall exploit the fact that this problem can be rewritten in the following canonical form: find

$$(u, \mathbf{v}, \boldsymbol{\sigma}) \in \mathscr{H}_A \times \mathscr{V}' \times \mathscr{G}_{A^*} \tag{1.59}$$

such that

$$\left.\begin{aligned} Au &= \mathbf{v} \qquad \gamma_1 u = g \\ E\mathbf{v} &= \boldsymbol{\sigma} \\ A^*\boldsymbol{\sigma} &= f \qquad \gamma_2^*\boldsymbol{\sigma} = s \end{aligned}\right\} \tag{1.60}$$

We would now like to construct a variational principle (i.e. a potential functional) corresponding to (1.60). Towards this end, we introduce the matrix operator

$$\begin{bmatrix} 0 & -1 & A & 0 & 0 \\ -1 & E & 0 & 0 & 0 \\ A^* & 0 & 0 & 0 & 0 \\ 0 & 0 & 0 & 0 & \pi_1 \\ 0 & 0 & 0 & \pi_2^* & 0 \end{bmatrix} \begin{bmatrix} \boldsymbol{\sigma} \\ \mathbf{v} \\ u \\ \gamma^*\boldsymbol{\sigma} \\ \gamma u \end{bmatrix} = \begin{bmatrix} 0 \\ 0 \\ f \\ g \\ s \end{bmatrix} \tag{1.61}$$

or

$$\mathscr{P}(\boldsymbol{\lambda}) = \mathbf{f} \tag{1.62}$$

where $\mathscr{P}$ is the coefficient matrix of operators in (1.61),

$$\boldsymbol{\lambda}^T = (\boldsymbol{\sigma}, \mathbf{v}, u, \gamma^*\boldsymbol{\sigma}, \gamma u) \in \mathscr{G}_{A^*} \times \mathscr{V}' \times \mathscr{H}_A \times \partial\mathscr{G} \times \partial\mathscr{H}$$
$$\mathbf{f}^T = (0, 0, f, g, s) \in \mathscr{V}' \times \mathscr{V} \times \mathscr{U}' \times \partial\mathscr{H}_1 \times \partial\mathscr{G}_2$$

If

$$\mathscr{W} = \mathscr{V} \times \mathscr{V}' \times \mathscr{U} \times \partial\mathscr{H}' \times \partial\mathscr{G}'$$

and

$$\langle\cdot,\cdot\rangle_{\mathscr{W}} \equiv \langle\cdot,\cdot\rangle_{\mathscr{V}} + \langle\cdot,\cdot\rangle_{\mathscr{V}'} + \langle\cdot,\cdot\rangle_{\mathscr{U}} + \langle\cdot,\cdot\rangle_{\partial\mathscr{H}_1'} + \langle\cdot,\cdot\rangle_{\partial\mathscr{G}_2'}$$

We compute easily the functional,

$$\begin{aligned} L(u, \mathbf{v}, \boldsymbol{\sigma}) &= \int_0^1 \langle \mathscr{P}(s\boldsymbol{\lambda}) - \mathbf{f}, \hat{\boldsymbol{\lambda}} \rangle_{\mathscr{W}}\, \mathrm{d}s \\ &= \tfrac{1}{2}\langle \mathscr{P}(\boldsymbol{\lambda}), \hat{\boldsymbol{\lambda}} \rangle_{\mathscr{W}} - \langle f, \hat{\boldsymbol{\lambda}} \rangle_{\mathscr{W}} \\ &= \tfrac{1}{2}\langle -\mathbf{v} + Au, \boldsymbol{\sigma} \rangle_{\mathscr{V}} + \tfrac{1}{2}\langle -\boldsymbol{\sigma} + E\mathbf{v}, \mathbf{v} \rangle_{\mathscr{V}'} + \tfrac{1}{2}\langle A^*\boldsymbol{\sigma}, u \rangle_{\mathscr{U}} \\ &\quad + \tfrac{1}{2}\langle \pi_1 \gamma u, \delta^*\boldsymbol{\sigma} \rangle_{\partial\mathscr{H}_1'} + \tfrac{1}{2}\langle \pi_2^* \gamma^*\boldsymbol{\sigma}, \delta u \rangle_{\partial\mathscr{G}_2'} \\ &\quad - \langle f, u \rangle_{\mathscr{U}} - \langle g, \delta^*\boldsymbol{\sigma} \rangle_{\partial\mathscr{H}_1'} - \langle s, \delta u \rangle_{\partial\mathscr{G}_2'} \end{aligned}$$

wherein

$$\hat{\boldsymbol{\lambda}}^T = (\boldsymbol{\sigma}, \mathbf{v}, u, \delta^*\boldsymbol{\sigma}, \delta u) \in \mathcal{V} \times \mathcal{V}' \times \mathcal{U} \times \partial\mathcal{H}' \times \partial\mathcal{G}'$$

Applying Green's formula to the term $\frac{1}{2}\langle A^*\boldsymbol{\sigma}, u\rangle_{\mathcal{U}}$ gives

I.
$$L(u, \mathbf{v}, \boldsymbol{\sigma}) = \tfrac{1}{2}\langle E\mathbf{v}, \mathbf{v}\rangle_{\mathcal{V}'} + \langle Au - \mathbf{v}, \boldsymbol{\sigma}\rangle_{\mathcal{V}} - \langle f, u\rangle_{\mathcal{U}} + \langle \gamma_1 u - g, \delta^*\boldsymbol{\sigma}\rangle_{\partial\mathcal{H}_1'} - \langle s, \delta u\rangle_{\partial\mathcal{G}_2'} \tag{1.63}$$

The Euler equations are (1.60). Indeed,

$$\begin{aligned}\delta L(u, \mathbf{v}, \boldsymbol{\sigma}; \bar{u}, \bar{\mathbf{v}}, \bar{\boldsymbol{\sigma}}) &= \langle E\mathbf{v}, \bar{\mathbf{v}}\rangle_{\mathcal{V}'} + \langle \mathrm{Au} - \mathbf{v}, \bar{\boldsymbol{\sigma}}\rangle_{\mathcal{V}} - \langle f, \bar{u}\rangle_{\mathcal{U}} + \langle \gamma_1 u - g, \delta^*\bar{\boldsymbol{\sigma}}\rangle_{\partial\mathcal{H}_1'} \\ &\quad - \langle s, \delta\bar{u}\rangle_{\partial\mathcal{G}_2'} + \langle A\bar{u} - \bar{\mathbf{v}}, \boldsymbol{\sigma}\rangle_{\mathcal{V}} + \langle \gamma_1\bar{u}, \delta^*\boldsymbol{\sigma}\rangle_{\partial\mathcal{H}_1'} \\ &= \langle E\mathbf{v} - \boldsymbol{\sigma}, \bar{\mathbf{v}}\rangle_{\mathcal{V}'} + \langle Au - \mathbf{v}, \bar{\boldsymbol{\sigma}}\rangle_{\mathcal{V}} + \langle A^*\boldsymbol{\sigma} - f, \bar{u}\rangle_{\mathcal{U}} \\ &\quad + \langle \gamma_1 u - g, \delta^*\bar{\boldsymbol{\sigma}}\rangle_{\partial\mathcal{H}_1'} + \langle \gamma_2^*\boldsymbol{\sigma} - s, \delta\bar{u}\rangle_{\partial\mathcal{G}_2'}\end{aligned}$$

where we have applied Green's formula into $\langle A\bar{u}, \boldsymbol{\sigma}\rangle_{\mathcal{V}}$.

Next, we list a number of additonal principles that can be derived directly from the functional L:

II. Constraint:

$$\mathbf{v} = E^{-1}\boldsymbol{\sigma} \tag{1.64}$$

$$\begin{aligned}R_\sigma(u, \boldsymbol{\sigma}) &= L(u, \mathbf{v}(\boldsymbol{\sigma}), \boldsymbol{\sigma}) \\ &= -\tfrac{1}{2}\langle E^{-1}\boldsymbol{\sigma}, \boldsymbol{\sigma}\rangle_{\mathcal{V}} + Au, \boldsymbol{\sigma}\rangle_{\mathcal{V}} - \langle f, u\rangle_{\mathcal{U}} \\ &\quad + \langle \delta^*\boldsymbol{\sigma}, \gamma_1 u - g\rangle_{\partial\mathcal{H}_1} - \langle \delta u, s\rangle_{\partial\mathcal{G}_2}\end{aligned} \tag{1.65}$$

Euler equations:

$$\begin{aligned} Au &= E^{-1}\boldsymbol{\sigma} \qquad & \gamma_1 u &= g \\ A^*\boldsymbol{\sigma} &= f & \gamma_2^*\boldsymbol{\sigma} &= s \end{aligned} \tag{1.66}$$

Constraint:

$$\hat{\boldsymbol{\sigma}} = \mathrm{Ev} \tag{1.67}$$

$$\begin{aligned}R_v(u, \mathbf{v}) &= L(u, \mathbf{v}, \boldsymbol{\sigma}(\mathbf{v})) \\ &= -\tfrac{1}{2}\langle \mathbf{v}, E\mathbf{v}\rangle_{\mathcal{V}} + \langle Au, E\mathbf{v}\rangle_{\mathcal{V}} - \langle f, u\rangle_{\mathcal{U}} \\ &\quad + \langle \delta^*\boldsymbol{\sigma}, \gamma_1 u - g\rangle_{\partial\mathcal{H}_1} - \langle \delta u, s\rangle_{\partial\mathcal{G}_2}\end{aligned} \tag{1.68}$$

Euler equations:

$$\begin{aligned} Au &= \mathbf{v} \qquad & \gamma_1 u &= g \\ A^*E\mathbf{v} &= f & \gamma_2^*E\mathbf{v} &= s \end{aligned} \tag{1.69}$$

III. Constraints:

$$\mathbf{v} = Au \qquad \gamma_1 u = g \qquad \boldsymbol{\sigma} = EAu \tag{1.70}$$

$$I(u) = L(u, \mathbf{v}(u), \boldsymbol{\sigma}(u)) = \tfrac{1}{2}\langle Au, EAu\rangle_{\mathcal{V}} - \langle f, u\rangle_{\mathcal{U}} - \langle \delta u, s\rangle_{\partial\mathcal{G}_2} \tag{1.71}$$

Euler equations:

$$A^*EAu=f \qquad \gamma_2^*EAu=s \tag{1.72}$$

$$\begin{aligned}\delta I(u,\bar{u}) &= \langle A\bar{u}, EAu\rangle_{\mathscr{V}} - \langle f,\bar{u}\rangle_{\mathscr{U}} - \langle \delta\bar{u}, s\rangle_{\partial\mathscr{G}_2}\\ &= \langle A^*EAu,\bar{u}\rangle_{\mathscr{U}} - \langle \delta^*Eau, \gamma_1\bar{u}\rangle_{\partial\mathscr{H}_1} + \langle \delta\bar{u}, \gamma_2^*EAu\rangle_{\partial\mathscr{G}_2}\\ &\quad -\langle f,\bar{u}\rangle_{\mathscr{U}} - \langle \delta\bar{u}, s\rangle_{\partial\mathscr{G}_2}\\ &= \langle A^*EAu-f,\bar{u}\rangle_{\mathscr{U}} + \langle \delta\bar{u}, \gamma_2^*EAu-s\rangle_{\partial\mathscr{G}_2}; \qquad \bar{u}\in\ker\gamma_1\end{aligned}$$

IV. Constraints:

$$\mathbf{v}=E^{-1}\boldsymbol{\sigma} \qquad A^*\boldsymbol{\sigma}=f \qquad \gamma_2^*\boldsymbol{\sigma}=s \tag{1.73}$$

$$\begin{aligned}J(\boldsymbol{\sigma}) &= L(u(\boldsymbol{\sigma}),\mathbf{v}(\boldsymbol{\sigma}),\boldsymbol{\sigma})\\ &= -\tfrac{1}{2}\langle E^{-1}\boldsymbol{\sigma},\boldsymbol{\sigma}\rangle_{\mathscr{V}} - \langle \delta^*\boldsymbol{\sigma}, g\rangle_{\partial\mathscr{H}_1}\end{aligned} \tag{1.74}$$

Euler equations:

$$Au=E^{-1}\boldsymbol{\sigma} \qquad \gamma_1 u=g \tag{1.75}$$

$$\begin{aligned}\delta J(\boldsymbol{\sigma},\bar{\boldsymbol{\sigma}}) &= -\langle E^{-1}\boldsymbol{\sigma},\bar{\boldsymbol{\sigma}}\rangle_{\mathscr{V}} - \langle \delta^*\bar{\boldsymbol{\sigma}}, g\rangle_{\partial\mathscr{H}_1}\\ &\quad +\langle Au,\bar{\boldsymbol{\sigma}}\rangle_{\mathscr{V}} + \langle \delta^*\bar{\boldsymbol{\sigma}}, \gamma_1 u\rangle_{\partial\mathscr{H}_1} - \langle \delta u, \gamma_2^*\bar{\boldsymbol{\sigma}}\rangle_{\partial\mathscr{G}_2} - \langle A^*\bar{\boldsymbol{\sigma}}, u\rangle_{\mathscr{U}}\\ &= \langle Au-E^{-1}\boldsymbol{\sigma},\bar{\boldsymbol{\sigma}}\rangle_{\mathscr{V}} + \langle \delta^*\bar{\boldsymbol{\sigma}}, \gamma_1 u-g\rangle_{\partial\mathscr{H}_1}; \qquad \bar{\boldsymbol{\sigma}}\in\ker A^*\cap\ker\gamma_2^*\end{aligned}$$

V. Constraints:

$$\begin{aligned}Au&=\mathbf{v} \qquad \gamma_1 u=g\\ A^*\boldsymbol{\sigma}&=f \qquad \gamma_2^*\boldsymbol{\sigma}=s\end{aligned} \tag{1.76}$$

$$K(\mathbf{v})=\tfrac{1}{2}\langle \mathbf{v}, E\mathbf{v}\rangle_{\mathscr{V}} - \langle \mathbf{v},\hat{\boldsymbol{\sigma}}\rangle_{\mathscr{V}}, \qquad \hat{\boldsymbol{\sigma}}\in\{\boldsymbol{\sigma}: A^*\boldsymbol{\sigma}=f,\ \gamma_2^*\boldsymbol{\sigma}=s\} \tag{1.77}$$

Euler equation:

$$E\mathbf{v}=\boldsymbol{\sigma} \tag{1.78}$$

VI. Constraints:

$$Au=\mathbf{v} \qquad \gamma_1 u=g \tag{1.79}$$

$$\begin{aligned}M(u,\boldsymbol{\sigma}) &= \langle Au, EAu\rangle_{\mathscr{V}} + \tfrac{1}{2}\langle E^{-1}\boldsymbol{\sigma},\boldsymbol{\sigma}\rangle_{\mathscr{V}} - \langle f,u\rangle_{\mathscr{U}}\\ &\quad -\langle Au,\boldsymbol{\sigma}\rangle_{\mathscr{V}} - \langle \delta u, s\rangle_{\partial\mathscr{G}_2}\end{aligned} \tag{1.80}$$

Euler equations:

$$\begin{aligned}&Au=E^{-1}\boldsymbol{\sigma}\\ &A^*(2EAu-\boldsymbol{\sigma})=f \qquad \gamma_2^*(2EAu-\boldsymbol{\sigma})=s\end{aligned} \tag{1.81}$$

VII. Constraints:

$$A^*\boldsymbol{\sigma}=f \qquad \gamma_2^*\boldsymbol{\sigma}=s \tag{1.82}$$

$$N(\mathbf{v},\boldsymbol{\sigma})=\tfrac{1}{2}\langle \mathbf{v}, E\mathbf{v}\rangle_{\mathscr{V}}-\langle \mathbf{v},\boldsymbol{\sigma}\rangle_{\mathscr{V}}+\langle \delta^*\boldsymbol{\sigma}, g\rangle_{\partial\mathscr{H}_1} \tag{1.83}$$

Euler equations:

$$\mathbf{v}=Au \qquad \gamma_1 u=g \qquad \boldsymbol{\sigma}=E\mathbf{v} \tag{1.84}$$

$$\begin{aligned}\delta N(\mathbf{v},\boldsymbol{\sigma};\bar{\mathbf{v}},\bar{\boldsymbol{\sigma}})&=\langle \bar{\mathbf{v}}, E\mathbf{v}-\boldsymbol{\sigma}\rangle_{\mathscr{V}}+\langle \mathbf{v},\bar{\boldsymbol{\sigma}}\rangle_{\mathscr{V}}+\langle \delta^*\bar{\boldsymbol{\sigma}}, g\rangle_{\partial\mathscr{H}_1}\\ &\quad+\langle A^*\bar{\boldsymbol{\sigma}}, u\rangle_{\mathscr{U}}-\langle Au,\bar{\boldsymbol{\sigma}}\rangle_{\mathscr{V}}+\langle \delta u, \gamma_2^*\bar{\boldsymbol{\sigma}}\rangle_{\partial\mathscr{G}_2}-\langle \delta^*\bar{\boldsymbol{\sigma}}, \gamma_1 u\rangle_{\partial\mathscr{H}_1}\\ &=\langle \bar{\mathbf{v}}, E\mathbf{v}-\boldsymbol{\sigma}\rangle_{\mathscr{V}}+\langle \mathbf{v}-Au,\bar{\boldsymbol{\sigma}}\rangle_{\mathscr{V}}+\langle \delta^*\bar{\boldsymbol{\sigma}}, g-\gamma_1 u\rangle_{\partial\mathscr{H}_1}\\ &\quad\bar{\boldsymbol{\sigma}}\in\ker A^*\cap\ker\gamma_2^*\end{aligned}$$

We summarize these results in Table 1.1.

Table 1.1

	Functional	Definition	Constraints	Euler equations
I	$L(u,\mathbf{v},\boldsymbol{\sigma})$	(1.63)	—	$Au=\mathbf{v};\ \gamma_1 u=g$ $E\mathbf{v}=\boldsymbol{\sigma}$ $A^*\boldsymbol{\sigma}=f;\ \gamma_2^*\boldsymbol{\sigma}=s$
II	$R_\sigma(u,\boldsymbol{\sigma})$ $R_v(u,\mathbf{v})$	(1.65) (1.68)	$v=E^{-1}\boldsymbol{\sigma}$ $\boldsymbol{\sigma}=E\mathbf{v}$	$Au=E^{-1}\boldsymbol{\sigma};\ \gamma_1 u=g$ $A^*\boldsymbol{\sigma}=f;\ \gamma_2^*\boldsymbol{\sigma}=s$ $Au=\mathbf{v},\ \gamma_1 u=g$ $A^*E\mathbf{v}=f;\ \gamma_2^*E\mathbf{v}=s$
III	$I(u)$	(1.71)	$\mathbf{v}=Au;\ \gamma_1 u=g$ $\boldsymbol{\sigma}=EAu$	$A^*EAu=f;\ \gamma_2^*EAu=s$
IV	$J(\boldsymbol{\sigma})$	(1.74)	$\mathbf{v}=E^{-1}\boldsymbol{\sigma}$ $A^*\boldsymbol{\sigma}=f;\ \gamma_2^*\boldsymbol{\sigma}=s$	$Au=E^{-1}\boldsymbol{\sigma};\ \gamma_1 u=g$
V	$K(\mathbf{v})$	(1.77)	$Au=\mathbf{v};\ \gamma_1 u=g$ $A^*\boldsymbol{\sigma}=f;\ \gamma_2^*\boldsymbol{\sigma}=s$	$E\mathbf{v}=\boldsymbol{\sigma}$
VI	$M(u,\boldsymbol{\sigma})$	(1.80)	$Au=\mathbf{v};\ \gamma_1 u=g$	$Au=E^{-1}\boldsymbol{\sigma}$ $A^*(2EAu-\boldsymbol{\sigma})=f$ $\gamma^*(2EAu-\boldsymbol{\sigma})=s$
VII	$N(\mathbf{v},\boldsymbol{\sigma})$	(1.83)	$A^*\boldsymbol{\sigma}=f;\ \gamma_2^*\boldsymbol{\sigma}=s$	$\mathbf{v}=Au;\ \gamma_1 u=g$ $\boldsymbol{\sigma}=E\mathbf{v}$

1.7 APPLICATIONS IN ELASTICITY

We now apply the theory in the previous section to the construction of seven variational principles of linear elastostatics.

The Lamé–Navier equations of linear elasticity are

$$\left.\begin{aligned}(A^*EA\mathbf{u})^j &\equiv -(E^{ijrs}u_{r,s})_{,i} = f^j \quad \text{in} \quad \Omega\\ (\gamma_1\mathbf{u})_i &\equiv u_i = \hat{u}_i \quad \text{on} \quad \partial\Omega_1\\ (A_2^*EA\mathbf{u})^j &\equiv n_i(E^{ijrs}u_{r,s}) = \hat{T}^j \quad \text{on} \quad \partial\Omega_2\end{aligned}\right\} \tag{1.85}$$

which correspond to equations (1.58). The canonical equations (1.60) are the strain-displacement equations, kinematical boundary conditions, constitutive equations, and equilibrium equations:

$$\left.\begin{aligned}&(A\mathbf{u})_{ij} \equiv \tfrac{1}{2}(u_{i,j}+u_{j,i}) = \varepsilon_{ij} \quad \text{in} \quad \Omega; \qquad (\gamma_1\mathbf{u})_i \equiv u_i = \hat{u}_i \quad \text{on} \quad \partial\Omega_1\\ &(E\boldsymbol{\varepsilon})^{ij} \equiv E^{ijrs}\varepsilon_{rs} = \sigma^{ij} \quad \text{in} \quad \Omega\\ &(A^*\boldsymbol{\sigma})^j \equiv -\sigma^{ij}_{,i} = f^j \quad \text{in} \quad \Omega; \qquad (\gamma_2^*\boldsymbol{\sigma})^j \equiv n_i\sigma^{ij} = \hat{T}^j \quad \text{on} \quad \partial\Omega_2\end{aligned}\right\} \tag{1.86}$$

In this case,

$$\left.\begin{aligned}\mathscr{U} &= \mathbf{L}_2(\Omega) = \{\mathbf{u} : u_1 \in L_2(\Omega),\ i = 1, 2, 3\}\\ \mathscr{V} &= \mathbf{L}_2(\Omega) = \{\boldsymbol{\sigma} : \sigma_{ij} \in L_2(\Omega),\ i, j = 1, 2, 3\}\end{aligned}\right\} \tag{1.87}$$

and

$$\left.\begin{aligned}\mathscr{H} &= \mathbf{H}^1(\Omega) = \{\mathbf{u} : u_i \in H^1(\Omega),\ i = 1, 2, 3\}\\ \mathscr{G} &= \mathbf{H}^1(\Omega) = \{\boldsymbol{\sigma} : \sigma_{ij} \in H^1(\Omega),\ i, j = 1, 2, 3\}\end{aligned}\right\} \tag{1.88}$$

Hence, the canonical problem of the elasticity can be expressed as follows: given

$$(\mathbf{f}, \hat{\mathbf{u}}, \hat{\mathbf{T}}) \in \mathscr{U}' \times \partial\mathscr{H}_1 \times \partial\mathscr{G}_2 = (L_2(\Omega))' \times \mathbf{H}^{1/2}(\partial\Omega_1) \times \mathbf{H}^{-1/2}(\partial\Omega_2) \tag{1.89}$$

find

$$(\mathbf{u}, \boldsymbol{\varepsilon}, \boldsymbol{\sigma}) \in \mathscr{H}_A \times \mathscr{V}' \times \mathscr{G}_{A^*} = \mathbf{H}^1(A, \Omega) \times (\mathbf{L}_2(\Omega))' \times \mathbf{H}^1(A^*, \Omega) \tag{1.90}$$

such that equations (1.86) are satisfied. Here

$$\left.\begin{aligned}\mathscr{H}_A &= \mathbf{H}^1(A, \Omega) = \{\mathbf{u} \in \mathbf{H}^1(\Omega) : A\mathbf{u} \in \mathbf{L}_2(\Omega)\\ \mathscr{G}_{A^*} &= \mathbf{H}^1(A^*, \Omega) = \{\boldsymbol{\sigma} \in \mathbf{H}^1(\Omega) : A^*\boldsymbol{\sigma} \in \mathbf{L}_2(\Omega)\}\end{aligned}\right\} \tag{1.91}$$

The Green's formula (1.57) becomes, for this problem,

$$\int_\Omega -\sigma^{ij}_{,i}u_j \, dx = \int_\Omega u_{j,i}\sigma_{ij} \, d\Omega + \int_{\partial\Omega_1} -n_i\sigma^{ij}u_j \, ds - \int_{\partial\Omega_2} u_j n_i \sigma^{ij} \, ds$$

$$\forall\mathbf{u} \in \mathbf{H}^1(A, \Omega), \qquad \forall\boldsymbol{\sigma} \in \mathbf{H}^1(A^*, \Omega) \tag{1.92}$$

from which we can identify the operators $\delta \in \mathscr{L}(\mathbf{H}^1(A, \Omega), \mathscr{G}^{-1/2}(\partial\Omega))$ and

$\delta^* \in \mathscr{L}(\mathbf{H}^1(A^*, \Omega), H^{-1/2}(\partial\Omega))$:

$$\left.\begin{aligned}(\delta \mathbf{u})_j &\equiv u_j && \text{on} \quad \partial\Omega_2\\ (\delta^* \boldsymbol{\sigma})^{ij} &\equiv -n_i\sigma^{ij} && \text{on} \quad \partial\Omega_1\end{aligned}\right\} \tag{1.93}$$

Now we are able to construct the variational theory for linear elastostatics. According to Table 1.1, we obtain:

I. *The Hu–Washizu principle*

$$L(\mathbf{u}, \boldsymbol{\varepsilon}, \boldsymbol{\sigma}) = \int_\Omega \{\tfrac{1}{2}\varepsilon_{ij}E^{ijrs}\varepsilon_{rs} + [\tfrac{1}{2}(u_{i,j}+u_{j,i}) - \varepsilon_{ij}]\sigma^{ij} - f^i u_i\}\,\mathrm{d}x$$
$$- \int_{\partial\Omega_1} n_i\sigma^{ij}(u_j - \hat{u}_j)\,\mathrm{d}s - \int_{\partial\Omega_2} u_j\hat{T}^j\,\mathrm{d}s \tag{1.94}$$

Euler equations:

$$\left.\begin{aligned}\tfrac{1}{2}(u_{i,j}+u_{j,i}) &= \varepsilon_{ij} \quad \text{in} \quad \Omega; && u_i = \hat{u}_i \quad \text{on} \quad \partial\Omega_1\\ E^{ijrs}\varepsilon_{rs} &= \sigma^{ij} \quad \text{in} \quad \Omega && \\ -\sigma^{ij}_{,i} &= f^j \quad \text{in} \quad \Omega; && n_i\sigma^{ij} = \hat{T}^j \quad \text{on} \quad \partial\Omega_2\end{aligned}\right\} \tag{1.95}$$

II. *The Hellinger–Reissner principle*

(i) Constraints:

$$\varepsilon_{rs} = C_{ijrs}\sigma^{ij}\,(C_{ijrs} \equiv (E^{ijrs})^{-1}) \tag{1.96}$$

$$R_\sigma(\mathbf{u}, \boldsymbol{\sigma}) = \int_\Omega [-\tfrac{1}{2}C_{ijrs}\sigma^{ij}\sigma^{rs} + \tfrac{1}{2}(u_{i,j}+u_{j,i})\sigma^{ij} - f^i u_i]\,\mathrm{d}x$$
$$- \int_{\partial\Omega_1} n_i\sigma^{ij}(u_j - \hat{u}_j)\,\mathrm{d}s - \int_{\partial\Omega_2} u_j\hat{T}^j\,\mathrm{d}s \tag{1.97}$$

Euler equations:

$$\left.\begin{aligned}\tfrac{1}{2}(u_{i,j}+u_{j,i}) &= C_{rsij}\sigma^{rs} \quad \text{in} \quad \Omega; && u_i = \hat{u}_i \quad \text{on} \quad \partial\Omega_2\\ -\sigma^{ij}_{,i} &= f^j \quad \text{in} \quad \Omega; && n_i\sigma^{ij} = \hat{T}^j \quad \text{on} \quad \partial\Omega_2\end{aligned}\right\} \tag{1.98}$$

(ii) Constraints:

$$\sigma^{ij} = E^{ijrs}\varepsilon_{rs} \quad \text{in} \quad \Omega \tag{1.99}$$

$$R_\varepsilon(\mathbf{u}, \boldsymbol{\varepsilon}) = \int_\Omega [-\tfrac{1}{2}\varepsilon_{ij}E^{ijrs}\varepsilon_{rs} + \tfrac{1}{2}(u_{i,j}+u_{j,i})E^{ijrs}\varepsilon_{rs} - f^i u_i]\,\mathrm{d}x$$
$$- \int_{\partial\Omega_1} n_i\sigma^{ij}(u_j - \hat{u}_j)\,\mathrm{d}s - \int_{\partial\Omega_2} u_j\hat{T}^j\,\mathrm{d}s \tag{1.100}$$

Euler equations:

$$\left.\begin{aligned} \tfrac{1}{2}(u_{i,j}+u_{j,i}) = \varepsilon_{ij} \quad \text{in} \quad \Omega; \qquad & u_i = \hat{u}_i \quad \text{on} \quad \partial\Omega_1 \\ -(E^{ijrs}\varepsilon_{rs}) = f^j \quad \text{in} \quad \Omega; \quad n_i E^{ijrs}\varepsilon_{rs} & = \hat{T}^j \quad \text{on} \quad \partial\Omega_2 \end{aligned}\right\} \tag{1.101}$$

III. *Potential energy principle*

Constraints:

$$\left.\begin{aligned} &\varepsilon_{ij} = \tfrac{1}{2}(u_{i,j}+u_{j,i}) \quad \text{in} \quad \Omega; \qquad u_i = \hat{u}_i \quad \text{on} \quad \partial\Omega_1 \\ &\sigma^{ij} = \tfrac{1}{2}E^{ijrs}(u_{r,s}+u_{s,r}) \quad \text{in} \quad \Omega \end{aligned}\right\} \tag{1.102}$$

$$I(\mathbf{u}) = \int_{\Omega} (\tfrac{1}{2}u_{i,j}E^{ijrs}u_{r,s} - f^i u_i)\,\mathrm{d}x - \int_{\partial\Omega_2} u_j \hat{T}^j\,\mathrm{d}s \tag{1.103}$$

Euler equations:

$$-(E^{ijrs}u_{r,s})_{,i} = f^j \quad \text{in} \quad \Omega; \qquad n_i E^{ijrs}u_{r,s} = \hat{T}^j \quad \text{on} \quad \partial\Omega_2 \tag{1.104}$$

IV. *Complementary energy principle*

Constraints:

$$\left.\begin{aligned} \varepsilon_{rs} = C_{ijrs}\sigma^{ij} \quad \text{in} \quad \Omega \qquad & \\ -\sigma^{ij}_{,i} = f^j \quad \text{in} \quad \Omega; \quad n_i\sigma^{ij} = \hat{T}^j \quad \text{on} \quad \partial\Omega_2 & \end{aligned}\right\} \tag{1.105}$$

$$J(\boldsymbol{\sigma}) = -\int_{\Omega} \tfrac{1}{2}C_{ijrs}\sigma^{ij}\sigma^{rs}\,\mathrm{d}x + \int_{\partial\Omega_1} n_i\sigma^{ij}\hat{u}_j\,\mathrm{d}s \tag{1.106}$$

Euler equations:

$$\tfrac{1}{2}(u_{r,s}+u_{s,r}) = C_{ijrs}\sigma^{ij} \quad \text{in} \quad \Omega; \qquad u_i = \hat{u}_i \quad \text{on} \quad \partial\Omega_1 \tag{1.107}$$

V. *A constitutive variational principle*

Constraints:

$$\left.\begin{aligned} \tfrac{1}{2}(u_{i,j}+u_{j,i}) = \varepsilon_{ij} \quad \text{in} \quad \Omega; \qquad & u_i = \hat{u}_i \quad \text{on} \quad \partial\Omega_1 \\ -\sigma^{ij}_{,i} = f^j \quad \text{in} \quad \Omega; \qquad & n_i\sigma^{ij} = \hat{T}^j \quad \text{on} \quad \partial\Omega_2 \end{aligned}\right\} \tag{1.108}$$

$$K(\boldsymbol{\varepsilon}) = \int (\tfrac{1}{2}\varepsilon_{ij}E^{ijrs}\varepsilon_{rs} - \varepsilon_{ij}\hat{\sigma}^{ij})\,\mathrm{d}x \tag{1.109}$$

$$\hat{\sigma}^{ij} \in \{\sigma^{ij} : -\sigma^{ij}_{,i} = f^j \text{ in } \Omega,\ n_i\sigma^{ij} = \hat{T}^j \text{ on } \partial\Omega_2\}$$

Euler equations:

$$E^{ijrs}\varepsilon_{rs} = \sigma^{ij} \quad \text{in} \quad \Omega \tag{1.110}$$

VI. *A constitutive-potential energy principle*

Constraints:

$$\tfrac{1}{2}(u_{i,j}+u_{j,i})=\varepsilon_{ij} \quad \text{in} \quad \Omega; \qquad u_i=\hat{u}_i \quad \text{on} \quad \partial\Omega_1 \tag{1.111}$$

$$M(\mathbf{u},\boldsymbol{\sigma})=\int_\Omega [u_{i,j}E^{ijrs}u_{r,s}+\tfrac{1}{2}C_{ijrs}\sigma^{ij}\sigma^{rs} \\ -\tfrac{1}{2}(u_{i,j}+u_{j,i})\sigma^{ij}-f^iu_i]\,\mathrm{d}x-\int_{\partial\Omega} u_i\hat{T}^i\,\mathrm{d}s \tag{1.112}$$

Euler equations:

$$\begin{aligned} &\tfrac{1}{2}(u_{r,s}+u_{s,r})=C_{ijrs}\sigma^{ij} \quad \text{in} \quad \Omega \\ &-\{E^{ijrs}(u_{r,s}+u_{s,r})-\sigma^{ij}\}_{,i}=f^j \quad \text{in} \quad \Omega \\ &n^i\{E^{ijrs}(u_{r,s}+u_{s,r})-\sigma^{ij}\}=\hat{T}^j \quad \text{on} \quad \partial\Omega_2 \end{aligned} \tag{1.113}$$

VII. *A compatibility-constitutive variational principle*

Constraints:

$$-\sigma^{ij}_{,i}=f^j \quad \text{in} \quad \Omega; \qquad n_i\sigma^{ij}=\hat{T}^j \quad \text{on} \quad \partial\Omega_2 \tag{1.114}$$

$$N(\boldsymbol{\varepsilon},\boldsymbol{\sigma})=\int_\Omega (\tfrac{1}{2}\varepsilon_{ij}E^{ijrs}\varepsilon_{rs}-\varepsilon_{ij}\sigma^{ij})\,\mathrm{d}x-\int_{\partial\Omega_1} n_i\sigma^{ij}\hat{u}_j\,\mathrm{d}s \tag{1.115}$$

Euler equations:

$$\left.\begin{aligned} &\varepsilon_{ij}=\tfrac{1}{2}(i_{i,j}+u_{j,i}) \quad \text{in} \quad \Omega; \qquad u_i=\hat{u}_i \quad \text{on} \quad \partial\Omega_1 \\ &\sigma^{ij}=E^{ijrs}\varepsilon_{rs} \quad \text{in} \quad \Omega \end{aligned}\right\} \tag{1.116}$$

1.8 DUAL PRINCIPLES

1.8.1 The dual problem

It was pointed out by Oden and Reddy[15] (see also Reference 14) that a parallel collection of so-called *dual* variational principles can be constructed in one-to-one correspondence with the variational principles described in the previous section. While we shall not elaborate on the detailed features of these dual functionals, we shall outline briefly the essential concepts for the sake of completeness.

To construct the dual principles, we consider a new Hilbert space $\mathscr{S}$ and a linear operator

$$C:\mathscr{S}\to\mathscr{G} \tag{1.117}$$

such that

$$\left.\begin{aligned}\mathcal{R}(C) \supset \mathcal{N}(A^*)\\ \mathcal{N}(C^*) \subset \mathcal{R}(A)\end{aligned}\right\} \tag{1.118}$$

Next, we consider the *dual problem* of finding $\varphi \in \mathcal{S}$ such that

$$\left.\begin{aligned}C^*E^{-1}C\varphi &= \eta\\ \gamma_1^{(C)}\varphi &= p\\ \gamma_2^{(C)*}E^{-1}C\varphi &= q\end{aligned}\right\} \tag{1.119}$$

where

$$\eta \in \mathcal{S}', \qquad p \in \partial\mathcal{G}_1', \qquad q \in \partial\mathcal{S}_2' \tag{1.120}$$

Owing to the similarity of (1.119) and (1.58), we can immediately construct the following dual functionals:

I′.
$$\begin{aligned}\mathcal{L}(\varphi, \sigma, \upsilon) = \tfrac{1}{2}\langle E^{-1}\sigma, \sigma\rangle_{\mathcal{V}'} + \langle \upsilon, C\varphi - \sigma\rangle_{\mathcal{V}'} - \langle \eta, \varphi\rangle_{\mathcal{S}}\\ + \langle \delta^{(C)*}\upsilon, \gamma_1^{(C)}\varphi - p\rangle_{\partial\mathcal{G}_1'} - \langle q, \delta^{(C)}\varphi\rangle_{\partial\mathcal{S}_2}\end{aligned} \tag{1.121}$$

II′.
$$\begin{aligned}\mathcal{R}_\upsilon(\varphi, \upsilon) = -\tfrac{1}{2}\langle \upsilon, E\upsilon\rangle_{\mathcal{V}'} + \langle \upsilon, C\varphi\rangle_{\mathcal{V}'} - \langle \eta, \varphi\rangle_{\mathcal{S}}\\ + \langle \delta^{(C)*}\upsilon, \gamma_1^{(C)}\varphi - p\rangle_{\partial\mathcal{G}_1'} - \langle q, \delta^{(C)}\varphi\rangle_{\partial\mathcal{S}_2}\end{aligned} \tag{1.122}$$

or

$$\begin{aligned}\mathcal{R}_\sigma(\varphi, \sigma) = -\tfrac{1}{2}\langle E^{-1}\sigma, \sigma\rangle_{\mathcal{V}'} + \langle E^{-1}\sigma, C\varphi\rangle_{\mathcal{V}'} - \langle \eta, \varphi\rangle_{\mathcal{S}}\\ + \langle \delta^{(C)*}E^{-1}\sigma, \gamma_1^{(C)}\varphi - p\rangle_{\partial\mathcal{G}_1'} - \langle q, \delta^{(C)}\varphi\rangle_{\partial\mathcal{S}_2}\end{aligned} \tag{1.123}$$

III′.
$$\mathcal{I}(\varphi) = \tfrac{1}{2}\langle E^{-1}C\varphi, C\varphi\rangle_{\mathcal{V}'} - \langle \eta, \varphi\rangle_{\mathcal{S}} - \langle q, \delta^{(C)}\varphi\rangle_{\partial\mathcal{S}_2} \tag{1.124}$$

IV′.
$$\mathcal{T}(\upsilon) = -\tfrac{1}{2}\langle \upsilon, E\upsilon\rangle_{\mathcal{V}'} - \langle \delta^{(C)*}\upsilon, p\rangle_{\partial\mathcal{G}_1'} \tag{1.125}$$

V′.
$$\mathcal{K}(\sigma) = \tfrac{1}{2}\langle E^{-1}\sigma, \sigma\rangle_{\mathcal{V}'} - \langle \upsilon, \sigma\rangle_{\mathcal{V}'} \tag{1.126}$$

VI′.
$$\begin{aligned}\mathcal{M}(\varphi, \upsilon) = \tfrac{1}{2}\langle E^{-1}C\varphi, C\varphi\rangle_{\mathcal{V}'} + \tfrac{1}{2}\langle \upsilon, E\upsilon\rangle_{\mathcal{V}'} - \langle \eta, \varphi\rangle_{\mathcal{S}}\\ - \langle \upsilon, C\varphi\rangle_{\mathcal{V}'} - \langle q, \delta^{(C)}\varphi\rangle_{\partial\mathcal{S}_2}\end{aligned} \tag{1.127}$$

VII′.
$$\mathcal{N}(\sigma, \upsilon) = \tfrac{1}{2}\langle E^{-1}\sigma, \sigma\rangle_{\mathcal{V}'} - \langle \upsilon, \sigma\rangle_{\mathcal{V}'} - \langle \delta^{(C)*}\upsilon, p\rangle_{\partial\mathcal{G}_1'} \tag{1.128}$$

Clearly, a table similar to Table 1.1 can also be constructed directly from (1.121)–(1.128) by using the following correspondences:

$$\left.\begin{matrix}\mathcal{G}' \sim \mathcal{G}, & \mathcal{S} \sim \mathcal{H}, & \varphi \sim \mu, & C \sim A,\\ C^* \sim A^*, & E^{-1} \sim E, & \sigma \sim \upsilon, & \upsilon \sim \sigma,\\ \gamma_1^{(C)} \sim \gamma_2^*, & \gamma_2^{(C)*} \sim \gamma_1, & q \sim g, & p \sim s\end{matrix}\right\} \tag{1.129}$$

1.8.2 Application to elastostatics

In the case of linear elasticity, we set

$$\begin{aligned}
\varphi = \varphi_{ij} &= \text{tensor of stress functions}\\
\sigma \sim \sigma^{ij} &= \text{stress tensor}\\
v \sim \varepsilon_{ij} &= \text{strain tensor}\\
\eta \sim \eta_{ij} &= \text{dislocation tensor}
\end{aligned}$$

Then (1.121)–(1.128) assume the following specific forms:

$$\begin{aligned}
\mathscr{L}(\varphi_{ij}, \sigma^{ij}, \varepsilon_{ij}) = \int_\Omega & [\tfrac{1}{2}\sigma^{ij}C_{ijrs}\sigma^{rs} - \varepsilon_{ij}(\sigma^{ij} - \mathrm{e}^{ims}\mathrm{e}^{jkr}\varphi_{rs,km})\\
& - \eta^{ij}\varphi_{ij}]\,\mathrm{d}x + \langle \delta^*\varepsilon_{ij}, \gamma(\varphi_{ij}) - p\rangle_{\partial\Omega_1}\\
& - \langle q, \delta(\varphi_{ij})\rangle_{\partial\Omega_2}
\end{aligned} \tag{1.130}$$

where, for compactness in notation, we have denoted

$$\begin{aligned}
\langle \delta^*\varepsilon_{ij}, \gamma(\varphi_{ij}) - p\rangle_{\partial\Omega_1} = \int_{\partial\Omega_1} & [(f^{imjnptqu}n_n\varphi_{tu} - p_1^{imjpq})\varepsilon_{im,jpq}\\
& - (F^{imjnptqu}n_p\varphi_{tu,m} - p_2^{ijnq})\varepsilon_{ij,nq}]\,\mathrm{d}s\\
\langle q, \delta(\varphi_{ij})\rangle_{\partial\Omega_2} = \int_{\partial\Omega_2} & (q_1^{ijq}\varphi_{ij,q} - q_2^{ij}\varphi_{ij})\,\mathrm{d}s
\end{aligned}$$

wherein

$$F^{imjnptqu} = \mathrm{e}^{imr}\mathrm{e}^{jns}C_{rslk}\mathrm{e}^{lpt}\mathrm{e}^{kqu} \qquad i, j, k, l, m, n, p, q, r, s, t, u = 1, 2, 3$$

$$\left.\begin{aligned}
F^{imjnptqu}n_n\varphi_{tu} &= p_1^{imjpq}\\
F^{imjnptqu}n_p\varphi_{tu,m} &= p_2^{ijnq}
\end{aligned}\right\} \quad \text{on} \quad \partial\Omega_1$$

$$\left.\begin{aligned}
F^{imjnptqu}n_p\varphi_{tu,mn} &= q_1^{ijq}\\
F^{imjnptqu}n_q\varphi_{tu,mnp} &= q_2^{ij}
\end{aligned}\right\} \quad \text{on} \quad \partial\Omega_2$$

$$\begin{aligned}
\mathscr{R}_\varepsilon(\varphi_{ij}, \varepsilon_{ij}) = \int_\Omega & (-\tfrac{1}{2}\varepsilon_{ij}E^{ijrs}\varepsilon_{rs} + \varepsilon_{ij}\mathrm{e}^{ims}\mathrm{e}^{jkr}\varphi_{rs,km} - \eta^{ij}\varphi_{ij})\,\mathrm{d}x\\
& + \langle \delta^*\varepsilon_{ij}, \gamma(\varphi_{ij}) - p\rangle_{\partial\Omega_1} - \langle q, \delta(\varphi_{ij})\rangle_{\partial\Omega_2}
\end{aligned} \tag{1.131}$$

$$\begin{aligned}
\mathscr{R}_\sigma(\varphi_{ij}, \sigma^{ij}) = \int_\Omega & [-\tfrac{1}{2}\sigma^{ij}C_{ijrs}\sigma^{rs} + C_{ijrs}\sigma^{ij}\mathrm{e}^{rmp}\mathrm{e}^{skq}\varphi_{pq,mk}\\
& - \eta^{ij}\varphi_{ij}]\,\mathrm{d}x + \langle \delta^*(C_{ijrs}\sigma^{rs}), \gamma(\varphi_{ij}) - p\rangle_{\partial\Omega_1}\\
& - \langle q, \delta(\varphi_{ij})\rangle_{\partial\Omega_2}
\end{aligned} \tag{1.132}$$

$$\mathscr{I}(\varphi_{ij}) = \int_\Omega [\tfrac{1}{2}C_{ijrs}e^{imp}e^{jqu}e^{rta}e^{sbk}\varphi_{mn,pq}y_{ak,tb} - \eta^{ij}\varphi_{ij}]\,dx$$
$$-\langle q, \delta(\varphi_{ij})\rangle_{\partial\Omega_2} \tag{1.133}$$

$$\mathscr{J}(\varepsilon_{ij}) = \int_\Omega (-\tfrac{1}{2}\varepsilon^{ijrs}\varepsilon_{ij}\varepsilon_{rs})\,dx - \langle \delta^*\varepsilon_{ij}, p\rangle_{\partial\Omega_1} \tag{1.134}$$

$$\mathscr{K}(\sigma^{ij}) = \int_\Omega (\tfrac{1}{2}C_{ijrs}\sigma^{rs}\sigma^{ij} - \sigma^{ij}\varepsilon_{ij})\,dx \tag{1.135}$$

$$\mathscr{M}(\varphi_{ij}, \varepsilon_{ij}) = \int_\Omega [\tfrac{1}{2}C_{ijrs}e^{ipm}e^{jqu}e^{rta}e^{sbk}\varphi_{mn,pq}y_{ak,tb}$$
$$+E^{ijrs}\varepsilon_{ij}\varepsilon_{rs} - 2\varepsilon_{ij}e^{irs}e^{jmk}\varphi_{sk,rm} - \eta^{ij}\varphi_{ij}]\,dx$$
$$-\langle q, \delta(\varphi_{ij})\rangle_{\partial\Omega_2} \tag{1.136}$$

$$\mathscr{N}(\sigma^{ij}, \varepsilon_{ij}) = \int_\Omega -\sigma^{ij}\varepsilon_{ij}\,dx + \int_{\partial\Omega_2} \varepsilon_{ij}\varphi^{ij}\,ds - \langle \delta^*\varepsilon_{ij}, p\rangle_{\partial\Omega_1} \tag{1.137}$$

ACKNOWLEDGEMENT

The support of this work by Grant 74–2660 from the U.S. Air Force Office of Scientific Research is gratefully acknowledged. I also wish to make a special note of thanks to Gonzalo Alduncin, who read the entire manuscript and made many helpful suggestions.

REFERENCES

1. Lions, J. L., and Magenes, E., *Non-Homogeneous Boundary-Value Problems and Applications,* Vols. I, II, III. Translated from a revision of the 1968 French edition by P. Kenneth, Springer-Verlag, New York, 1972.
2. Lions, J. L., *Quelques Méthodes de Résolution des Problèmes aux Limites Non-Linéaires,* Dunod, Paris, 1969.
3. Brezis, H., 'Equations et inéquations non-linéaires dans les espaces vectoriels en dualité', *Ann. Inst. Fourier,* **18,** 115–175 (1968).
4. Browder, F. G., 'Nonlinear elliptic boundary-value problems', *Bull. Amer. Math. Soc.,* **69,** 6, 862–874 (1963).
5. Duvaut, G., and Lions, J. L., *Les Inéquations en Mécanique et en Physique,* Dunod, Paris, 1972.
6. Lions, J. L., and Stampacchia G., 'Inequations variationelles noncoercives', *C. R. Acad. Sci. Paris,* **261,** 25–27 (1965).
7. Vainberg, M. M., *Variational Methods for the Study of Nonlinear Operators,* (translated from the 1956 Russian monograph), Holden-Day, San Francisco, 1964.
8. Nashed, M. Z., 'Differentiability and related properties of non-linear operators: some aspects of the role of differentials in nonlinear functional analysis', in *Non-linear Functional Analysis and Applications,* edited by L. B. Rall, Academic Press, New York, pp. 103–309.

9. Noble, B., 'Complementary variational principles for boundary-value problems I, Basic principles', *Report 473*, Mathematics research Center, University of Wisconsin, Madison, 1966.
10. Noble, B., and Sewell, M. J., 'On dual extremum principles in applied mathematics', *J. Inst. Math. Applics.*, **9,** 123–193 (1972).
11. Arthurs, A. M., *Complementary Variational Principles*, Oxford University Press, London, 1970.
12. Tonti, E., 'On the mathematical structure of a large class of physical theories', *Acad. Naz. dai Lincei*, III, **LII,** 48–56 (1972).
13. Tonti, E., 'A mathematical model for physical theories', *Acad. Naz. dei Lincei*, VIII, **LII,** 175–181, 350–356 (1972).
14. Oden, J. T., and Reddy, J. N., 'On dual complementary variational principles in mathematical physics', *Int'l. J. Engr'g. Sci.*, **12,** 1–29 (1974).
15. Oden, J. T., and Reddy, J. N., *Variational Methods in Theoretical Mechanics*, Springer-Verlag, Heidelberg, 1976.
16. Dunford, N., and Schwartz, J. T., *Linear Operators*, Part I: *General Theory*, Interscience, New York, 1958.
17. Nečas, J., *Les Méthodes Directes in Théorie des Équations Elliptiques*, Masson, Paris, 1967.
18. Adams, R. A., *Sobolev Spaces*, Academic Press, New York, 1975.
19. Oden, J. T., and Reddy, J. N., *An Introduction to the Mathematical Theory of Finite Elements*, Wiley–Interscience, New York, 1976.
20. Aubin, J. P., *Approximation of Elliptic Boundary-Value Problems*, Wiley–Interscience, New York, 1972.
21. Taylor, A. E., *Introduction to Functional Analysis*, Wiley, New York, 1958.
22. Oden, J. T., *Applied Functional Analysis*, Prentice-Hall, Englewood Cliffs, N.J., 1977.
23. Babuška, I., 'Errors bounds for the finite element method', *Numer. Math.*, **16,** 322–333 (1971).

Chapter 2

Approximation of Three-Dimensional Models by Two-Dimensional Models in Plate Theory

P. G. Ciarlet and P. Destuynder

2.1 INTRODUCTION

Usually, a plate theory is derived from a three-dimensional model by making *a priori* assumptions on the form of some of the unknowns (the components of the displacement vector and the stresses), so as to take into account as best possible the fact that a plate is a body with 'small' thickness. Various developments along these lines are found in the books of Duvaut and Lions,[1] Fraeijs de Veubeke,[2] Landau and Lifchitz,[3] Timoshenko and Woinowsky-Krieger,[4] Washizu,[5] Wempner.[6]

In this paper, we develop what seems to be a new approach (to the authors' best knowledge), in which *no* a priori *assumption is needed.* The basic ideas are the following. After posing the plate problem as a standard three-dimensional problem of (linear) elasticity, we transform the problem into a problem posed over a fixed domain, i.e. one which does not depend on the thickness of the plate. We next use the same technique as in Lions[7] in order to construct a formal series of 'approximations', and when we go back to the volume occupied by the actual plate, we are able to show that the two first terms of the series are precisely the solutions found in standard two-dimensional linear plate models. In addition, standard *a priori* assumptions concerning the variations of the stresses and of the components of the displacement vector across the thickness of the plate are automatically derived from our approach to the problem.

Complete proofs of these results are found in Ciarlet and Destuynder.[8,9] As we shall show in Destuynder,[10] this approach is also particularly well-adapted to the *error analysis,* i.e. to the comparison between the three-dimensional and the two-dimensional solutions.

For further developments, notably regarding the computation of the subsequent terms in the formal series (with the corresponding error analysis), the case of a simply supported plate (or, more generally, different boundary conditions—we restrict ourselves here to the case of a clamped plate), and other linear plate models (such as models with 'rotations'), see Destuynder.[12]

Let us also point out that the domain of applicability of the present method is not restricted to plates and linear models. With methods based on the same ideas, we are able to obtain (cf. Ciarlet[11]) nonlinear plate models, such as the von Kármán model, and also (cf. Destuynder[12]) shell models.

We shall use the following notation: The usual partial derivatives will be denoted

$$\partial_i v = \frac{\partial v}{\partial x_i}, \qquad \partial_{\alpha\beta} v = \frac{\partial^2 v}{\partial x_\alpha \partial x_\beta}, \quad \text{etc.} \ldots$$

If θ is an open subset of $\mathbb{R}^n$, we define the norms

$$|v|_\theta = \left(\int_\theta |v|^2 \, dx\right)^{1/2}$$

$$\|v\|_{1,\theta} = \left(|v|_\theta^2 + \sum_{i=1}^n |\partial_i v|_\theta^2\right)^{1/2}.$$

For brevity we shall usually write integrals in the form

$$\int_\theta v \, dx = \int_\theta v$$

i.e. we shall omit the symbol dx.

As a rule, Greek indices, $\alpha, \beta, \mu, \ldots$, take their values in the set $\{1, 2\}$, while Latin indices, $i, j, p, \ldots$, takes their values in the set $\{1, 2, 3\}$.

Finally, we shall use the repeated index convention combined with the above rule. Thus, for instance,

$$\gamma_{pp} = \gamma_{11} + \gamma_{22} + \gamma_{33}$$

$$\partial_\alpha \sigma_{\alpha\beta} = \partial_1 \sigma_{1\beta} + \partial_2 \sigma_{2\beta} \ldots, \qquad \beta = 1, 2$$

2.2 THE THREE-DIMENSIONAL PROBLEM

Let ω be a bounded open subset of the plane, with boundary γ (later on we shall make specific, but mild, assumptions on the regularity of the boundary γ). Given a constant $\varepsilon > 0$, we let

$$\Omega^\varepsilon = \omega \times]-\varepsilon, \varepsilon[$$

$$\Gamma_0^\varepsilon = \gamma \times [-\varepsilon, \varepsilon]$$

$$\Gamma_+^\varepsilon = \omega \times \{\varepsilon\}$$

$$\Gamma_-^\varepsilon = \omega \times \{-\varepsilon\}$$

so that the boundary of the set Ω^ε is partitioned into the sets Γ_0^ε, Γ_+^ε, and Γ_-^ε.

We are concerned with the problem of finding the displacement vector $u=(u_i)$ and the stress tensor $\sigma=(\sigma_{ij})$ of a three-dimensional body which occupies the set $\bar{\Omega}^\varepsilon$ in the absence of exterior forces. Because the thickness 2ε is thought of as being 'small' compared to the dimensions of the set ω, we shall henceforth refer to the body (identified with the set) $\bar{\Omega}^\varepsilon$ as a plate. The plate is assumed to be *clamped*, in the sense that the displacement u is imposed to vanish on the lateral surface Γ_0^ε.

The applied forces are accounted for by a volumic density $f=(f_i)$ defined over the set Ω^ε and by a superficial density $g=(g_i)$ defined over the upper and lower faces Γ_+^ε and Γ_-^ε.

As a mathematical model for this problem, we shall use the linear, three-dimensional displacement-stress model known as the Hellinger–Reissner variational principle (see in particular Oden and Reddy[13] and Oden[14]).

To describe this model we introduce the Hilbert spaces

$$\mathbf{V}^\varepsilon=\{v=(v_i)\in(H^1(\Omega^\varepsilon))^3;\ v=0 \text{ on } \Gamma_0^\varepsilon\} \tag{2.1}$$

$$\mathbf{\Sigma}^\varepsilon=\{\tau=(\tau_{ij})\in(L^2(\Omega^\varepsilon))^9;\ \tau_{ij}=\tau_{ji}\} \tag{2.2}$$

The space $\mathbf{V}^\varepsilon$ is equipped with the product norm $\left(\sum_i \|v_i\|_{1,\Omega^\varepsilon}^2\right)^{1/2}$, while the space $\mathbf{\Sigma}^\varepsilon$ is equipped with the norm

$$|\tau|_{\Omega^\varepsilon}=\left(\sum_{i,j}|\tau_{ij}|_{\Omega^\varepsilon}^2\right)^{1/2} \tag{2.3}$$

Also, we shall have to consider mappings which map the space of 3×3 tensors with two indices into itself. More specifically, with each tensor $X=(X_{ij})$, we associate the tensor $Y=AX=(Y_{ij})$ defined by

$$Y_{ij}=(AX)_{ij}=\left(\frac{1+\nu}{E}\right)X_{ij}-\frac{\nu}{E}X_{pp}\delta_{ij} \tag{2.4}$$

where the constants E and ν are respectively the Young's modulus and the Poisson coefficient of the constitutive material of the plate. We note that the inequalities

$$E>0, \qquad 0<\nu<\tfrac{1}{2} \tag{2.5}$$

hold (for physical reasons).

In the following easily proved lemma, we record some useful properties of the mapping A defined by relations (2.4).

Lemma 2.2.1 If inequalities (2.5) *are satisfied, relations* (2.4) *can be inverted through*

$$X_{ij}=(A^{-1}Y)_{ij}=\frac{E}{1+\nu}Y_{ij}+\frac{E\nu}{(1+\nu)(1-2\nu)}Y_{pp}\delta_{ij} \tag{2.6}$$

Besides,

$$\exists c_0 = c_0(\nu, E) > 0, \qquad \forall \tau \in \mathbf{\Sigma}^\varepsilon, \qquad \int_{\Omega^\varepsilon} (A\tau)_{ij}\tau_{ij} \geq c_0 |\tau|^2_{\Omega^\varepsilon} \tag{2.7}$$

$$\exists c_1 = c_1(\nu, E) > 0, \qquad \forall \tau \in \mathbf{\Sigma}^\varepsilon, \qquad \int_{\Omega^\varepsilon} (A^{-1}\tau)_{ij}\tau_{ij} \geq c_1 |\tau|^2_{\Omega^\varepsilon} \tag{2.8}$$

$$\exists c_2 = c_2(\nu, E) > 0, \qquad \forall \tau \in \mathbf{\Sigma}^\varepsilon, \qquad |\tau|_{\Omega^\varepsilon} \geq c_2 |A^{-1}\tau|_{\Omega^\varepsilon} \tag{2.9}$$ ■

The Hellinger–Reissner variational principle consists in finding an element $(\sigma, u) \in \mathbf{\Sigma}^\varepsilon \times \mathbf{V}^\varepsilon$ which satisfies

$$\forall \tau \in \mathbf{\Sigma}^\varepsilon, \qquad \int_{\Omega^\varepsilon} (A\sigma)_{ij}\tau_{ij} - \int_{\Omega^\varepsilon} \tau_{ij}\gamma_{ij}(u) = 0 \tag{2.10}$$

$$\forall v \in \mathbf{V}^\varepsilon, \qquad \int_{\Omega^\varepsilon} \sigma_{ij}\gamma_{ij}(v) = \int_{\Omega^\varepsilon} f_i v_i + \int_{\Gamma_+^\varepsilon \cup \Gamma_-^\varepsilon} g_i v_i \tag{2.11}$$

where the operator A is that defined in (2.4) and where we let

$$\gamma_{ij}(v) = \tfrac{1}{2}(\partial_i v_j + \partial_j v_i) \tag{2.12}$$

This problem can be written in the more condensed form: Find $(\sigma, u) \in \mathbf{\Sigma}^\varepsilon \times \mathbf{V}^\varepsilon$ such that

$$\forall \tau \in \mathbf{\Sigma}^\varepsilon, \qquad \mathscr{A}^\varepsilon(\sigma, \tau) + \mathscr{B}^\varepsilon(\tau, u) = 0 \tag{2.13}$$

$$\forall v \in \mathbf{V}^\varepsilon, \qquad \mathscr{B}^\varepsilon(\sigma, v) = \mathscr{F}^\varepsilon(v) \tag{2.14}$$

where the bilinear forms $\mathscr{A}^\varepsilon$, $\mathscr{B}^\varepsilon$, and the linear form $\mathscr{F}^\varepsilon$ are respectively given by

$$\mathscr{A}^\varepsilon : (\sigma, \tau) \in \mathbf{\Sigma}^\varepsilon \times \mathbf{\Sigma}^\varepsilon \to \mathscr{A}^\varepsilon(\sigma, \tau) = \int_{\Omega^\varepsilon} (A\sigma)_{ij}\tau_{ij} \tag{2.15}$$

$$\mathscr{B}^\varepsilon : (\tau, v) \in \mathbf{\Sigma}^\varepsilon \times \mathbf{V}^\varepsilon \to \mathscr{B}^\varepsilon(\tau, v) = -\int_{\Omega^\varepsilon} \tau_{ij}\gamma_{ij}(v) \tag{2.16}$$

$$\mathscr{F}^\varepsilon : v \in \mathbf{V}^\varepsilon \to \mathscr{F}^\varepsilon(v) = -\int_{\Omega^\varepsilon} f_i v_i - \int_{\Gamma_+^\varepsilon \cup \Gamma_-^\varepsilon} g_i v_i \tag{2.17}$$

In order that the linear form $\mathscr{F}^\varepsilon$ be defined and continuous over the space $\mathbf{V}^\varepsilon$, we shall henceforth assume that the densities of the applied forces satisfy:

$$f_i \in L^2(\Omega^\varepsilon), \qquad g_i \in L^2(\Gamma_+^\varepsilon \cup \Gamma_-^\varepsilon) \tag{2.18}$$

To prove existence and uniqueness of the solution of problem (2.13)–(2.14), we shall call upon the following result, due to Brezzi[15] and also to Babuška[16] in a weaker form.

Theorem 2.2.1 *Let V and Σ be two Hilbert spaces with norms $\|\cdot\|_V$ and $\|\cdot\|_\Sigma$ respectively, and let $\mathscr{A}:\Sigma\times\Sigma\to\mathbf{R}$, $\mathscr{B}:\Sigma\times V\to\mathbf{R}$, $\mathscr{F}: V\to\mathbf{R}$, and $\mathscr{G}:\Sigma\to\mathbf{R}$ be continuous bilinear and linear forms, respectively. We define the space*

$$\Xi=\{\tau\in\Sigma; \forall v\in V, \mathscr{B}(\tau, v)=0\} \tag{2.19}$$

and we assume that the bilinear form $\mathscr{A}$ is Ξ-elliptic in the sense that

$$\exists a>0, \qquad \forall\tau\in\Xi, \qquad \mathscr{A}(\tau,\tau)\geq a\|\tau\|_\Sigma^2 \tag{2.20}$$

Finally we assume that

$$\exists b>0, \qquad \forall v\in V, \qquad \sup_{\tau\in\Sigma}\frac{|\mathscr{B}(\tau, v)|}{\|\tau\|_\Sigma}\geq b\|v\|_V \tag{2.21}$$

Then there exists a unique element $(\sigma, u)\in\Sigma\times V$ such that

$$\forall\tau\in\Sigma, \qquad \mathscr{A}(\sigma,\tau)+\mathscr{B}(\tau, u)=\mathscr{G}(\tau) \tag{2.22}$$

$$\forall v\in V, \qquad \mathscr{B}(\sigma, v)=\mathscr{F}(v) \tag{2.23}$$ ■

Theorem 2.2.2 *Problem* (2.10)–(2.11) *has one and only one solution.*

Proof: The continuity of the forms $\mathscr{A}^\varepsilon$, $\mathscr{B}^\varepsilon$ and $\mathscr{F}^\varepsilon$ is clear. Since, by Lemma 2.2.1,

$$\exists c_0>0, \qquad \forall\tau\in\mathbf{\Sigma}^\varepsilon, \qquad \mathscr{A}^\varepsilon(\tau,\tau)=\int_{\Omega^\varepsilon}(A\tau)_{ij}\tau_{ij}\geq c_0|\tau|_{\Omega^\varepsilon}^2$$

property (2.20) holds *a fortiori*. It therefore sufficies to verify assumption (2.21). For a given $v\in V$, let

$$\tau=-A^{-1}\gamma(v)$$

where the operator A^{-1} is defined as in (2.6) and where

$$\gamma(v)=(\gamma_{ij}(v))$$

With this choice for the tensor τ, we find that

$$\sup_{\tau\in\Sigma}\frac{|\mathscr{B}^\varepsilon(\tau, v)|}{\|\tau\|_\Sigma}\geq\frac{\int_{\Omega^\varepsilon}(A^{-1}\gamma(v))_{ij}\gamma_{ij}(v)}{|A^{-1}\gamma(v)|_{\Omega^\varepsilon}}\geq c_1c_2|\gamma(v)|_{\Omega^\varepsilon}$$

by Lemma 2.2.1. Using *Korn's inequality* (for a proof, see Duvaut and Lions[1] (Chapter 3, §3.3) or Fichera[17] (Sec. 12)), one can show that the mapping

$$v\in\mathbf{V}^\varepsilon\to|\gamma(v)|_{\Omega^\varepsilon}$$

is a norm over the space $\mathbf{V}^\varepsilon$ equivalent to the product norm $\left(\sum_i \|v_i\|^2_{1,\Omega}\right)^{1/2}$, as long as the functions in $\mathbf{V}^\varepsilon$ vanish on a set of strictly positive superficial measure, as is the case here. Therefore, the proof is complete. ■

Remark 2.2.1 It is easily realized that solving problem (2.10)–(2.11) amounts to solving the following boundary-value problem:

$$\gamma_{ij}(u) = (A\sigma)_{ij} = \left(\frac{1+\nu}{E}\right)\sigma_{ij} - \frac{\nu}{E}\sigma_{pp}\delta_{ij} \quad \text{in} \quad \Omega^\varepsilon \tag{2.24}$$

$$-\partial_j \sigma_{ij} = f_i \quad \text{in} \quad \Omega^\varepsilon \tag{2.25}$$

$$u = 0 \quad \text{on} \quad \Gamma_0^\varepsilon \tag{2.26}$$

$$\sigma_{i3} = \begin{cases} g_i & \text{on} \quad \Gamma_+^\varepsilon \\ -g_i & \text{on} \quad \Gamma_-^\varepsilon \end{cases} \tag{2.27}$$

■

2.3 FORMAL CONSTRUCTION OF APPROXIMATIONS

Our first aim is to define a problem equivalent to problem (2.10)–(2.11), but now posed on a domain which does *not* depend on ε. Accordingly, we shall successively define appropriate changes of *variables* and changes of *functions.* We let

$$\begin{aligned} \bar{\Omega} &= \bar{\omega} \times [-1, 1] \\ \Gamma_0 &= \gamma \times [-1, 1] \\ \Gamma_+ &= \omega \times \{1\} \\ \Gamma_- &= \omega \times \{-1\} \end{aligned} \tag{2.28}$$

and, for each $\varepsilon > 0$, we define the mapping

$$F^\varepsilon : X = (x_1, x_2, x_3) \in \bar{\Omega} \rightarrow F^\varepsilon(X) = (x_1, x_2, \varepsilon x_3) \in \bar{\Omega}^\varepsilon \tag{2.29}$$

As regards the changes of functions, we shall associate with the four fields

$$v, \sigma, f, g : \Omega^\varepsilon \rightarrow \mathbf{R}$$

the four fields

$$v^\varepsilon, \sigma^\varepsilon, f^\varepsilon, g^\varepsilon : \Omega \rightarrow \mathbf{R}$$

defined as follows:

$$v_\alpha^\varepsilon = v_\alpha \cdot F^\varepsilon, \qquad v_3^\varepsilon = \varepsilon v_3 \cdot F^\varepsilon \tag{2.30}$$

$$\sigma_{\alpha\beta}^\varepsilon = \sigma_{\alpha\beta} \cdot F^\varepsilon, \qquad \sigma_{\alpha 3}^\varepsilon = \sigma_{3\alpha}^\varepsilon = \varepsilon^{-1}\sigma_{\alpha 3} \cdot F^\varepsilon, \qquad \sigma_{33}^\varepsilon = \varepsilon^{-2}\sigma_{33} \cdot F^\varepsilon \tag{2.31}$$

$$f_i^\varepsilon = f_i \cdot F^\varepsilon, \qquad g_i^\varepsilon = g_i \cdot F^\varepsilon \tag{2.32}$$

Remark 2.3.1 Although we shall not elaborate on this aspect, the above changes of functions may be completely justified as follows: The applied force field and the displacement field are in duality with respect to the inner-product $\int_{\Omega^\varepsilon} f_i v_i + \int_{\Gamma_+^\varepsilon \cup \Gamma_-^\varepsilon} g_i v_i$, while the stress tensor field and the strain tensor field are in duality with respect to the inner-product $\int_{\Omega^\varepsilon} \tau_{ij}\gamma_{ij}(v)$. Then the change of variables $X^\varepsilon \in \bar{\Omega}^\varepsilon \to X = (F^\varepsilon)^{-1}(X^\varepsilon) \in \bar{\Omega}$ may be interpreted as a change of parametrization of the *same* manifold $\bar{\Omega}^\varepsilon$. As such, it induces well-determined transformations of the four fields under considerations, which are precisely those given in (2.30), (2.31), and (2.32). ■

We define the spaces

$$\mathbf{V} = \{v = (v_i) \in (H^1(\Omega))^3;\ v = 0 \text{ on } \Gamma_0\} \tag{2.33}$$

$$\mathbf{\Sigma} = \{\tau = (\tau_{ij}) \in (L^2(\Omega))^9;\ \tau_{ij} = \tau_{ji}\} \tag{2.34}$$

We also define the mapping A^t which transforms any 2×2 tensor $X = (X_{\alpha\beta})$ into a tensor $Y = A^t X = (Y_{\alpha\beta})$ through the expressions:

$$Y_{\alpha\beta} = (A^t X)_{\alpha\beta} = \left(\frac{1+\nu}{E}\right) X_{\alpha\beta} - \frac{\nu}{E} X_{\mu\mu}\delta_{\alpha\beta} \tag{2.35}$$

where the constants E and ν have the same meaning as before (cf. (2.5)). Then we have:

Theorem 2.3.1 Let $(\sigma^\varepsilon, u^\varepsilon) \in \mathbf{\Sigma} \times \mathbf{V}$ *be the element constructed from the solution* $(\sigma, u) \in \mathbf{\Sigma}^\varepsilon \times \mathbf{V}^\varepsilon$ *of problem* (2.10)–(2.11) *through formulae* (2.30) –(2.31). *Then this element satisfies*

$$\forall \tau \in \mathbf{\Sigma}, \qquad \mathscr{A}_0(\sigma^\varepsilon, \tau) + \varepsilon^2 \mathscr{A}_2(\sigma^\varepsilon, \tau) + \varepsilon^4 \mathscr{A}_4(\sigma^\varepsilon, \tau) + \mathscr{B}(\tau, u^\varepsilon) = 0 \tag{2.36}$$

$$\forall v \in \mathbf{V}, \qquad \mathscr{B}(\sigma^\varepsilon, v) = \varepsilon^{-2}\mathscr{F}_{-2}(v) + \varepsilon^{-1}\mathscr{F}_{-1}(v) + \mathscr{F}_0(v) \tag{2.37}$$

where, for arbitrary elements $\sigma, \tau \in \mathbf{\Sigma}$ and $v \in \mathbf{V}$,

$$\mathscr{A}_0(\sigma, \tau) = \int_\Omega (A^t\sigma)_{\alpha\beta}\tau_{\alpha\beta} \tag{2.38}$$

$$\mathscr{A}_2(\sigma, \tau) = \int_\Omega \left\{2\left(\frac{1+\nu}{E}\right)\sigma_{\alpha 3}\tau_{\alpha 3} - \frac{\nu}{E}(\sigma_{33}\tau_{\mu\mu} + \sigma_{\mu\mu}\tau_{33})\right\} \tag{2.39}$$

$$\mathscr{A}_4(\sigma, \tau) = \int_\Omega \frac{1}{E}\sigma_{33}\tau_{33} \tag{2.40}$$

$$\mathscr{B}(\tau, v) = -\int_\Omega \tau_{ij}\gamma_{ij}(v) \tag{2.41}$$

$$\mathscr{F}_{-2}(v) = -\int_{\Gamma_+ \cup \Gamma_-} g_3^\varepsilon v_3 \tag{2.42}$$

$$\mathscr{F}_{-1}(v) = -\int_\Omega f_3^\varepsilon v_3 - \int_{\Gamma_+ \cup \Gamma_-} g_\alpha^\varepsilon v_\alpha \tag{2.43}$$

$$\mathscr{F}_0(v) = -\int_\Omega f_\alpha^\varepsilon v_\alpha \tag{2.44}$$ ■

Notice that the bilinear forms $\mathscr{A}_0$, $\mathscr{A}_2$, $\mathscr{A}_4$, and $\mathscr{B}$ are now *independent* of ε, since the functions in the spaces $\mathbf{\Sigma}$ and $\mathbf{V}$ are defined on a fixed domain and since the coefficients which occur in these bilinear forms are also independent of ε. Following a technique of Lions,[7] this observation, together with the fact that ε may be thought of as a 'small' parameter, suggests that we construct a formal power series expansion of the solution $(\sigma^\varepsilon, u^\varepsilon)$ in the form

$$\sigma^\varepsilon = \varepsilon^{-2}\sigma^{-2} + \varepsilon^{-1}\sigma^{-1} + \sigma^0 + \varepsilon\sigma^1 + \cdots \tag{2.45}$$

$$u^\varepsilon = \varepsilon^{-2}u^{-2} + \varepsilon^{-1}u^{-1} + u^0 + \varepsilon u^1 + \cdots \tag{2.46}$$

where the elements $(\sigma^p, u^p) \in \mathbf{\Sigma} \times \mathbf{V}$, $p \geqslant -2$, are obtained by equating to zero the coefficients of ε^p, $p \geqslant -2$, in the resulting series. In this fashion, we find that the successive elements (σ^p, u^p) should satisfy the following equations, for arbitrary elements $\tau \in \mathbf{\Sigma}$, $v \in \mathbf{V}$:

$$\left.\begin{aligned} \mathscr{A}_0(\sigma^{-2}, \tau) + \mathscr{B}(\tau, u^{-2}) &= 0 \\ \mathscr{B}(\sigma^{-2}, v) &= \mathscr{F}_{-2}(v) \end{aligned}\right\} \tag{2.47}$$

$$\left.\begin{aligned} \mathscr{A}_0(\sigma^{-1}, \tau) + \mathscr{B}(\tau, u^{-1}) &= 0 \\ \mathscr{B}(\sigma^{-1}, v) &= \mathscr{F}_{-1}(v) \end{aligned}\right\} \tag{2.48}$$

$$\left.\begin{aligned}\mathscr{A}_0(\sigma^0, \tau)+\mathscr{B}(\tau, u^0)&=-\mathscr{A}_2(\sigma^{-2}, \tau)\\ \mathscr{B}(\sigma^0, v)&=\mathscr{F}_0(v)\end{aligned}\right\} \tag{2.49}$$

$$\left.\begin{aligned}\mathscr{A}_0(\sigma^1, \tau)+\mathscr{B}(\tau, u^1)&=-\mathscr{A}_2(\sigma^{-1}, \tau)\\ \mathscr{B}(\sigma^1, v)&=0\end{aligned}\right\} \tag{2.50}$$

$$\vdots$$

$$\left.\begin{aligned}\mathscr{A}_0(\sigma^p, \tau)+\mathscr{B}(\tau, u^p)&=-\mathscr{A}_2(\sigma^{p-2}, \tau)-\mathscr{A}_4(\sigma^{p-2}, \tau)\\ \mathscr{B}(\sigma^p, v)&=0, \qquad p\geq 2\end{aligned}\right\} \tag{2.51}$$

$$\vdots$$

Several comments are in order about this formal expansion:

(i) The reason we start out with a factor of ε^{-2} in (2.45)–(2.46) is clear from (2.37).

(ii) At this stage, nothing guarantees that the successive equations (2.47)–(2.51) have solutions, of course. In particular, Theorem 2.2.1 does *not* apply to a problem of the form: Find $(\sigma, u)\in\mathbf{\Sigma}\times\mathbf{V}$ such that

$$\left.\begin{aligned}\forall\tau\in\mathbf{\Sigma} \qquad \mathscr{A}_0(\sigma, \tau)+\mathscr{B}(\tau, u)&=\mathscr{G}(\tau)\\ \forall v\in\mathbf{V}, \qquad \mathscr{B}(\sigma, v)&=\mathscr{F}(v)\end{aligned}\right\} \tag{2.52}$$

which is precisely the type of problem found at each stage of the computation of the elements (σ^p, u^p), for appropriate linear forms $\mathscr{G}$ and $\mathscr{F}$. In fact, such a problem does not have a solution in general, except for special forms $\mathscr{G}$ and $\mathscr{F}$, such as those which correspond to the terms (σ^{-2}, u^{-2}) and (σ^{-1}, u^{-1}) (cf. the next two sections). Notice, however, that we can show that if there is a solution to a problem of the form (2.52), then it is necessarily unique.

(iii) Relations (2.47)–(2.51) are equivalent to two *independent* sets of relations, which define successively the terms with even and odd index, respectively.

2.4 COMPUTATION OF THE TERM (σ^{-2}, u^{-2})

We define the space

$$V=\{v\in H^1(\Omega);\, v=0 \text{ on } \Gamma_0\} \tag{2.53}$$

In view of the specific form of relations (2.38), (2.41), (2.42), and (2.47), the element $(\sigma^{-2}, u^{-2}) \in \mathbf{\Sigma} \times \mathbf{V}$ must satisfy the following set of equations:

$$\forall \tau_{\alpha\beta} \in L^2(\Omega), \qquad \tau_{12} = \tau_{21},$$

$$\int_\Omega (A^t\sigma^{-2})_{\alpha\beta}\tau_{\alpha\beta} - \int_\Omega \tau_{\alpha\beta}\partial_\alpha u_\beta^{-2} = 0 \tag{2.54}$$

$$\forall \tau_{\alpha 3} \in L^2(\Omega), \qquad \int_\Omega \tau_{\alpha 3}(\partial_\alpha u_3^{-2} + \partial_3 u_\alpha^{-2}) = 0 \tag{2.55}$$

$$\forall \tau_{33} \in L^2(\Omega), \qquad \int_\Omega \tau_{33}\partial_3 u_3^{-2} = 0 \tag{2.56}$$

$$\forall v_\beta \in V, \qquad \int_\Omega \{\sigma_{\alpha\beta}^{-2}\partial_\alpha v_\beta + \sigma_{3\beta}^{-2}\partial_3 v_\beta\} = 0 \tag{2.57}$$

$$\forall v_3 \in V, \qquad \int_\Omega \{\sigma_{\alpha 3}^{-2}\partial_\alpha v_3 + \sigma_{33}^{-2}\partial_3 v_3\} = \int_{\Gamma_+\cup\Gamma_-} g_3 v_3 \tag{2.58}$$

We let

$$\left.\begin{aligned} g_3^+(x_1, x_2) &= g_3(x_1, x_2, \varepsilon) = g_3^\varepsilon(x_1, x_2, 1) \\ g_3^-(x_1, x_2) &= g_3(x_1, x_2, -\varepsilon) = g_3^\varepsilon(x_1, x_2, -1) \end{aligned}\right\} \tag{2.59}$$

and, whenever necessary, we shall identify the functions g_3^+ and g_3^- with functions defined over the set ω.

Remarkably, we can explicitly construct the (unique) solution of problem (2.54)–(2.58) (it is henceforth assumed that the boundary γ is sufficiently smooth, so that the solution of the variational problem (2.60) below is in the space $H^3(\omega)$):

Theorem 2.4.1 Problem (2.54)–(2.58) has one and only one solution $(\sigma^{-2}, u^{-2}) \in \mathbf{\Sigma} \times \mathbf{V}$, which is obtained as follows:

The function u_3^{-2}, which is independent of the variable x_3, is identified with the solution $u_3^{-2} \in H_0^2(\omega)$ of the variational problem:

$$\forall v \in H_0^2(\omega), \qquad \frac{2E}{3(1-\nu^2)}\int_\omega \Delta u_3^{-2}\Delta v = \int_\omega (g_3^+ + g_3^-)v \tag{2.60}$$

The other unknowns are then given by:

$$u_\alpha^{-2} = -x_3\partial_\alpha u_3^{-2} \tag{2.61}$$

$$\sigma_{\alpha\beta}^{-2} = -\frac{Ex_3}{(1-\nu^2)}\{(1-\nu)\partial_{\alpha\beta}u_3^{-2} + \nu\Delta u_3^{-2}\delta_{\alpha\beta}\} \tag{2.62}$$

$$\sigma_{3\beta}^{-2} = -\frac{E(1-x_3^2)}{2(1-\nu^2)} \partial_\beta \Delta u_3^{-2} \tag{2.63}$$

$$\sigma_{33}^{-2} = \frac{x_3(3-x_3^2)}{4}(g_3^+ + g_3^-) + \frac{g_3^+ - g_3^-}{2} \tag{2.64}$$ ■

2.5 COMPUTATION OF THE TERM (σ^{-1}, u^{-1})

In view of the specific form of relations (2.38), (2.41), (2.43), and (2.48) the element $(\sigma^{-1}, u^{-1}) \in \mathbf{\Sigma} \times \mathbf{V}$ must satisfy the following set of equations:

$\forall \tau_{\alpha\beta} \in L^2(\Omega), \qquad \tau_{12} = \tau_{21},$

$$\int_\Omega (\mathrm{A}^t \sigma^{-1})_{\alpha\beta} \tau_{\alpha\beta} - \int_\Omega \tau_{\alpha\beta} \partial_\alpha u_\beta^{-1} = 0 \tag{2.65}$$

$$\forall \tau_{\alpha 3} \in L^2(\Omega), \qquad \int_\Omega \tau_{\alpha 3}(\partial_\alpha u_3^{-2} + \partial_3 u_\alpha^{-2}) = 0 \tag{2.66}$$

$$\forall \tau_{33} \in L^2(\Omega), \qquad \int_\Omega \tau_{33} \partial_3 u_3^{-1} = 0 \tag{2.67}$$

$$\forall v_\beta \in V, \qquad \int_\Omega \{\sigma_{\alpha\beta}^{-1} \partial_\alpha v_\beta + \sigma_{3\beta}^{-1} \partial_3 v_\beta\} = \int_{\Gamma_+ \cup \Gamma_-} g_\alpha^\varepsilon v_\alpha \tag{2.68}$$

$$\forall v_3 \in V, \qquad \int_\Omega \{\sigma_{\alpha 3}^{-1} \partial_\alpha v_3 + \sigma_{33}^{-1} \partial_3 v_3\} = \int_\Omega f_3^\varepsilon v_3 \tag{2.69}$$

We introduce the functions $g_\alpha^+, g_\alpha^- : \omega \to \mathbf{R}$ defined by

$$g_\alpha^+(x_1, x_2) = g_\alpha(x_1, x_2, \varepsilon) = g_\alpha^\varepsilon(x_1, x_2, 1) \tag{2.70}$$

$$g_\alpha^-(x_1, x_2) = g_\alpha(x_1, x_2, -\varepsilon) = g_\alpha^\varepsilon(x_1, x_2, -1) \tag{2.71}$$

Again, we can explicitly construct the (unique) solution of problem (2.65)–(2.69).

Theorem 2.5.1 Assume that the functions g_α^+ *and* g_α^-*, as defined in* (2.70)–(2.71), *are in the space* $H^1(\omega)$. *Then problem* (2.65)–(2.69) *has one and only one solution* $(\sigma^{-1}, u^{-1}) \in \mathbf{\Sigma} \times \mathbf{V}$, *which is obtained as follows:*

The function u_3^{-1}*, which is independent of the variable* x_3*, is identified with the solution* $u_3^{-1} \in H_0^2(\omega)$ *of the variational problem:*

$$\forall v \in H_0^2(\omega), \qquad \frac{2E}{3(1-\nu^2)} \int_\omega \Delta u_3^{-1} \Delta v = \int_\omega \left(\int_{-1}^{1} f_3 \, \mathrm{d}x_3\right) v - \int_\omega (g_\alpha^+ - g_\alpha^-) \partial_\alpha v \tag{2.72}$$

Next, let $(n_{\alpha\beta}^{-1}, (u_{\alpha}^{0})^{-1})$ *be the unique solution of the mixed variational problem* $(n^{-1} = (n_{\alpha\beta}^{-1}))$:

$$\forall \tau_{\alpha\beta}^{0} \in L^2(\omega), \qquad \int_{\omega} (A^{t} n^{-1})_{\alpha\beta} \tau_{\alpha\beta}^{0} - 2 \int_{\omega} \tau_{\alpha\beta}^{0} \partial_{\alpha} (u_{\beta}^{0})^{-1} = 0 \tag{2.73}$$

$$\forall v_{\alpha}^{0} \in H_0^1(\omega), \qquad \int_{\omega} n_{\alpha\beta}^{-1} \partial_{\alpha} v_{\beta}^{0} = \int_{\omega} (g_{\alpha}^{+} + g_{\alpha}^{-}) v_{\alpha}^{0} \tag{2.74}$$

The other unknowns are then given by

$$u_{\alpha}^{-1} = (u_{\alpha}^{0})^{-1} - x_3 \partial_{\alpha} u_3^{-1} \tag{2.75}$$

$$\sigma_{\alpha\beta}^{-1} = \frac{n_{\alpha\beta}^{-1}}{2} - \frac{E x_3}{(1-\nu^2)} \{(1-\nu) \partial_{\alpha\beta} u_3^{-1} + \nu \Delta u_3^{-1} \, \delta_{\alpha\beta}\} \tag{2.76}$$

$$\sigma_{3\beta}^{-1} = \frac{g_{\beta}^{+} - g_{\beta}^{-}}{2} + \frac{g_{\beta}^{+} + g_{\beta}^{-}}{2} x_3 - \frac{E(1-x_3^2)}{2(1-\nu^2)} \partial_{\beta} \Delta u_3 \tag{2.77}$$

$$\sigma_{33}^{-1} = \frac{x_3(1-x_3^2)}{4} \int_{-1}^{1} f_3^{\varepsilon} \, dx_3 + \left(\frac{(x_3+1)}{2} \int_{-1}^{1} f_3^{\varepsilon} \, dx_3 - \int_{-1}^{x_3} f_3^{\varepsilon} \, dx_3 \right) + \frac{(1-x_3^2)}{4} \partial_{\alpha} (g_{\alpha}^{+} + g_{\alpha}^{-}) + \frac{x_3(1-x_3^2)}{4} \partial_{\alpha} (g_{\alpha}^{+} - g_{\alpha}^{-}) \tag{2.78}$$ ■

2.6 CONCLUSIONS

After returning to the set occupied by the actual plate, it is an easy matter to write down the equations, analogous to those found in Theorems 2.4.1 and 2.5.1, but now posed over the set Ω^{ε} (this simply amounts to multiplying some of the terms by appropriate powers of ε). It is then realized that, without any *a priori* assumption (whether of a geometrical nature or of a mechanical nature), we have *simultaneously* found (i) the standard biharmonic model for a clamped plate (cf. (2.60) or (2.72)), *and* (ii) specific expressions (cf. (2.61)–(2.64) or (2.75)–(2.78)) for the other unknowns, which are precisely the *assumed* expressions found in various places in the literature (see, for example, Duvaut and Lions[1] (pp. 199, 202), Timoshenko and Woinowsky-Krieger[4] (p. 81), and Green and Zerna[18] (eqns. (7.7.3) and (7.7.4)).

ACKNOWLEDGEMENTS

Part of this work was completed while the first author was visiting the Texas Institute for Computational Mechanics, University of Texas at Austin. In this respect, the first author expresses his particular gratitude to J. Tinsley

Oden who had invited him. The financial support of the U.S. National Science Foundation (Grant ENG 75-07846) is also gratefully acknowledged.

REFERENCES

1. Duvaut, G., Lions, J. L., *Les Inéquations en Mécanique et en Physique*, Dunod, Paris, 1972.
2. Fraeijs de Veubeke, B., *Cours d'Elasticité* (to appear).
3. Landau, L., and Lifchitz, E., *Théorie de l'Elasticité*, Mir, Moscow, 1967.
4. Timoshenko, S., and Woinowsky-Krieger, S., *Theory of Plates and Shells*, McGraw-Hill, New York, 1959.
5. Washizu, K., *Variational Methods in Elasticity and Plasticity*, Pergamon Press, Oxford, 1968.
6. Wempner, G., *Mechanics of Solids*, McGraw-Hill, New York, 1973.
7. Lions, J. L., *Perturbations Singulières dans les Problèmes aux Limites et en Contrôle Optimal*, Lecture Notes in Mathematics, Vol. 323, Springer-Verlag, Berlin, 1973.
8. Ciarlet, P. G., and Destuynder, P., 'A justification of the two-dimensional linear plate model—Part I: Derivation of the two-dimensional model from the three-dimensional model', *TICOM Report 77-9*, The Texas Institute for Computational Mechanics, The University of Texas at Austin, 1977.
9. Ciarlet, P. G., and Destuynder, P., A justification of the two-dimensional linear plate model (to appear).
10. Destuynder, P., A comparison between the two-dimensional linear plate model and the three-dimensional elasticity model (to appear).
11. Ciarlet, P. G., A justification of the von Kármán plate model (to appear). model (to appear).
12. Destuynder, P., Doctoral dissertation (to appear).
13. Oden, J. T., and Reddy, C. T., *Variational Methods in Theoretical Mechanics*, Springer-Verlag, Heidelberg, 1976.
14. Oden, J. T., 'The classical variational principles of mechanics', Chapter 1 of this book.
15. Brezzi, F., 'On the existence, uniqueness and approximation of saddle-point problems arising from Lagrangian multipliers', *Rev. Française Automat. Informat. Recherche Opérationnelle, Sér. Rouge Anal. Numér.*, **R-2**, 129–151 (1974).
16. Babuška, I., 'Error-bounds for finite element method', *Numer. Math.*, **16,** 322–333 (1971).
17. Fichera, G., 'Existence theorems in elasticity—Boundary value problems of elasticity with unilateral constraints', in *Encyclopedia of Physics* (S. Flügge, Chief Ed.), Vol. VI a/2, *Mechanics of Solids* II (C. Truesdell, Ed.), pp. 347–424, Springer-Verlag, Berlin, 1972.
18. Green, A. E., and Zerna, W., *Theoretical Elasticity*, 2nd edn, Oxford University Press, London, 1968.

Chapter 3

The Global Constant Strain Condition and the Patch Test

A. Samuelsson

3.1 INTRODUCTION

Structural engineers have since Rayleigh's, Ritz's, and Galerkin's time used variational principles to find approximate solutions to problems in linear elasticity. During the same time engineers have also used more intuitive approaches supported by engineering judgement. Regardless of the method used, final confidence in the result has been gained first after a series of numerical experiments.

Finite element methods based on the principle of minimum potential energy and Ritz–Galerkin procedure are theoretically simple and easy to analyse with regard to existence and uniqueness of the approximate solution and convergence of a series of approximate solutions towards the exact solution.[1–4] Typical of the methods of this type is that the approximate solution is obtained as the minimum of the potential energy over a space V_h of functions that is a subspace of the space V of functions in which the exact solution can be found, $V_h \subset V$.

Alternatively, finite element methods can be based on the principle of minimum complementary energy and Ritz–Galerkin method. This type of method has been thoroughly studied by Fraeijs de Veubeke and his co-workers and also been successfully implicated.[5,6] In this case too the minimum is sought over a space V_h that is a subspace of the space V of functions in which the exact solution can be found.

Many of the intuitively invented finite element methods can, however, be obtained neither as a minimum of the potential energy nor as a minimum of the complementary energy. Let us call such methods *non-conventional.* Typical for these methods is that $V_h \not\subset V$. Well-known examples of elements of such methods are the rectangle for 2D-elasticity proposed by Turner *et al.*[7] and the triangle for plates in bending proposed by Zienkiewicz.[8] In Turner's and Zienkiewicz's methods displacements are unknowns, in other non-conventional methods displacements and moments are primary unknowns, thus being classified as mixed methods.

Non-conventional methods have been deduced in many different ways.

One way is to make the deduction in exactly the same way as used for conventional methods but with trial functions from a space $V_h \not\subset V$.[9] Another way is to use numerical instead of exact integration which violates the inclusion $V_h \subset V$.[10]

A third way is to use a variational principle of the saddle-point type such as the Hellinger–Reissner principle,[11] to some extent also used by engineers in connection with solution methods other than the finite element method. This principle has been applied to finite element methods in two slightly different ways by Herrmann[12] and by Pian.[13] Herrmann arrives at methods that are called mixed methods. Thus, the plate bending element by Hellan mentioned above has been derived by Herrmann from Hellinger–Reissner's principle. Pian's methods are called hybrid methods since they are derived from saddle-point principles but appear with only one type of variable—displacements or stresses—in the final algebraic equations. The method by Turner *et al.* mentioned above can be derived as a hybrid method.[14]

The non-conventional methods are not as easily analysed with respect to existence, uniqueness, and convergence as the conventional ones. An intuitive but correct understanding of the condition for convergence of these methods was, however, given by Irons.[15] He advised a test for convergence which he called 'the patch test'. The first strict proof of convergence of a non-conventional finite element method was—to my knowledge—given by Johnson.[16] He analysed a mixed method proposed by Bäcklund.[17] Shortly afterwards other non-conventional methods were analysed by Strang and Fix,[18] Ciarlet,[19] and Brezzi.[20]

3.2 THE KIRCHHOFF PLATE PROBLEM

The Kirchhoff plate problem gives rise to a fourth-order differential equation. Conventional finite elements are then either of the high order polynomial type or of the super-element type. Simple elements are all non-conventional, which is why this problem is used for demonstration.

The variables in the Kirchhoff plate problem are the curvatures $\kappa(x, y) = \{\kappa_x, \kappa_y, \kappa_{xy}\}$, the moments $M(x, y) = \{M_x, M_y, M_{xy}\}$, the deflection $w(x, y)$ and the transverse load intensity $W(x, y)$ here all referred to a Cartesian coordinate system. The sets of these variables are related according to the diagram.

$$\begin{array}{ccc} \{M\} & \xrightarrow{\nabla_2^T} & \{W\} \\ {\scriptstyle S^d}\uparrow & & \\ \{\kappa\} & \xleftarrow{\nabla_2} & \{w\} \end{array} \tag{3.1}$$

where

$$\nabla_2 = -\begin{bmatrix} \partial^2/\partial x^2 \\ \partial^2/\partial y^2 \\ 2\partial^2/\partial x\,\partial y \end{bmatrix}, \qquad S^d = D\begin{bmatrix} 1 & \nu & 0 \\ \nu & 1 & 0 \\ 0 & 0 & (1-\nu)/2 \end{bmatrix} \tag{3.2}$$

and D is the plate modulus.

Elimination of M and κ in (3.1) gives the Kirchhoff plate equation:

$$\Delta\Delta w = W/D \tag{3.3}$$

For simplicity, we assume that the boundary $\partial\Omega$ of the region Ω is polygonal and that

$$w = \partial w/\partial n = 0 \quad \text{on} \quad \partial\Omega \tag{3.4}$$

The variational formulation (virtual work equation) of (3.3), (3.4) is: Find that $x \in V$ for which

$$\int_\Omega M^T(w)\nabla_2 w'\,dx\,dy = \int_\Omega Ww'\,dx\,dy \quad \text{for all} \quad w' \in V \tag{3.5}$$

where V is the space $H_0^2(\Omega)$ of functions w, whose derivatives are square integrable up to the second order and that satisfy (3.4). The left-hand side of (3.5) can with use of (3.1) be written:

$$a(w, w') = D\int_\Omega \{\Delta w \Delta w' - (1-\nu)(w_{xx}w'_{yy} + w_{yy}w'_{xx} - 2w_{xy}w'_{xy})\}\,dx\,dy \tag{3.6}$$

where indices denote differentiation.

By use of Green's formula, (3.6) can alternatively be written

$$a(w, w') = D\left\{\int_\Omega \Delta\Delta w \cdot w'\,dx\,dy - \oint_{\partial\Omega} (\Delta w)_n w'\,ds + \oint_{\partial\Omega} \Delta w \cdot w'_n\,ds + (1-\nu)\oint_{\partial\Omega} w_{nt}w'_t - w_{tt}w'_n)\,ds\right\} \tag{3.7}$$

where n and t are normal and tangential directions at the boundary.

Finally, (3.5) can be written

$$a(w, w') = (W, w') \quad \text{for all} \quad w' \in V \tag{3.8}$$

3.3 THE GLOBAL CONSTANT STRAIN CONDITION

An approximate solution to the problem (3.5) can be obtained by restricting the space of functions to an inclusion $V_h \subset V = H_0^2(\Omega)$ where V_h is a finite-dimensional space. The approximate problem is then: Find a function $w_h \in V_h$ for which

$$a(w_h, w'_h) = (W, w'_h) \quad \text{for all} \quad w'_h \in V_h \tag{3.9}$$

Now also

$$a(w, w'_h) = (W, w'_h) \quad \text{for all} \quad w'_h \in V_h \tag{3.10}$$

so

$$a(w - w_h, w'_h) = 0 \quad \text{for all} \quad w'_h \in V_h \tag{3.11}$$

From this orthogonality in energy between the error $w - w_h$ and the space V_h it follows that the energy error is limited according to

$$a(w - w_h, w - w_h) \leqslant a(w - w'_h, w - w'_h) \quad \text{for all} \quad w'_h \in V_h \tag{3.12}$$

A finite element approximation of this problem consists of three steps:

(1) A triangulation T (all or some triangles may be paired to quadrilaterals).
(2) A polynomial approximation over each element $E: w_{h|E} \in P_E \subset P_k$, where P_k is the space of all polynomials of degree $\leqslant k$. Since for the plate problem $V = H_0^2(\Omega)$ a minimum requirement is that P_E contains all second order polynomials of degree $\leqslant k$, that is $P_2 \subset P_E$. This minimum inclusion corresponds to constant curvatures, see (3.1). The curvatures can be regarded as generalized strains so the requirement can be called *the local constant strain condition*, see Bazeley *et al.*[21]
(3) A choice of basis for the element polynomial expressed in parameters at the boundary of the element so that continuity of the deflection w and its normal derivatives w_n can be achieved by equalization of the parameters at the interelement boundaries. When w and w_n is continuous, that is $w \subset C^1(\bar{\Omega})$, the element is said to be *conforming*.

In short, for a conventional finite element method for the plate problem the minimum conditions are:

$$\left.\begin{array}{l} P_2 \subset P_E \quad \text{for all} \quad E \\ V_h \subset C^1(\bar{\Omega}) \end{array}\right\} \Rightarrow V_h \subset H^2(\Omega) \tag{3.13}$$

It is quite natural that (3.13) is satisfied if P_2 is reproducible in the whole region Ω. This condition can be called the *global constant strain condition.* This has been expressed by Fraeijs de Veubeke[22] by saying that *strong conformity* is replaced by *weak conformity*. The method is then 'variational' also along the interelement boundaries—local continuity is replaced by average continuity.

For a non-conforming method satisfying the global constant strain condition (3.10) is not satisfied because $V_h \not\subset V$. Even if the left-hand side of (3.10) is not integrable in ordinary way, it can be formally calculated as a sum of integrals $a(w, w_h')$ over each element. Let this sum be denoted by $a_h(w, w_h')$. Then the global constant strain condition says that

$$a_h(w, w_h') = (W, w_h') = 0 \quad \text{for all} \quad w \in P_2, \qquad w_h' \in V_h \tag{3.14}$$

where

$$a_h(w, w_h') = \sum_{E \in T} a(w, w_h')|_E \tag{3.15}$$

According to (3.7) after partial integration

$$a(w, w_h')\,|_E = D\Big\{\int_\Omega \Delta\Delta w \cdot w_h' \, dx\, dy + \oint_{\partial E} \{\partial^3 w/\partial n^3 + \partial^3 w/\partial t^3$$

$$+ (1-\nu)\, \partial^3 w/\partial n\, \partial t^2\} w_h' \, ds - \sum_{\text{sides}} [(1-\nu)(\partial^2 w/\partial n\, \partial t) w_h']_{\text{end 1}}^{\text{end 2}}$$

$$+ \oint_{\partial E} (-\Delta w + (1-\nu)\, \partial^2 w/\partial t^2)(w_h')_n \, ds\Big\}$$

With $w \in P_2$ the two first terms are identically zero. It remains

$$a(w, w_h')\,|_E = \sum_{\text{sides}} [(M_{nt}(w) w_h')_1 - (M_{nt}(w) w_h')_2] + \oint_{\partial E} M_n(w)(w_h')_n \, ds \tag{3.16}$$

where M_{nt} and M_n are the twisting and bending moments along the element boundaries.

The global constant strain condition is then satisfied if for all $w \in P_2$, $w_h' \in V_h$

$$\begin{aligned} &\sum_{\text{elements}} \Big\{\sum_{\text{sides}} [(M_{nt}(w) w_h')_1 - (M_{nt}(w) w_h')_2]\Big\} = 0, \\ &\sum_{\text{elements}} \Big\{\oint_{\partial E} (w_h')_n \, ds\Big\} = 0. \end{aligned} \tag{3.17}$$

We observe that, since $M_{nt}(w)$ is continuous, the first equation (3.17) is satisfied if w_h' is continuous at the vertices.

3.4 THE PATCH TEST

A direct numerical test of the global constant strain condition

$$a_h(w, w_h') = 0 \quad \text{for all} \quad w \in P_2, \qquad w_h' \in V_h \tag{3.18}$$

has been proposed by Irons *et al.* and has been called the *patch test.* The name is motivated by that the test is performed on representative 'patches' of the region.

Two alternative procedures have been proposed. In one procedure the finite element set of equations corresponding to $a_h(w_h, w_h') = 0$ is solved with prescribed 'unknowns' at the patch boundary calculated from a given second order polynomial P_2 defined over the patch. The test is passed if the unknowns at the inner nodes coincide (up to round-off errors) with those obtained directly from P_2.

In the other procedure all unknowns are given values corresponding to P_2. The test is then passed if the finite element set of equations is satisfied (up to round-off errors).

Both procedures are easily programmed and are of special value in cases when (3.17) is difficult to apply. Three examples for which application of (3.17) presents no difficulties will be demonstrated here.

3.5 EXAMPLES

3.5.1 Example 1

The first example is the constant curvature triangle, often called the Morley triangle.[23] A mixed version of this element was derived in 1967 independently by Hellan[9] and Herrmann.[12] It was shown by Allman[24] that a solution obtained from this element represented a minimum of complementary energy. Bäcklund proved[17] that the stiffness matrix for the Morley triangle could be derived from the Hellan–Herrmann triangle so the two methods were essentially the same. Finally, Johnson[25] gave a comprehensive mathematical analysis of the mixed version.

The deflection $w(x, y)$, Figure 3.1(a), can be expressed in the vertex deflections w^1, w^2, w^3 and the midside slopes m_b^1, m_b^2, m_b^3 measured from a plane through w^1, w^2, w^3 thus excluding slopes caused by rigid-body rotations:

$$w = [1-(x+y)]w^1 + yw^2 + xw^3 + x(1-x)m_b^1 + (1/\sqrt{2})(x+y)[1-(x+y)]m_b^2 + y(1-y)m_b^3 \quad (3.19)$$

The normal slope at the side $x = 0$ is

$$w_x = w_n = -w^1 + w^3 + m_b^1 + (1/\sqrt{2})(1-2y)m_b^2 \quad (3.20)$$

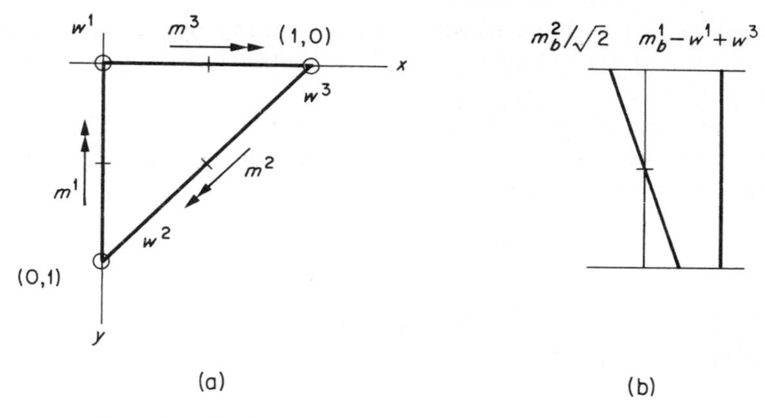

Figure 3.1 (a) The constant curvature triangle. (b) w_n at $x=0$

At the midside the normal slope is thus $m^1 = -w^1 + w^3 + m_b^1$. The two first terms come from rigid-body rotations. The three midside rotations are then

$$m_e = \begin{bmatrix} m^1 \\ m^2 \\ m^3 \end{bmatrix} = \begin{bmatrix} -1 & 0 & 1 \\ \sqrt{2} & -1/\sqrt{2} & -1/\sqrt{2} \\ -1 & 1 & 0 \end{bmatrix} \begin{bmatrix} w^1 \\ w^2 \\ w^3 \end{bmatrix} + \begin{bmatrix} m_b^1 \\ m_b^2 \\ m_b^3 \end{bmatrix} = Aw_e + m_b \tag{3.21}$$

The curvatures are obtained from (3.19) as

$$\begin{bmatrix} \kappa_x \\ \kappa_y \\ \kappa_{xy} \end{bmatrix} = \begin{bmatrix} -w_{xx} \\ -w_{yy} \\ -2w_{xy} \end{bmatrix} = \begin{bmatrix} 2 & \sqrt{2} & 0 \\ 0 & \sqrt{2} & 2 \\ 0 & 2\sqrt{2} & 0 \end{bmatrix} \begin{bmatrix} m_b^1 \\ m_b^2 \\ m_b^3 \end{bmatrix} = B_b m_b \tag{3.22}$$

The stiffness matrix for the element can now be obtained from the diagram:

$$\begin{array}{ccccc} M & \xrightarrow{\int B^T M \, dx \, dy} & & & \begin{bmatrix} M_e \\ W_e \end{bmatrix} \\ S^d \uparrow & & & & \\ \kappa & \xleftarrow{B_b} & m_b & \xleftarrow{[I,-A]} & \begin{bmatrix} m_e \\ w_e \end{bmatrix} \\ & & \xleftarrow{\quad B \quad} & & \end{array} \tag{3.23}$$

where $M_e = \{M^1, M^2, M^3\}$ are moment loads on the interelement lines and $W_e = \{W^1, W^2, W^3\}$ concentrated loads on the nodes. The structure matrix is finally obtained after identification of node deflections w_e and midside nodes m_e.

According to the global constant strain condition the method converges towards the correct solution if (3.17) is satisfied. In this case the first condition of (3.17) is satisfied directly.

Regarding condition (3.18) the constant part of $(w_h')_n$ is conforming so contributions to (3.18) cancel. It remains integrals for each triangle side of the type, see Figure 3.1(b),

$$\int_0^1 (1/\sqrt{2})(1-2y)m_b^2\,dy \tag{3.24}$$

valid for the side $x=0$. All three integrals are zero so (3.17) is satisfied and the method is convergent.

3.5.2 Example 2

The second example is a rectangular element proposed in a mixed version by Bäcklund[17] and in a consistent (non-mixed) version by Pettersson.[26] A mathematical treatment of the mixed version has been performed by Johnson.[16] Variables in the latter version are the four corner deflections and the four midside slopes (Figure 3.2).

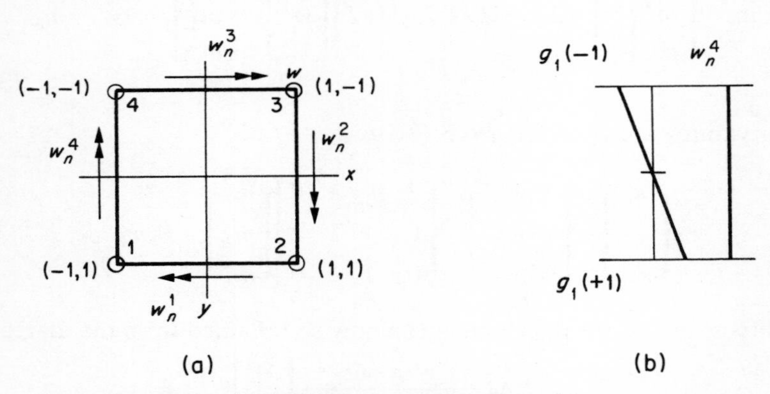

Figure 3.2 (a) The linear curvature rectangle. (b) w_n at $x=1$

With the corner and midside node co-ordinates denoted by (x_i, y_i) the assumed deflection can be written

$$w=\sum_1^4\left[\frac{1}{4}(1+xx_i)(1+yy_i)+\frac{1}{8}xx_i(1-x^2)+\frac{1}{8}yy_i(1-y^2)\right]w^i$$
$$-\sum_1^4\left[\frac{1}{4}x_i(1-x^2)(1+xx_i)+\frac{1}{4}y_i(1-y^2)(1+yy_i)\right]w_n^i \tag{3.25}$$

The deflection and slope along the side $x = -1$ is

$$w = (1/2)(1+y)w^1 + (1/2)(1-y)w^4 + f_1(y)$$
$$w_n = w_n^4 + g_1(y)$$

where

$$f_1(y) = (1/8)(1-y^2)\{y(w^1+w^2-w^3-w^4) - 2(1+y)w_n^1 + 2(1-y)w_n^3\}$$

The condition (3.17) for convergence is satisfied since the corner deflections and the constant part of the slope is continuous and the remaining integrals over the sides, as for example

$$\int_{-1}^{1} g_1(y)\,dy \tag{3.26}$$

are zero for each side.

3.5.3 Example 3

The last example is the Hermitian rectangle, Figure 3.3, with twelve variables, namely w, w_x, and w_y at each corner. It was first proposed by Adini and Clough[27] and was mathematically analysed by Ciarlet.[19]

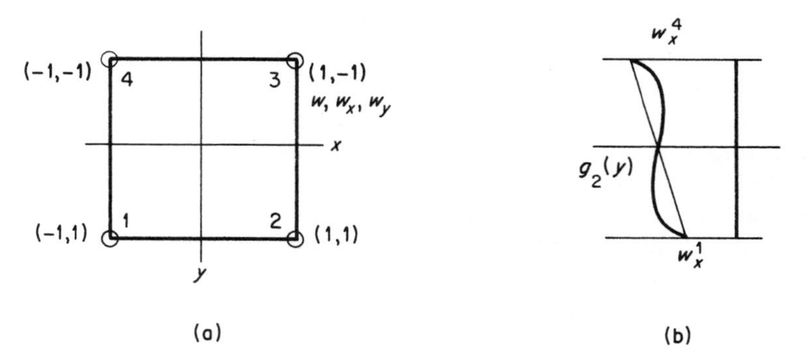

Figure 3.3 (a) The Hermitian rectangle. (b) w_n at $x = -1$

The assumed deflection can be written

$$w = \sum_1^4 [\tfrac{1}{4}(1+xx_i)(1+yy_i) + \tfrac{1}{8}xx_i(1-x^2)(1+yy_i)$$
$$+ \tfrac{1}{8}yy_i(1-y^2)(1+xx_i)]w^i - \sum_1^4 [\tfrac{1}{8}x_i(1-x^2)(1+xx_i)(1+yy_i)]w_x^i$$
$$- \sum_1^4 [\tfrac{1}{8}y_i(1-y^2)(1+xx_i)(1+yy_i)]w_y^i \tag{3.27}$$

The deflection and slope along the side $x=-1$ is

$$w=(1/2)(1+y)w^1+(1/2)(1-y)w^4+f_2(y)$$
$$w_x=(1/2)(1+y)w_x^1+(1/2)(1-y)w_x^4+g_2(y)$$

where

$$f_2(y)=(1/8)(1-y^2)\{2y(w^1-w^4)-2(1+y)w_y^1+2(1-y)w_y^4\}$$
$$g_2(y)=(1/8)y(1-y^2)(-w^1+w^2-w^3+w^4)$$

Regarding condition (3.17) the corner deflections and the linear part of the slope is continous. There remain integrals over the sides of the type

$$\int_{-1}^{1} g_2(y)\,dy \tag{3.28}$$

which all are zero, so the global constant strain condition is satisfied.

3.6 CONCLUDING REMARKS

In this chapter an attempt has been made to explain the concept of 'global constant strain condition' for convergence of non-conforming finite element methods. It has in its numerical version been called 'the patch test'. It has been shown that for the plate bending problem this condition is satisfied if the deflections at the nodes are unique (pointwise continuity) and if a line integral of slopes around each element is zero. This is, of course, a sufficient but not a necessary condition for convergence.

For illustration three plate bending elements have been tested. The first element is of the minimum of complementary energy type—is 'strongly diffusive' in Fraeijs de Veubeke's nomenclature. It has here been tested for convergence as a 'weakly conforming' element. The second element has been tested by Johnson[25] for convergence as a mixed 'weakly diffusive' element. It has here been tested as a 'weakly conforming' element. Also the third element is tested as a 'weakly conforming' element, see Ciarlet.[19]

All three elements satisfy the slope integral condition in such a way that the integral is zero for each single side of the element. This is certainly not the only way to satisfy the condition. For example, for rectangular elements, integrals on opposite sides may cancel each other.[18,25]

Fraeijs de Veubeke[22] explained the connection between 'the patch test' and hybrid elements. At the construction of such elements a condition of the type given by the second equation (3.17) is added to variational equations with coefficients of area integral type.

REFERENCES

1. N. M. Johnson and R. N. McLay, 'Convergence of the finite element method in the theory of elasticity', *J. Appl. Mech. Trans. Am. Soc. Mech. Eng.*, 274–278 (1968).
2. R. N. Clough, 'The finite element method in structural mechanics', Chapter 7 of *Stress Analysis* (ed. O. C. Zienkiewicz), Holister, 1965.
3. M. Zlámal, 'On the finite element method', *Num. Math.*, **12,** 394–409 (1968).
4. P. G. Ciarlet and C. Wagschal, 'Multipoint Taylor formulas and applications to the finite element methods', *Num. Math.*, **17,** 84–100 (1971).
5. B. Fraeijs de Veubeke, 'Bending and stretching of plates', *Conf. Matrix Meth. Struct. Mech.*, Air Force Flight Dyn. Lab., Ohio, 863–885 (1965).
6. B. Fraeijs de Veubeke and O. C. Zienkiewicz, 'Strain energy bounds in finite element analysis by slab analogy', *J. Strain Analysis*, **2,** 265–271 (1967).
7. M. J. Turner, R. W. Clough, H. C. Martin, and L. J. Topp, 'Stiffness and deflection analysis of complex structures', *J. aeronaut. Sci.*, **23,** 805-823 (1956).
8. O. C. Zienkiewicz, *The Finite Element Method in Engineering Science*, McGraw-Hill, London, 1971.
9. K. Hellan, 'Analysis of elastic plates in flexure by a simplified finite element method', *Acta Polyt. Scandinavia*, **Ci46** (Trondheim), (1967).
10. A. Samuelsson, 'Non-conventional Finite Element Methods', in *Finite Elements in Fluids*, Vol. 3 (ed. R. H. Gallagher *et al.*), Wiley, London, 1978.
11. E. Reissner, 'On variational principles in elasticity', *Proc. Symposia Appl. Math.*, **8,** 1–6 (1958).
12. L. Herrmann, 'Finite element bending analysis for plates', *Proc ASCE*, **EM5** (1967).
13. T. H. H. Pian and P. Tong, Basis of finite element methods for solid continua, *Int. J. Num. Methods in Engineering*, **1,** 3–28 (1969).
14. M. Fröier, Lg. Nilsson, and A. Samuelsson, 'The rectangular, plane stress element by Turner, Pian and Wilson', *Int. J. Num. Meth. Engng*, **8,** 433–437 (1974).
15. B. M. Irons and A. Razzaquee, 'Experience with the patch test for convergence of finite element methods', *The Mathematical foundations of the finite element method*, Academic Press, New York, 1972.
16. C. Johnson, 'Convergence of Another Mixed Finite-element Method for Plate Bending Problems', *Publ. Department of Math.*, Chalmers University of Technology, Göteborg, Sweden, No. 27, 1972.
17. J. Bäcklund, 'Mixed Finite Element Analysis of elastic and Elasto-plastic Plates in Bending', *Publ. Depart. of Struct. Mech.*, Chalmers University of Technology, Göteborg, Sweden, No. 71:1.
18. G. Strang and G. Fix, *An analysis of the finite element method*, Prentice-Hall, New York, 1973.
19. P. G. Ciarlet, 'Conforming and non-conforming finite element methods for solving the plate problem', paper at the *Conf. on the Numerical Solution of Diff. Eqs.*, University of Dundee, 1973.
20. F. Brezzi, 'Sur la méthode des elements finis hybrides pour le problème biharmonique', *Num. Math.*, **24** (1975).
21. G. P. Bazeley, Y. K. Cheung, B. M. Irons, and O. C. Zienkiewicz, 'Triangular elements in bending—conforming and non-conforming solutions', *Proc. Conf. on Matrix Methods*, Wright-Patterson Air Force Base, Ohio, 1965.

22. B. Fraeijs de Veubeke, 'Variational principles and the patch test', *Int. J. Num. Meth. Engng,* **8,** 783–801 (1974).
23. L. S. D. Morley, 'The triangular equilibrium element in the solution of plate bending problems', *Aero. Quart.,* **19,** 149–169 (1968).
24. D. J. Allman, 'Triangular finite elements for plate bending with constant and varying bending moments', *Proc. IUTAM,* Liège, (1970).
25. C. Johnson, 'On the convergence of some mixed element methods in plate bending problems', *Num. math.,* **21** (1973).
26. H. Pettersson, 'A rectangular mixed element for plate bending', *Publ. Dept. of Building Construction, Chalmers University of Technology.,* Göteborg, Sweden, 1972.
27. A. Adini and R. W. Clough, 'Analysis of plate bending by the finite element method', *NSF Report* G, 7337 (1961).

Chapter 4

The Influence of the Finite Element Method on Developments in Structural Dynamics

G. B. Warburton

4.1 INTRODUCTION

During the last twenty years many problems in aeronautical, civil, mechanical, and nuclear engineering, which required the determination of dynamic response, have been solved by the finite element method. In some areas there was also a concurrent growing realization that proper allowance for dynamic, as opposed to static, loading should be made, whereas in others, e.g. aeronautical and mechanical engineering, although dynamic stress analysis has a longer history, the development of the finite element method has increased the range and complexity of problems that can be solved. In the context of structural dynamics the dominant characteristic of the finite element method is to replace the actual continuous system, which has theoretically an infinite number of degrees of freedom, by an approximate multi-degree-of-freedom system. It is usual, but not essential, for this number of degrees of freedom to be large. Thus recent emphasis in structural dynamics has been on the determination of natural frequencies and response for various types of dynamic loads applied to large multi-degree-of-freedom systems. Formal solutions in terms of matrix algebra are not affected in principle by the number of degrees of freedom, but computational problems and cost grow rapidly as this number increases. Thus approximations in the method of solution and improvements in technique, which reduce the computation time for large problems, are of great value.

It should be mentioned that other topics in structural dynamics have been studied concurrently, including the development and use of other approximate methods and the determination of exact solutions of the governing equations of motion for relatively simple geometries. However, the dynamic matrix equations for a linear system, derived by the finite element method in its conventional displacement approach, the Rayleigh–Ritz method, and the finite difference method if based on an energy or variational approach, are of identical form, so that advances in structural dynamics associated with the finite element method are applicable to these well-known approximate methods. Although exact solutions are limited to simple geometries, they are useful for parametric studies and for improving our understanding of the dynamic behaviour of simple components.

The development of the finite element method has had a unifying effect on structural dynamics, as the method is general, being applicable to one-, two- and three-dimensional structures, and can accommodate changes in geometry, boundary conditions, and loading. In contrast, the form, or even the existence, of closed-form exact solutions is very sensitive to boundary conditions; for example, consider the effects on the method of solution if the boundary conditions are changed from simply supported to clamped for a rectangular plate or thin cylindrical shell. In the past, approximate methods were developed for particular classes of problem. For instance, torsional vibrations of non-uniform shafts—an important problem in power transmission—generated many approximate methods, of which the best known and most versatile was the Holzer method; later this was extended to flexural vibrations of non-uniform beams,[1] but the soluble problems remained one-dimensional.

4.2 MATRIX EQUATION FOR LINEAR STRUCTURES

4.2.1 *General equation*

The finite element model for the dynamic response of linear elastic structures to specific excitations is represented by the matrix equation

$$\mathbf{M}\ddot{\mathbf{x}} + \mathbf{C}\dot{\mathbf{x}} + \mathbf{K}\mathbf{x} = \mathbf{p}(t) \tag{4.1}$$

where the vector $\mathbf{x}$ lists the displacements, linear or angular, associated with the degrees of freedom of the finite element model; $\mathbf{M}$, $\mathbf{C}$, and $\mathbf{K}$ are the square and symmetric matrices for the mass, damping and stiffness of the structure; and $\mathbf{p}(t)$ is the vector of generalized excitation forces, corresponding to the nodal displacements, the velocity vector $\dot{\mathbf{x}} = \mathrm{d}\mathbf{x}/\mathrm{d}t$ and the acceleration vector $\ddot{\mathbf{x}} = \mathrm{d}^2\mathbf{x}/\mathrm{d}t^2$. The order of the matrices in equation (4.1) equals the number of degrees of freedom of the constrained structure. The form of the equation is not altered substantially, if the excitation consists of prescribed accelerations or displacements at the base or foundation of the structure, instead of applied forces. Thus the solution of equation (4.1) gives the response of the structure to prescribed excitation and from these displacements the dynamic stresses can be obtained.

For a particular structure the same stiffness matrix $\mathbf{K}$ is required whether a static or a dynamic problem is being considered. Thus considering the modelling of complex structures by finite elements, a survey of the generation of stiffness matrices is effectively a history of the finite element method and will not be attempted here. For a specified type of deformation various elements exist and have been described in several textbooks, for example, References 2 and 3. Extensive bibliographies of recent publications exist.[4,5]

4.2.2 Damping matrix C

Generation of the damping matrix presents special difficulties, due to lack of information regarding damping mechanisms and damping levels in structures. If the damping mechanisms are limited to internal, or material, damping and the structure is made from one homogeneous material, the damping matrix is proportional to the stiffness matrix. For structures built from two or more homogeneous materials or in interaction problems (i.e. dynamic interaction between a structure and the ground or between a structure and surrounding fluid), each component damping matrix may be proportional to the corresponding stiffness matrix, but as the constants of proportionality differ, the system damping matrix will not be proportional to the system stiffness matrix. The matrices for structures, which have sliding or yielding joints or constraints or incorporate absorbing mechanisms to limit vibrations, also exhibit this non-proportionality. Non-proportionally damped systems may require special consideration when determining response (Section 4.5.2).

Often there is insufficient information to formulate the complete damping matrix. If the normal mode method of determining response is used (Section 4.5.1), an alternative procedure is to use estimates of the modal damping factors for all significant modes. In some response calculations damping is neglected. This is permissible in problems, where the damping is light and the dynamic magnification factor is small, and can have computational advantages (Section 4.6).

4.2.3 Mass matrix M

In standard applications of the finite element method a consistent mass matrix is determined for each element and assembly leads to a symmetric mass matrix. With this procedure and the use of conforming elements natural frequencies determined from an eigenvalue solution are upper bounds. However, there can be computational advantages (Section 4.6.1) if the mass matrix is diagonal. This type of matrix is always obtained with the lumped mass procedure, as used in the standard finite difference approach. Special improved diagonal mass matrices, which avoid the inaccuracies of the lumped mass procedure, have been developed by Key and Beisinger[6] for their doubly curved quadilateral shell element and by Hinton, Rock, and Zienkiewicz[7] for parabolic isoparametric elements.

4.2.4 Force vector $\mathbf{p}(t)$

The consistent force vector for an element is derived as in static problems. One classification of dynamic response problems depends upon how the elements of the force vector $\mathbf{p}(t)$ vary with time. This leads to the following categories: zero; harmonic; periodic; aperiodic or transient; and random. Free undamped vibration is considered in Section 4.3. If the elements of $\mathbf{p}$ are periodic, but not harmonic, the excitation can be decomposed into a sum of harmonic terms by a Fourier series and the resulting dynamic response found by superposition for linear systems; thus this case does not require separate consideration. Methods of determining response when the time variation is harmonic, aperiodic, and random are considered in Sections 4.4, 4.5, and 4.6.

4.3 NATURAL FREQUENCIES AND REDUCTION METHODS

Before considering methods of determining response from equation (4.1), we consider the associated free vibration problem, represented by

$$\mathbf{M}\ddot{\mathbf{x}}+\mathbf{K}\mathbf{x}=\mathbf{0} \tag{4.2}$$

The natural frequencies ω_r are found from the determinant

$$\det|\mathbf{K}-\omega^2\mathbf{M}|=0 \tag{4.3}$$

For a particular natural frequency ω_r the associated modal vector ${}_r\mathbf{x}$ can be found from equation (4.2). If equations (4.1) to (4.3) are of order n, i.e. n degrees of freedom, n natural frequencies can be obtained from equation (4.3). Although these frequencies can be determined to any desired accuracy from standard computer programs for eigenvalues and eigenvectors, they are the frequencies of the approximate system. With a reasonable number of appropriate elements some lower fraction of these n frequencies should be good approximations to the frequencies of the real structure. In general the higher frequencies found from equation (4.3) have no physical significance. Several methods of computing natural frequencies and modal vectors from equation (4.2) exist. The following factors affect their suitability and efficiency; the size of the matrices $\mathbf{K}$ and $\mathbf{M}$, their bandwidths; whether all, or only the lower, frequencies are required. Bathe and Wilson[8] review four methods: Householder—QR—inverse iteration; generalized Jacobi iteration; determinant search; and subspace iteration. Of these the last two determine the lower frequencies, while the first two yield all the frequencies. Subspace iteration, developed by Bathe and Wilson,[9] is particularly suitable for very large systems with large bandwidth.

If the size of the matrix equation (4.2) is too large, the technique of eigenvalue economizers or reduction can be used to reduce the order of the equations by eliminating degrees of freedom associated with small contributions to the kinetic energy of the structure.[2,10] The vector $\mathbf{x}$ and matrices $\mathbf{K}$ and $\mathbf{M}$ in equation (4.2) are written in partitioned form as

$$\mathbf{x} = \left[\begin{array}{c} \mathbf{x}_m \\ \hline \mathbf{x}_s \end{array}\right], \qquad \mathbf{K} = \left[\begin{array}{c|c} \mathbf{K}_{mm} & \mathbf{K}_{ms} \\ \hline \mathbf{K}_{sm} & \mathbf{K}_{ss} \end{array}\right], \text{ etc.} \tag{4.4}$$

where the subscripts m and s refer to master and slave respectively. It is assumed that the slave displacements $\mathbf{x}_s$ can be expressed in terms of the master displacements $\mathbf{x}_m$ by expanding the equation below the partition and assuming that inertia effects in this relationship are negligible, leading to

$$\mathbf{x}_s = -\mathbf{K}_{ss}^{-1}\mathbf{K}_{sm}\mathbf{x}_m \tag{4.5}$$

Using equations (4.4) and (4.5) the following reduced equation is obtained:

$$\mathbf{M}^*\ddot{\mathbf{x}}_m + \mathbf{K}^*\mathbf{x}_m = \mathbf{0} \tag{4.6}$$

where the order of the reduced matrices $\mathbf{M}^*$ and $\mathbf{K}^*$ equals the number of master displacements and $\mathbf{M}^*$ and $\mathbf{K}^*$ can be defined in terms of the partitioned matrices of $\mathbf{K}$ and $\mathbf{M}$.

In practice the slave displacements are eliminated progressively as the matrices are assembled so that the overall system matrices are never set up. If the system has n degrees of freedom and n^* masters are chosen, the lower natural frequencies from equation (4.6) will have small errors compared to those from equation (4.2), but in general the former will be more accurate than frequencies found from a coarser finite element idealization, which has n^* degrees of freedom initially and to which no reduction has been applied. Obviously the slave degrees of freedom must be chosen carefully. From the assumption that the degrees of freedom are associated with small contributions to the kinetic energy, rotations in bending problems and in-plane components of displacement in coupled bending-membrane problems tend to be eliminated first. Recently a simple technique has been developed by Henshell and Ong[11] for selecting automatically these degrees of freedom. When the number of degrees of freedom assembled exceeds the specified number of masters, the degree of freedom associated with the largest value of the ratio

$$\omega_j^2 = K_{jj}/M_{jj}$$

is eliminated first. K_{jj} and M_{jj} are corresponding diagonal values from the matrices $\mathbf{K}_{ss}$ and $\mathbf{M}_{ss}$ respectively. Examples of varying complexity demonstrate its usefulness, i.e. for a given (not small) number n^*, automatic selection gives lower errors in the natural frequencies of interest than manual selection.

Although reduction has been discussed in the context of frequency determination, which is its conventional application, it can be applied to the full matrix equation (4.1) in order to reduce its order and then the response is determined by one of the standard methods (see following sections). In this case it must be ensured that when the force vector $\mathbf{p}$ is partitioned, all non-zero entries in $\mathbf{p}$ are maintained in $\mathbf{p}_m$, i.e. $\mathbf{p}_s$ consists solely of zeros. In many problems $\mathbf{p}$ contains a large number of zero entries, so this is not unduly restrictive.

The method of substructures or component modal synthesis, introduced by Hurty[12] (see References 13 and 14 for recent reviews and contributions), avoids the necessity of dealing with the large number of degrees of freedom associated with the complete structure by considering appropriate components and their modes and assembling global matrices by applying equilibrium and compatibility conditions at the interface or master degrees of freedom.

4.4 RESPONSE

As mentioned in Section 4.2.4, the time variation of the force vector $\mathbf{p}(t)$ may be harmonic, aperiodic, or random; this affects the method chosen to determine response.

If the elements of $\mathbf{p}(t)$ vary harmonically with time, the steady-state response of the structure can be found by the frequency response method or by the normal mode method (Section 4.5). (In theory the numerical integration methods of Section 4.6 could be used, but this would be highly uneconomic computationally. The steady-state response is obtained after a time interval which is sufficiently large for the starting transient to decay to negligible proportions; with numerical integration this would require an excessive number of time steps. However, numerical integration methods can be used to investigate the initial response, but this is usually of less importance.) The frequency response method, which is limited to the determination of the steady-state response for harmonic excitation of frequency ω, leads to

$$\mathbf{x} = \mathrm{Re}[\mathbf{J}^{-1}\mathbf{p}\exp(i\omega t)] \tag{4.7}$$

where $\mathbf{J} = \mathbf{K} - \omega^2\mathbf{M} + i\omega\mathbf{C}$, $\mathbf{J}^{-1}$ is the inverse of $\mathbf{J}$, $i = (-1)^{1/2}$, the vector $\mathbf{p}$ consists of constants (complex if there are phase differences between different excitation forces) and Re signifies that the real part of the complete complex expression is taken. If the structure has hysteretic damping with constants forming the matrix $\mathbf{H}$, $\mathbf{H}\dot{\mathbf{x}}/\omega$ replaces $\mathbf{C}\dot{\mathbf{x}}$ in equation (4.1) and $\mathbf{H}$ replaces $\omega\mathbf{C}$ in the definition of $\mathbf{J}$. Evaluation of the response from equation (4.7) requires the inversion of the complex matrix $\mathbf{J}$ at each excitation frequency of interest.

For aperiodic excitation the normal mode method (Section 4.5) and numerical integration methods (Section 4.6) can be used to determine the response. For random excitation determination of receptances or transfer functions, which can be obtained from equation (4.7), is an essential prerequisite to the evaluation of mean square response.

Other methods of determining response exist. Apart from the use of Fourier transforms for interaction problems (Section 4.7), these will not be considered.

4.5 THE NORMAL MODE METHOD

4.5.1 Classically or proportionally damped structures

The natural frequencies ω_r and the associated normalized eigenvectors ${}_r\mathbf{z}$ which satisfy

$$ {}_r\mathbf{z}^{\mathrm{T}}\mathbf{M}{}_r\mathbf{z} = 1 \tag{4.8} $$

must be known and are found from the solution of equation (4.2); r is an integer defining the mode number. Also the eigenvectors satisfy the orthogonality conditions

$$ \begin{aligned} {}_r\mathbf{z}^{\mathrm{T}}\mathbf{M}{}_s\mathbf{z} &= 0 \\ {}_r\mathbf{z}^{\mathrm{T}}\mathbf{K}{}_s\mathbf{z} &= 0 \end{aligned} \qquad r \neq s \tag{4.9} $$

where r and s relate to any two modes with different natural frequencies. We define the transformation

$$ \mathbf{x} = \mathbf{Zq} \tag{4.10} $$

where $\mathbf{q}^{\mathrm{T}}, = [q_1, q_2, \ldots, q_n]$, is a vector of principal coordinates and the normalized vector ${}_r\mathbf{z}$, defined by equation (4.8), forms the rth column of the modal matrix $\mathbf{Z}$. Ideally substitution of the transformation (4.10) into equation (4.1), followed by pre-multiplication by $\mathbf{Z}^{\mathrm{T}}$, leads to a set of uncoupled equations in q_r. The orthogonality conditions (4.9) ensure that the stiffness and mass terms are uncoupled. The damping terms are uncoupled if $\mathbf{B}, = \mathbf{Z}^{\mathrm{T}}\mathbf{CZ}$, is a diagonal matrix. It can be shown that the necessary and sufficient condition for uncoupling is:[15]

$$ \mathbf{CM}^{-1}\mathbf{K} = \mathbf{KM}^{-1}\mathbf{C} \tag{4.11} $$

In general, this condition will not be satisfied, but various relations, which impose a restriction on the form of the damping matrix by expressing $\mathbf{C}$ in terms of $\mathbf{K}$ and $\mathbf{M}$, satisfy condition (4.11) and make $\mathbf{B}$ diagonal. A simple form is

$$ \mathbf{C} = \lambda_k\mathbf{K} + \lambda_m\mathbf{M} \tag{4.12} $$

where λ_k and λ_m are constants. More complex forms of equation (4.12) are obtained by adding terms such as $\lambda_1\mathbf{K}\mathbf{M}^{-1}\mathbf{K}$ and $\lambda_2\mathbf{M}\mathbf{K}^{-1}\mathbf{M}$. The process can be extended;[16] each additional term must individually satisfy condition (4.11).

In equation (4.12) the damping matrix is assumed to be a linear combination of the mass and stiffness matrices. For practical structures, where light internal damping in the elastic members must be represented approximately, the assumption that $\mathbf{C}$ is proportional to $\mathbf{K}$ is common and reasonable. Thus equation (4.12) is not so restrictive as it first appears.

With assumption (4.12) the resulting uncoupled equations are of the form

$$\ddot{q}_r + 2\gamma_r\omega_r\dot{q}_r + \omega_r^2 q_r = \sum_{j=1}^{n} {}_r z_j P_j(t), \qquad r = 1, 2, 3, \ldots, n \tag{4.13}$$

where ${}_r z_j$ and $P_j(t)$ are the jth entries in the column vectors ${}_r\mathbf{z}$ and $\mathbf{p}(t)$ and $2\gamma_r\omega_r = \lambda_m + \lambda_k\omega_r^2$ where γ_r is the modal damping parameter.

Evaluation of the principal coordinates q_r, using the Duhamel integral solution of equation (4.13)[10] or numerical integration, and use of transformation (4.10) gives the response x_s at any coordinate. In general, the response from the higher modes (larger values of r) is insignificant and it is necessary to compute q_r only for a limited range of values $r = 1, 2, \ldots, n_1$ with $n_1 \ll n$. No general relation between n_1 and n can be given, but it depends on the parameters of the structure, the nature of the excitation, and the response quantity of interest.

It must be recalled that equation (4.1) is a multi-degree-of-freedom approximation to the real structure. For accuracy it is essential that all the n_1 modes which contribute significantly to the response should be included in the n_2 modes of the idealization which are true modes of the real structure. How can this be verified for a practical problem for which only approximate solutions exist? The natural frequencies and mode shapes are determined from equations (4.2) and (4.3) for a particular finite element idealization. Then the mesh is refined, i.e. n is increased, and the frequencies and mode shapes determined again. Comparison of frequencies and mode shapes from the two idealizations will show that the frequencies of n_2 identifiable modes have not been changed significantly by the change in idealization. Thus, for these n_2 modes the frequencies are converged values and, in the absence of other information, can be assumed to be reasonable approximations to the true values. Although these n_2 modes will be at the lower end of the frequency spectrum, they will not usually be the lowest n_2 modes. Then the response quantities of interest for the specified excitation can be determined with various numbers of modes included in the series in order to obtain n_1.

If the modal damping parameters γ_r are assumed for the first few modes, $r=1, 2, \ldots, n^+$, instead of formulating the complete damping matrix **C** (Section 4.2.2), it is always possible to obtain uncoupled equations in q_r by taking n^+ terms with coefficients λ_k, λ_m, λ_1, λ_2 etc. in the relation linking **C** to **K** and **M**.[10,16]

Ewins[17] gives a simple method of adequate accuracy for estimating the response of a damped structure from the responses at frepuencies $\omega_r(1 \pm \frac{1}{2}\mu_r)$ of the corresponding undamped system, where μ_r is the hysteretic damping in mode r. The method depends upon natural frequencies being reasonably well separated and damping values being small. Although formulated in terms of the modal approach, it could be used in conjunction with the frequency response method of equation (4.7); in this case the matrix **J** will be real, instead of complex, with consequent savings in computational effort.

4.5.2 Non-classically damped structures

In many practical problems the matrix $\mathbf{B}(=\mathbf{Z}^{\mathrm{T}}\mathbf{C}\mathbf{Z})$ is not diagonal, e.g. in ground–structure interaction problems where the level of damping in the ground, mainly due to radiation, is considerably higher than that in the structure. In such cases use of the normal mode method leads to a matrix equation, which comprises n coupled equations in q_r. (In fluid–structure problems analysis in terms of the normal modes of the structure without the fluid present leads also to a set of coupled equations in the principal coordinates q_r, owing to the additional fluid inertia terms which contribute to a modified mass matrix.) These coupled equations can be solved by one of the methods of numerical integration (Section 4.6). At first sight this procedure is not advantageous, as the original n coupled equations in the physical coordinates, equation (4.2), have been exchanged for n coupled equations in terms of q_r. However, as only n_1 modes make a significant contribution to the response and $n_1 \ll n$, working in terms of the principal coordinates q_r requires numerical integration of significantly smaller matrix equations. This procedure is advocated and illustrated by Clough and Mojtahedi.[18]

Approximations that allow the normal mode method to be used when the matrix **B** is not diagonal are important in practice. Thomson *et al.*[19] suggest that the non-diagonal matrix **B** should be replaced by a diagonal matrix with the same diagonal terms as the original matrix, (i.e. the off-diagonal terms of **B** are replaced by zeros); then the standard normal mode procedure is followed. Although the literature contains some conflicting numerical evidence, it appears that, provided damping is light and natural frequencies are reasonably well spaced, this approximation introduces acceptable errors allowing for the uncertainties regarding damping values in real structures.

The author[20] has given a criterion, which should be satisfied in order that neglecting the off-diagonal terms in **B** does not lead to excessive errors in major response quantities. The criterion is:

$$\gamma_r \leqslant 0.05 \left| \frac{b_{rr}}{2b_{rs}} \left(\frac{\omega_s^2}{\omega_r^2} - 1 \right) \right|_{\text{min } s} \tag{4.14}$$

where γ_r is the damping ratio for the rth mode, ω_r and ω_s are natural frequencies, b_{rr} and b_{rs} are elements from the matrix **B** and the minimum of the expression $|\ldots.|$ with respect to s is taken; s may take any integer value between 1 and n other than r. In practice, equation (4.14) is applied only to the lower values of r, where significant resonant response may occur. For each of these values of r the right-hand side is obtained by considering a few values of s on either side of r. Further numerical evidence in support of criterion (4.14) has been obtained by the author[21] when considering coupled structure–ground–water vibrations for a simplified offshore structure.

4.5.3 Determination of maxima

In the normal mode method a typical displacement x_s is given by (equation (4.10))

$$x_s = \sum_r {}_r z_s q_r(t) \tag{4.15}$$

with $q_r(t)$ obtained from equation (4.13). As each of the principal coordinates q_r is a different function of time, determination of the maximum value of x_s requires evaluation of each significant principal coordinate for a large number of values of t. (Although equation (4.15) and the following discussion is in terms of displacements x_s, comparable expressions can be written down if maximum stress or maximum acceleration is required.) Now

$$x_s(\max) \leqslant \sum_r [{}_r z_s q_r(\max)] \tag{4.16}$$

where $q_r(\max)$ is the maximum value of $q_r(t)$ and can be determined from an appropriate response spectrum. For a single applied force $P_0 f_j(t)$ applied at coordinate j, where the maximum value of $f_j(t)$ is unity and P_0 is a constant equal to the maximum value of the force, equation (4.13) becomes

$$\ddot{q}_r + 2\gamma_r \omega_r \dot{q}_r + \omega_r^2 q_r = {}_r z_j P_0 f_j(t) \tag{4.17}$$

(Multi-point excitation can be introduced without major alteration of the following discussion if all the applied forces are the same function of time; this will occur when a single force is applied to the actual structure and is represented by several entries in the consistent force vector.) The equation of motion for the classical single-degree-of-freedom system is

$$\ddot{x}+2\gamma\omega_n\dot{x}+\omega_n^2 x=P_0 f(t)/m \tag{4.18}$$

and a response spectrum shows the variation of the dynamic magnification factor $kx(\max)/P_0$ with a period or frequency ratio, e.g. the ratio of some characteristic time T_0 of the applied force to the period $T(=2\pi/\omega_n)$ for a specified value of the damping factor γ. Noting the similarity between equations (4.17) and (4.18), $q_r(\max)$ may be found from the appropriate response spectrum. Using the equality sign in equation (4.16), i.e. taking the upper bound as an approximation,

$$x_s(\max)=P_0\sum_r\left|\frac{{}_rz_{s\,r}z_j}{\omega_r^2}(DMF)_r\right| \tag{4.19}$$

where $(DMF)_r$ is the dynamic magnification factor from the response spectrum for the force-time history $f_j(t)$.

For complex excitations, e.g. earthquakes, equation (4.19) may seriously overestimate the maximum response as the underlying assumption is not true. An alternative, empirical expression, which is often used, is

$$x_s(\max)=P_0\left[\sum_r\left\{\frac{{}_rz_{s\,r}z_j}{\omega_r^2}(DMF)_r\right\}^2\right]^{1/2} \tag{4.20}$$

Fung and Barton[22] suggest that when $f_j(t)$ is a single pulse of any shape and the rise time/$T_1>0.5$, the contributions from the various modes may be assumed to reach their maxima at the same time. With this assumption the algebraic sum of modal contributions, instead of the absolute sum in equation (4.19), is used to evaluate $x_s(\max)$, i.e.

$$x_s(\max)=P_0\sum_r\frac{{}_rz_{s\,r}z_j}{\omega_r^2}(DMF)_r \tag{4.21}$$

As dynamic magnification factors are available for standard simple force-time histories[23] or can be determined relatively simply for a given history, use of any of equations (4.19), (4.20), or (4.21) reduces the quantity of computation required to determine maximum displacements. However, these equations are approximate and resulting maxima may be unacceptably inaccurate. As an example, the author considered an undamped four degree-of-freedom-system of the chain type subjected to the excitation

$$\begin{aligned}P(t)&=P_0\sin\frac{2\pi t}{T_0} && 0\leq t\leq\tfrac{1}{2}NT_0\\ &=0 && t>\tfrac{1}{2}NT_0\end{aligned} \tag{4.22}$$

Considering $N=1, 2$, and 4 and $T_1/T_0=0.4$, 1.0, and 2.5, equation (4.19) could overestimate the maximum response of one of the masses by 42 per cent; errors in maximum response from equation (4.20) varied from −32 per cent to 15 per cent; equation (4.21) gave errors up to 10 per cent but was not applicable for $T_1/T_0=2.5$. Even for this simple example errors can be large and are sensitive to small changes in time history. Thus these approximate equations must currently be used with caution; it is hoped that future studies will illustrate how the many parameters affect the errors.

4.6 NUMERICAL INTEGRATION METHODS

4.6.1 Outline of methods

In these methods assumptions are made about the variation of the displacements or accelerations during small time intervals; for example, it may be assumed that during a small time interval the displacement is a cubic function of time or the acceleration varies linearly or is constant. With such assumptions the set of n second order differential equations (4.1) is replaced, in general, by n simultaneous equations. The solution of the latter gives the displacements at the end of the short time step for known conditions at the beginning. Successive application of this procedure gives the response.

Many methods exist: the established methods, which have been incorporated in finite element computer codes, include the Newmark β,[24] Houbolt,[25] Wilson θ[16] and central difference methods.[10] Methods which exhibit improved characteristics for test problems have been developed by Park,[26] by Melosh,[27] and by Medland[28] (effectively the same method was proposed independently by the two authors) and by Hilber, Hughes, and Taylor.[29] The latter authors and also Bathe and Wilson[30] compare some of the methods in current use.

In the Newmark method the following assumptions are made:[24]

$$\begin{aligned} \dot{\mathbf{x}}_{s+1} &= \dot{\mathbf{x}}_s + \tfrac{1}{2}\Delta t(\ddot{\mathbf{x}}_s + \ddot{\mathbf{x}}_{s+1}) \\ \mathbf{x}_{s+1} &= \mathbf{x}_s + \Delta t\dot{\mathbf{x}}_s + (\tfrac{1}{2} - \beta)(\Delta t)^2\ddot{\mathbf{x}}_s + \beta(\Delta t)^2\ddot{\mathbf{x}}_{s+1} \end{aligned} \tag{4.23}$$

where $\mathbf{x}_{s+1}$, $\mathbf{x}_s$ and $\mathbf{x}_{s-1}$ are the values of $\mathbf{x}$ at times t_{s+1}, t_s and t_{s-1} respectively, $t_{s+1}-t_s=t_s-t_{s-1}=\Delta t$, a small time interval, and velocity and acceleration vectors are defined similarly. The parameter β is often taken as $\frac{1}{4}$; in this case the physical interpretation of assumptions (4.23) is that the acceleration vector during the time interval t_s to t_{s+1} is constant and equal to the mean of $\ddot{\mathbf{x}}_s$ and $\ddot{\mathbf{x}}_{s+1}$. (Therefore it is sometimes called the average acceleration method.) Using these assumptions in equation (4.1) we obtain finally

$$
\begin{aligned}
[\mathbf{M}+\tfrac{1}{2}(\Delta t)\mathbf{C}+\beta(\Delta t)^2\mathbf{K}]\mathbf{x}_{s+1} &= (\Delta t)^2[\beta\mathbf{p}_{s+1}+(1-2\beta)\mathbf{p}_s+\beta\mathbf{p}_{s-1}] \\
&\quad +[2\mathbf{M}-(\Delta t)^2(1-2\beta)\mathbf{K}]\mathbf{x}_s \qquad (4.24) \\
&\quad -[\mathbf{M}-\tfrac{1}{2}(\Delta t)\mathbf{C}+\beta(\Delta t)^2\mathbf{K}]\mathbf{x}_{s-1}
\end{aligned}
$$

where $\mathbf{p}_{s+1}$, $\mathbf{p}_s$ and $\mathbf{p}_{s-1}$ are the values of the force vector $\mathbf{p}(t)$ at times t_{s+1}, t_s and t_{s-1} respectively. The solution of equation (4.24) gives the displacements at the end of the time interval. Equation (4.24) is applicable for $s = 1, 2, 3, \ldots$; a modified equation is required to evaluate $\mathbf{x}_1$. The central difference method of numerical integration, which is based on standard first and second order differences with regard to time, leads to equation (4.24) with $\beta = 0$.

A recurrence relation, similar to equation (4.24), can be obtained for other methods. Although the latter depend on different mathematical or physical assumptions, they can be shown to be special cases of general procedures. Using a difference operator formulation, Krieg and Key[31] show that the standard methods, and also that recently developed by Hilber, Hughes, and Taylor,[29] are special cases of a general equation containing four parameters. The Newmark recurrence relation (4.24) is obtained by substituting equations (4.23) in equation (4.1) evaluated at the three successive times t_{s+1}, t_s and t_{s-1}. In its most general form the Newmark method contains a second parameter γ, which has been assumed to be $\frac{1}{2}$ in the first of equations (4.23) and consequently in equation (4.24). By expressing the displacement vector in terms of shape functions (i.e. using a standard finite element prodedure with respect to time), and using weighted residuals, Zienkiewicz[32] obtains the generalized form of equation (4.24) when three consecutive time stations are used; the parameters β and γ are defined in terms of integrals containing the weighting functions. Further, using this procedure for four stations, a general recurrence relation containing three parameters is obtained; he shows that the Houbolt and Wilson algorithms correspond to particular values of these parameters. It is of interest to speculate whether one set of the values of these parameters, corresponding to a particular standard method, can be demonstrated to be the optimum.

In general, equation (4.24) (and similar equations obtained by other methods) represents n simultaneous equations, which must be solved at each time step. If $\mathbf{M}$ is diagonal, $\mathbf{C}$ is diagonal or zero and the central difference method is used, explicit expressions for $x_1, x_2, \ldots$ are obtained. For large systems and when a large number of time steps must be used, the existence of these explicit expressions reduces the quantity of computation significantly. Although primarily concerned here with linear problems, numerical integration methods, unlike the normal mode method, can be applied to nonlinear problems, for which the computational advantages of having explicit expressions are considerably greater as several iterative solutions of the set of equations may be required at each time step. The improved diagonal mass matrices, mentioned in Section 4.2.3, have been developed so that when damping is neglected explicit expressions for x_s are obtained.

However, the central difference method (and other explicit methods) is only conditionally stable, i.e. if too large a time step Δt is used, the computations become numerically unstable. The Houbolt method and the Newmark β method with $\beta \geqslant 1/4$ are unconditionally stable (hence the use of $\beta = 1/4$ for practical calculations with the Newmark method). For the central difference method the stability condition is $\Delta t/T_n \leqslant 1/\pi$ or $\omega_n \Delta t \leqslant 2$, where T_n and ω_n are the period and circular frequency of the nth or highest mode of the system. Numerical integration methods do not require the calculation of the natural frequencies, so some estimate of ω_n is required. Considering the response of shells, Underwood[33] gives $C_0 \Delta t \leqslant \Delta l$, where Δl is the smallest spatial mesh size and $C_0 (= \sqrt{E/\rho})$ is the velocity of extensional waves in the medium, as the stability criterion—i.e. the time step must not exceed the time for a wave to travel across the smallest element dimension. If a long one-dimensional mesh of simple bar elements is considered, the period of the highest mode T_n is $\pi \Delta l / C_0$ when lumped mass matrices are used, agreeing with Underwood's criterion. (A somewhat lower value of T_n, and a correspondingly lower value of Δt at the stability limit, is obtained with consistent mass matrices.) Comparable values for Δt for flexural deformations and for two-demensional meshes exist.[6,34] In general, mesh refinement (reduction of Δl) leads to a lower stability limit Δt.

4.6.2 Accuracy

Considering the response to comprise contributions from normal modes, the approximations in numerical integration methods usually cause artificial attenuation of the response and some error in the period of the mode predicted by the numerical solution. These two effects increase as $\Delta t/T_r$ increases, where T_r is the period of the mode whose contribution is being considered. However, for the Newmark method with $\beta = 1/4$ the period error occurs without the artificial attenuation. Bathe and Wilson[30] survey several methods and comment on accuracy. From some examples with the Newmark method ($\beta = 1/4$) they show that the percentage error in period is 1 per cent and 10 per cent approximately for $\Delta t/T_r$ equal to 0.06 and 0.18 respectively. Thus a possible criterion for reasonable accuracy is $\Delta t/T_j \leqslant 0.06$, where T_j is the period of the highest mode making a significant contribution to the response. However, an estimate for T_j in a practical problem is difficult. For large problems only a fraction of the modes make a significant contribution, i.e. j/n is small, thus T_j/T_n is large and the criterion $\Delta t/T_j \leq 0.06$ may lead to a larger value of Δt than that predicted by the stability criterion for the central difference method. For the Houbolt method Johnson and Greif[34] give $\Delta t/T_j \leq 0.02$ as an accuracy criterion in order to ensure that the jth modal contribution to the response is not artificially attenuated.

Hilber, Hughes, and Taylor,[29] and earlier Bathe and Wilson,[30] recommend that, as often only the response from the lower modes is of interest, it is advantageous for an algorithm to possess some artificial attenuation or numerical dissipation in order that the spurious response from higher modes is damped out. Controllable dissipation is included, together with unconditional stability and second order accuracy, in the desirable attributes of any numerical integration method by Hilber and Hughes.[62] They propose a family of algorithms, the α methods, in which the parameter α controls the numerical dissipation; when $\alpha = 0$, it is identical to the Newmark average acceleration method. They obtain the desired numerical dissipation, but as expected there is some degradation in accuracy as α departs from zero. Accepting their criteria, the α methods are demonstrated to be superior to the family based on the Wilson θ method and to the Houbolt and Park methods. However, their stipulation of unconditional stability rules out the central difference method, which has computational advantages, and the Newmark method with $\beta = 1/12$, which has third order accuracy. Also their comprehensive study is based on the matrix equation (4.1) without the viscous damping term $\mathbf{C}\dot{\mathbf{x}}$. Even when the idealization of equation (4.1) represents accurately all the modes which make a significant contribution to the response, it does not follow that the Newmark average acceleration method (i.e. $\beta = 1/4$ and hence no numerical dissipation) is necessarily the best algorithm, as the period errors will cause modal contributions to be combined with incorrect relative phase angles and thus the maximum response may be underestimated or overestimated. Nevertheless, several investigators[26,35,36] consider that this method is the most accurate for problems in linear structural dynamics. Recently Park[37] has shown that introduction of numerical dissipation is not advantageous for explicit schemes, including the central difference method.

The above comments refer mainly to errors caused by the numerical integration procedure. As this is applied to equation (4.1), which represents a finite element idealization of a complex structure, there may also be errors due to this idealization. Krieg and Key[31] consider both types of error and also accuracy of computation by studying one-dimensional meshes in extension and flexure. It is possible to chose well-matched methods, i.e. the errors due to the spatial and temporal approximations tend to cancel; in particular, the diagonal mass matrix (the improved version mentioned earlier)[6] and the central difference integration method form a good combination in terms of accuracy and economy. Better accuracy could be obtained with the same mesh of elements by combining a non-zero value of β and a non-diagonal mass matrix, but computing costs would be increased appreciably.

4.6.3 Nonlinear problems

Broadly there are two types: material nonlinearity, where stress is not proportional to strain; and geometric nonlinearity, where additional terms allowing for large deformations must be included in the analysis. The algorithms discussed in previous sections can be used for nonlinear problems, but new criteria for accuracy and stability of computation are required. The literature contains examples of numerical instability in response calculations for nonlinear structures, when an algorithm that is unconditionally stable for linear problems has been used.[38]

The computational advantages for nonlinear problems of the explicit central difference method have been mentioned (Section 4.6.1) and this method has been used in several investigations.[39] Tillerson and Stricklin[40] and Wu and Witmer[41] compared the Houbolt, Newmark average acceleration, and central difference methods for nonlinear materials problems and concluded that the first of these methods is probably the most efficient. McNamara[42] included the Wilson method in his comparison. Park[26] has developed an improved method with an apparently superior performance to that of Houbolt. Belytschko and Schoeberle[43] have introduced an energy criterion with an iterative procedure to be followed until this criterion is satisfied. The Newmark average acceleration method is unconditionally stable when this procedure is applied to problems with material nonlinearities. Recently Hughes[44] has extended this procedure to include geometric nonlinearities.

When determining the internal force matrix **F**, which replaces the linear term **Kx** in equation (4.1) and is a nonlinear function of the displacement vector **x**, either the fictitious force or the tangential stiffness method is normally used to obtain an effective stiffness matrix at each time step. Belytschko, Chiapetta, and Bartel[45] describe a direct method of determining the nodal force matrix without recourse to a stiffness matrix and show that for large meshes of simple elements and using explicit time integration this procedure results in an efficient computational scheme.

4.7 DYNAMIC INTERACTION PROBLEMS

The methods discussed in previous sections are applicable when the exciting forces are known and can be extended to deal with excitation by known displacements or accelerations of the foundation or base of the structure. However, there exists a growing class of important dynamic problems, where the excitation mechanism is affected by the properties of the surrounding or underlying medium. Thus a proper determination of structural response requires some dynamic analysis of the medium. Broadly there are two types, interaction between soil and structure and between fluid and structure. The former occurs when determining the response of structures to earthquakes, as the known input may be the acceleration of the free surface of the soil, i.e. without the structure present. It occurs also when the source of excitation is located in one structure and the response of neighbouring structures is required. The latter occurs when the coupled vibrations of water and dams, piers or offshore structures are considered. Indeed, both types of interaction may have to be considered, if, for example, the response of a dam to an earthquake is required. When aerodynamic effects are included, wind-induced oscillations of structures exhibit interaction effects. We will consider briefly the use of the finite element method in soil–structure interaction problems. (Some similar problems occur for fluid–structure interaction, particularly if fluid compressibility is allowed for in the analysis.)

There are two general approaches, depending upon whether a finite element or a continuum model is used for the ground; in both cases a finite element model is used for the structure. Hadjian, Luco, and Tsai[46] have compared the two approaches. In the first approach practical factors, including variation of soil properties with depth, strain-dependent soil properties, and irregularly shaped foundations, can be incorporated. Strictly these problems are three-dimensional and the ground should be represented by a mesh of three-dimensional finite elements. However, this leads to a system with a very large number of degrees of freedom and consequently to excessive computing costs. For long structures, e.g. dams, two-dimensional (plane strain) finite elements can be used; also axisymmetric elements are appropriate for some problems. Nevertheless, the use of two-dimensional elements for general three-dimensional situations, e.g. the interaction of soil with nuclear power plant, may lead to errors in response.[46] For any finite element model of an infinite domain the boundaries of the finite model can generate spurious reflected waves that obscure the true response; Lysmer and Kuhlemeyer[47] have devised a method for synthesizing non-reflecting boundaries by introducing viscous dampers at boundary nodes. A refinement of this method has been given recently by White, Valliappan, and Lee.[48] Weaver, Brandow, and Höeg[49] have used this method and their numerical results demonstrate the necessity to allow for interaction when determining the response of structures to earthquakes. The problem of representing an infinite domain by a finite number of elements exists in other fields; thus extension of the concepts of infinite elements[50,51] and boundary elements, based on boundary solution procedures,[52] to soil problems will be useful and is likely to be developed soon, Usually many degrees of freedom of the overall model will be associated with soil motion, although the response of the structure to prescribed excitation is of prime interest. Substructuring techniques, mentioned earlier, reduce the size of the matrices to be handled.

In seismic problems, when substructuring is not used, the input must be specified at the base of the finite element model, which ideally represents bedrock. However, available information is in the form of free-surface ground acceleration records. Deconvolution of this surface motion, using a one-dimensional model of shear waves propagating vertically, can be used to generate the corresponding motion of the bedrock;[53] an alternative procedure has been proposed by Clough and Penzien.[16] Nevertheless, controversy still exists regarding the proper procedure.[54]

In addition to demonstrating the advantages of using substructures in interaction problems, whether the soil is modelled by finite elements or a continuum, Gutierrez and Chopra[55] relax two assumptions, which have been common to most earlier work, by allowing for footing flexibility and for spatial variations in the motion along the soil–structure interface. No doubt these two important effects will be the subject of further studies.

Lee[56] has surveyed the continuum approach; recent work on layered media and embedded foundations has increased its scope. Only the degrees of freedom of the structure and its foundation have to be considered, but the force–displacement relation between a rigid foundation and the ground is required. In general, the stiffness and damping matrices, which occur in this relation, are frequency dependent. For earthquake or other transient excitation yielding an effective force vector $\mathbf{p}(t)$, the corresponding Fourier transform $\mathbf{P}(\omega)$ is formed, the Fourier transform $\mathbf{X}(\omega)$ of the response $\mathbf{x}(t)$ found by solving the standard matrix equation by the frequency response method and the inverse transform used to determine $\mathbf{x}(t)$.[10] Although the response vector $\mathbf{x}(t)$ is obtained from $\mathbf{X}(\omega)$, and also $\mathbf{P}(\omega)$ from $\mathbf{p}(t)$, using the fast Fourier transform algorithm, which is highly efficient,[57] $\mathbf{X}(\omega)$ has to be calculated from $\mathbf{P}(\omega)$, using equation (4.7) in the form $\mathbf{X}(\omega) = \mathbf{J}^{-1}\mathbf{P}(\omega)\exp(i\omega t)$ for a large number of values of ω. This requires considerable computation for a finite element model for which the order of the matrices is high. For many practical excitations, the response is limited to the first few modes; this has led to methods of solution which economize in computation.[55,58,59] Richart, Hall, and Woods[60] discuss approximate, frequency-independent forms for the stiffness and damping interaction matrices, and these have been used in practical problems.[61]

4.8 CONCLUDING REMARKS

Use of the finite element method in structural dynamics has emphasized the importance of the determination of natural frequencies and response for large multi-degree-of-freedom systems. Although standard methods of solution have existed for many years, considerable effort is being expended to increase the size of problems that can be solved with acceptable accuracy and efficiency of computation and to gain more knowledge of the effect of approximations and simplifications on accuracy.

There is increasing interest in nonlinear problems with earthquake engineering supplying many of the dynamic problems; this stimulates the current, significant research activity on numerical integration methods and their stability, accuracy, and efficiency of computation. Dynamic interaction between structures and soil or fluid is another important area where the use of finite elements in dynamic problems is increasing our current knowledge.

REFERENCES

1. J. P. Den Hartog, *Mechanical Vibrations*, 4th edn, McGraw-Hill, New York, 1956.
2. O. C. Zienkiewicz, *The Finite Element Method*, 3rd edn, McGraw-Hill, London, 1977.
3. C. A. Brebbia and J. J. Connor, *Fundamentals of Finite Element Techniques*, Butterworths, London, 1973.
4. D. H. Norrie and G. de Vries, *Finite Element Bibliography*, Plenum Press, New York, 1976.
5. J. R. Whiteman, *A Bibliography for Finite Elements*, Academic Press, London, 1975.
6. S. W. Key and Z. E. Beisinger, 'The transient dynamic analysis of thin shells by the finite element method', *Proceedings of the 3rd Conference on Matrix Methods in Structural Mechanics*, Dayton, 1971.
7. E. Hinton, T. Rock, and O. C. Zienkiewicz, 'A note on mass lumping and related processes', *Earthquake Engineering and Structural Dynamics*, **4,** 245–249 (1975).
8. K. J. Bathe and E. L. Wilson, 'Solution methods for eigenvalue problems in structural mechanics', *International Journal for Numerical Methods in Engineering*, **6,** 213–226 (1973).
9. K. J. Bathe and E. L. Wilson, 'Large eigenvalue problems in dynamic analysis', *Journal of the Engineering Mechanics Division, ASCE*, **98,** 1471–1485 (1972).
10. G. B. Warburton, 'The dynamical behaviour of structures', 2nd edn, Pergamon, Oxford, 1976.
11. R. D. Henshell and J. H. Ong, 'Automatic masters for eigenvalue economization', *Earthquake Engineering and Structural Dynamics*, **3,** 375–383 (1975).
12. W. C. Hurty, 'Dynamic analysis of structural systems using component modes', *AIAA Journal*, **3,** 678–685 (1965).

13. S. Rubin, 'Improved component mode representation for structural dynamic analysis', *AIAA Journal*, **13,** 995–1006 (1975).
14. R. M. Hintz, 'Analytical methods in component modal synthesis', *AIAA Journal*, **13,** 1007–1016 (1975).
15. T. K. Caughey and M. E. J. O'Kelly, 'Classical normal modes in damped linear dynamic systems', *Journal of Applied Mechanics*, **32,** 583–588 (1965).
16. R. W. Clough and J. Penzien, 'Dynamics of structures', McGraw-Hill, New York, 1975.
17. D. J. Ewins, 'Estimation of resonant peak amplitudes', *Journal of Sound and Vibration*, **43,** 595–605 (1975).
18. R. W. Clough and S. Mojtahedi, 'Earthquake response analysis considering non-proportional damping', *Earthquake Engineering and Structural Dynamics*, **4,** 489–496 (1976).
19. W. T. Thomson, T. Calkins, and P. Caravani, 'A numerical study of damping', *Earthquake Engineering and Structural Dynamics*, **3,** 97–103 (1974).
20. G. B. Warburton and S. R. Soni, 'Errors in response calculations for non-classically damped structures', *Earthquake Engineering and Structural Dynamics*, **5,** 365–376 (1977).
21. G. B. Warburton, 'Some effects of dynamic interaction on the response of off-shore structures', *Euromech Colloquium No. 96, Numerical Analysis of Dynamic Interaction of Structures with Fluids*, University College of Swansea, 1977.
22. Y. C. Fung and M. V. Barton, 'Some shock spectra characteristics and uses', *Journal of Applied Mechanics*, **25,** 365–372 (1958).
23. L. S. Jacobsen and R. S. Ayre, *Engineering Vibrations*, McGraw-Hill, New York, 1958.
24. N. M. Newmark, 'A method of computation for structural dynamics', *Journal of the Engineering Mechanics Division, ASCE*, **85,** No. EM3, 67–94 (1959).
25. J. C. Houbolt, 'A recurrence matrix method for the dynamic response of elastic aircraft', *Journal of the Aeronautical Sciences*, **17,** 540–550 (1950).
26. K. C. Park, 'An improved stiffly stable method for direct integration of nonlinear structural dynamic equations', *Journal of Applied Mechanics*, **42,** 464–470 (1972).
27. R. J. Melosh, 'Integration of linear equations of motion', *Journal of the Structural Division, ASCE*, **101,** 1551–1558 (1975).
28. A. J. Medland, 'The computation of the dynamic deflections of an impacted structure by a step-by-step finite element procedure', *Computing Developments in Experimental and Numerical Stress Analysis* (Ed. P. Stanley), Applied Science Publishers, London, 1976.
29. H. M. Hilber, T. J. R. Hughes, and R. L. Taylor, 'Improved numerical dissipation for time integration algorithms in structural dynamics', *Earthquake Engineering and Structural Dynamics*, **5,** 283–292 (1977).
30. K. J. Bathe and E. L. Wilson, 'Stability and accuracy analysis of direct integration methods', *Earthquake Engineering and Structural Dynamics*, **1,** 283–291 (1973).
31. R. D. Krieg and S. W. Key, 'Transient shell response by numerical time integration', *International Journal for Numerical Methods in Engineering*, **7,** 273–286 (1973).
32. O. C. Zienkiewicz, 'A new look at the Newmark, Houbolt and other time

stepping formulas, a weighted residual approach', *Earthquake Engineering and Structural Dynamics*, **5,** 413–418 (1977).
33. P. Underwood, 'Transient response of inelastic shells of revolution', *Computers and Structures*, **2,** 975–989 (1972).
34. D. E. Johnson and R. Greif, 'Dynamic response of a cylindrical shell: two numerical methods', *AIAA Journal*, **4,** 486–494 (1966).
35. R. E. Nickell, 'Direct integration methods in structural dynamics', *Journal of the Engineering Mechanics Division, ASCE*, **99,** 303–317 (1973).
36. J. M. Smith, 'Recent developments in numerical integration', *Journal of Dynamic Systems, Measurement and Control, Trans. ASME*, **96G,** 61–70 (1974).
37. K. C. Park, 'Practical aspects of numerical time integration', *Computers and Structures*, **7,** 343–353 (1977).
38. R. E. Ball, W. F. Hubka, N. J. Huffington Jr., P. Underwood, and W. A. Von Riesemann, 'A comparison of computer results for the dynamic response of the LMSC truncated cone', *Computers and Structures*, **4,** 485–498 (1974).
39. D. Shantaram, D. R. J. Owen, and O. C. Zienkiewicz, 'Dynamic transient behaviour of two- and three-dimensional structures including plasticity, large deformation effects and fluid interaction', *Earthquake Engineering and Structural Dynamics*, **4,** 561–578 (1976).
40. J. R. Tillerson and J. A. Stricklin, 'Numerical methods of integration applied in the nonlinear dynamic analysis of shells of revolution', NASA CR-108639, Washington DC, 1970.
41. R. W. H. Wu and E. A. Witmer, 'Nonlinear transient responses of structures by a spatial finite element method', *AIAA Journal*, **11,** 1110–1117 (1973).
42. J. F. McNamara, 'Solution schemes for problems of nonlinear structural dynamics', *Journal of Pressure Vessel Technology, Trans. ASME*, **96J,** 96–102 (1974).
43. T. Belytschko and D. F. Schoeberle, 'On the unconditional stability of an implicit algorithm for nonlinear structural dynamics', *Journal of Applied Mechanics*, **42,** 865–869 (1975).
44. T. J. R. Hughes, 'A note on the stability of Newmark's algorithm in nonlinear structural dynamics', *International Journal for Numerical Methods in Engineering*, **11,** 383–386 (1977).
45. T. Belytschko, R. L. Chiapetta, and H. D. Bartel, 'Efficient large scale nonlinear transient analysis by finite elements', *International Journal for Numerical Methods in Engineering*, **10,** 579–596 (1976).
46. A. H. Hadjian, J. E. Luco, and N. C. Tsai, 'Soil–structure interaction: continuum or finite element', *Nuclear Engineering and Design*, **31,** 151–167 (1974).
47. J. Lysmer and R. S. Kuhlemeyer, 'Finite dynamic model for infinite media', *Journal of the Engineering Mechanics Division, ASCE*, **95,** 859–877 (1969).
48. W. White, S. Valliappan, and I. K. Lee, 'Unified boundary for finite dynamic models', *Journal of the Engineering Mechanics Division, ASCE*, **103,** 949–964 (1977).
49. W. Weaver Jr., G. E. Brandow, and K. Höeg, 'Three-dimensional soil–structure response to earthquakes', *Bulletin of the Seismological Society of America*, **63,** 1041–1056 (1973).
50. P. Bettess, 'Infinite elements', *International Journal for Numerical Methods in Engineering*, **11,** 53–64 (1977).
51. S. S. Saini, P. Bettess, and O. C. Zienkiewicz, 'Coupled hydrodynamic response

of concrete gravity dams using finite and infinite elements', *Earthquake Engineering and Structural Dynamics*, **6,** 363–374 (1978).
52. O. C. Zienkiewicz, D. W. Kelly, and P. Bettess, 'The coupling of the finite element method and boundary solution procedures', *International Journal for Numerical Methods in Engineering,* **11,** 355–375 (1977).
53. I. M. Idriss and K. Sadigh, 'Seismic SSI of nuclear power plant structures', *Journal of the Geotechnical Engineering Division, ASCE,* **102,** 663–682 (1976).
54. E. L. Wilson, 'Finite elements for foundations, joints and fluids', Chapter 10 of *Finite Elements in Geomechanics* (Ed. G. Gudehus), Wiley, London, 1977.
55. J. A. Gutierrez and A. K. Chopra, 'A substructure method for earthquake analysis of structures including structure–soil interaction', *Earthquake Engineering and Structural Dynamics,* **6,** 51–69 (1978).
56. T. H. Lee, 'Soil–structure interaction—nuclear reactors. The continuum approach', *Shock and Vibration Digest,* **8,** No. 6, 15–23, (1976).
57. E. O. Brigham, *The Fast Fourier Transform,* Prentice-Hall, Englewood Cliffs, N.J., 1974.
58. P. C. Jennings and J. Bielak, 'Dynamics of building soil interaction', *Bulletin of the Seismological Society of America,* **63,** 9–48 (1973).
59. A. K. Chopra and J. A. Gutierrez, 'Earthquake response analysis of multistorey buildings including foundation interaction', *Earthquake Engineering and Structural Dynamics,* **3,** 65–77 (1974).
60. F. E. Richart Jr., J. R. Hall Jr., and R. D. Woods, '*Vibration of Soils and Foundations*, Prentice-Hall, Englewood Cliffs, N. J., 1970.
61. N. C. Tsai, D. Niehoff, M. Swatta, and A. H. Hadjian, 'The use of frequency-independent soil–structure interaction parameters', *Nuclear Engineering and Design,* **31,** 168–183 (1974).
62. H. M. Hilber and T. J. R. Hughes, 'Collocation, dissipation and "overshoot" for time integration schemes in structural dynamics', *Earthquake Engineering and Structural Dynamics,* **6,** 99–117 (1978).

Chapter 5

Marriage à la mode—*The Best of Both Worlds (Finite Elements and Boundary Integrals)*

O. C. Zienkiewicz, D. W. Kelly, and P. Bettess

5.1 INTRODUCTION

The finite element method and boundary solution (Trefftz) methods[1] are general approximation processes applicable to a wide variety of engineering problems. Both are discretization procedures based on trial functions and the main difference between the methods lies in the criteria used for selection of these functions. A different discretization of the domain follows as a result of the properties of the trial functions.

Exact solutions satisfying both the governing equations and the boundary conditions of a given boundary value problem are not in general available. The boundary solution procedures therefore choose trial functions which satisfy the governing equations *a priori* and attempt to satisfy the boundary conditions by a collocation, least-squares, or Galerkin procedure. The solution is therefore confined to the boundaries of the region.

The finite element method, on the other hand, does not require that the trial functions satisfy the governing equations but they retain the ability to satisfy many boundary conditions identically. The problem is then defined over the domain and variational principles form the most natural procedure for seeking approximate satisfaction of the governing equations.

In the finite element method the domain is divided into a discrete number of elements, over each of which the trial functions apply. The solution then generates both interior and surface values simultaneously. However, the boundary integral methods, which are a particular form of boundary solution, deal only with the boundaries of the region. These are invariably divided into surface elements over which interpolation functions for the surface behaviour are coupled to the solutions to the governing equations which apply over the domain. The resulting equation is then solved numerically for values on the boundary alone and values at interior points are calculated subsequently from the surface data.

Paper presented at the International Symposium on Innovative Numerical Analysis in Applied Engineering Science, Versailles (France), May 1977.

The Trefftz type trial functions are usually confined to regions in which the governing equations are linear and homogeneous so that the finite element methodology is better suited to problems in which the material properties are not homogeneous and nonlinearity occurs. On the other hand the finite element process is restricted to problems which are bounded and the boundary integral procedures are obviously better suited to problems in which the domain extends to infinity. The confinement of the boundary integral processes to the surface of the domain also reduces the order of the problem by one which is of particular relevance in three-dimensional problems if the volume to surface ratio is large.

The arguments relating to two procedures have been set out elsewhere[2,3] and it has been suggested that an optimal solution process (based on discretization and solution using trial functions) might result if the trial functions are not bound to either selection criterion over the entire domain. Ideas similar to these had already been expounded in the literature (for example, References 4, 5, 6, 7, and others) for exterior boundary solution with the nominal aim of producing the correct boundary conditions on the truncated finite element mesh. A more fundamental indication of the similarity between the procedures has more recently appeared in the formulations of boundary solution processes based on variational principles.[2,8] Loof[9] and others have also suggested the generation of macro-finite elements using solutions of the governing differential equations.

If a direct coupling strategy is used the equations based on the Trefftz form of trial functions can be directly linked to the finite element equations and one such procedure will be discussed here. However, the discretization generally used when formulating the approximation to the B. I. equation involves a collocation procedure and leads to fully coupled and non-symmetric matrices. In such a form the approximation is not generally suitable for a 'marriage' with a finite element discretization where, for self-adjoint problems, banded and symmetric equations arise.

Two quite general procedures for generating symmetric element matrices using the Trefftz procedures will therefore be discussed here. In the first the starting point of the *direct B. I. discretization* process follows the standard collocation procedure. From this an approximate 'energy' expression is generated and a symmetric coupling matrix derived. Here the comparative disadvantage of having to invert a matrix equal in order to the number of collocation points arises but this is compensated for by the possibility of using directly available programs for B.I. solution.

For the second process the B.I. of the direct type (i.e. not involving extraneous variables) is replaced by *an indirect form* in which an auxiliary variable such as the density of distributed sources on the boundary is used. This allows the independent interpolation of such auxiliary variables and retains these in the final solution thus avoiding the matrix inversion. This procedure has an additional merit of allowing a different number of degrees of freedom for the Trefftz 'element' to be used from that of the nodes connecting the interface boundary and the addition of supplementary solutions becomes a trivial extension. However, these advantages are counter-balanced by an additional complexity of integration.

The examples presented are mainly concerned with exterior domains where the advantage of B.I. elements is pronounced. However, for com-(pleteness some simple examples of the interior kind are presented and indicate the viability of the more general ideas proposed, thus encouraging continuing work in this area.

5.2 THE DISCRETIZATION PROCEDURES

We shall illustrate our processes of discretization on a simple Laplace equation and shall show later how the same method can be applied to the Helmholtz equation or elasticity problems. Here we thus seek solutions ϕ which satisfy

$$\nabla^2\phi = 0 \quad \text{in} \quad \Omega \tag{5.1}$$

with

$$\phi = \bar{\phi} \quad \text{on} \quad \Gamma_\phi \tag{5.2a}$$

and

$$\frac{\partial \phi}{\partial n} = \overline{\frac{\partial \phi}{\partial n}} \quad \text{on} \quad \Gamma_q \tag{5.2b}$$

where Ω is a region with boundary Γ.

For *solution by the F.E. method*[10] the region is divided into m elements. The solution to (5.1) satisfying (5.2b) will then minimize the functional

$$\pi = \sum_{l=1}^{m} \left[\int_{\Omega_l} \tfrac{1}{2}\left(\left(\frac{\partial \phi}{\partial x}\right)^2 + \left(\frac{\partial \phi}{\partial y}\right)^2 \right) d\Omega - \int_{\Gamma_{ql}} \phi \overline{\frac{\partial \phi}{\partial n}} \, d\Gamma \right] \tag{5.3}$$

Solution follows by selecting trial functions to specify ϕ on each element in the form

$$\phi(x, y) = \sum N_i(x, y)\phi_i = \mathbf{N}(x, y)\hat{\boldsymbol{\phi}} \tag{5.4}$$

where the $\hat{\boldsymbol{\phi}}$ are nodal values. Requiring $\partial\pi/\partial\hat{\boldsymbol{\phi}} = 0$ for the minimum leads to the matrix equation

$$\mathbf{K}\hat{\boldsymbol{\phi}} + \mathbf{F} = \mathbf{0} \tag{5.5}$$

with $\mathbf{K} = \sum_{i=1}^{m} \mathbf{K}_i$ being a symmetric matrix and $\mathbf{F} = \sum_{i=1}^{m} \mathbf{F}_i$. Boundary conditions (5.2a) are automatically included in the solution of (5.5).

In contrast the *direct B.I. formulation* is based on a reciprocal relation which for Laplace's equation is given by Green's Identity. This identity states that for all functions ψ

$$\int_\Omega (\phi\nabla^2\psi - \psi\nabla^2\phi)\,d\Omega = \oint_\Gamma \left(\phi\frac{\partial\psi}{\partial n} - \psi\frac{\partial\phi}{\partial n}\right) d\Gamma \tag{5.6}$$

leading, after selection of ψ to be an appropriate singular Green's function $G(p, \Gamma)$, to

$$\alpha\phi(p) = \oint_\Gamma \frac{\partial G(p,\Gamma)}{\partial n_\Gamma}\,\phi(\Gamma)\,d\Gamma - \oint_\Gamma G(p,\Gamma)\frac{\partial\phi}{\partial n_\Gamma}(\Gamma)\,d\Gamma \tag{5.7}$$

where the subscript Γ denotes the position at which the normal derivative is taken, α is a constant and p defines the position at which ϕ is sampled. Note here that once $\phi(\Gamma)$ is defined $\partial\phi/\partial n(\Gamma)$ is uniquely specified but solution proceeds as an approximation assuming independent interpolation functions for the variation of both ϕ and $\partial\phi/\partial n$ on boundary segments in the form

$$\phi(\Gamma) = \sum N_i(\Gamma)\phi_i = \mathbf{N}(\Gamma)\hat{\boldsymbol{\phi}}$$

and

$$\frac{\partial\phi}{\partial n}(\Gamma) = \sum \mathrm{M}_i(\Gamma)\left(\frac{\partial\phi}{\partial n}\right)_i = \mathbf{M}(\Gamma)\widehat{\frac{\partial\boldsymbol{\phi}}{\partial\mathbf{n}}} \tag{5.8}$$

with (5.7) being then imposed at a correct number of collocating points. Substitution in (5.7) leads to a set of linear relations

$$\mathbf{A}\hat{\boldsymbol{\phi}} = \mathbf{B}\widehat{\frac{\partial\boldsymbol{\phi}}{\partial\mathbf{n}}} \tag{5.9}$$

which can be solved for the unknown values of either $\hat{\boldsymbol{\phi}}$ or $\widehat{\partial\boldsymbol{\phi}/\partial\mathbf{n}}$.

Global equilibrium in the form

$$\oint \frac{\partial\phi}{\partial n}\,d\Gamma = 0 \tag{5.10}$$

is satisfied automatically by the exact solution to the integral equation (5.7). However, there appears no guarantee that the solution of the discretized form represented by (5.9) will satisfy this condition.

The *indirect formulation of the B.I. equation* on the other hand recognizes that the solution can be simulated by a surface distribution of point solutions giving

$$\phi(p) = \oint G(p,\Gamma)\mu(\Gamma)\,d\Gamma \tag{5.11}$$

and

$$\frac{\partial \phi}{\partial n}(p) = \oint \frac{\partial G}{\partial n_p}(p, \Gamma)\mu(\Gamma)\, d\Gamma + \beta\mu(p) \tag{5.12}$$

where β is a constant on the boundary and zero internally. Here the source distribution defined on the boundary is alone interpolated as

$$\mu(\Gamma) = \sum M_i(\Gamma)\hat{\mu}_i = \mathbf{M}(\Gamma)\hat{\boldsymbol{\mu}} \tag{5.13}$$

and the determination of the discrete nodal values of this distribution using boundary values of flux or potential at specified collocation points leads to the solution of the problem.

The matrices in (5.9) and those derived after substitution of (5.13) into (5.11) and (5.12) are in general non-symmetric and fully populated. In contrast the matrix **K** in (5.5) is symmetric and banded. A detailed comparison of the direct and indirect boundary integral formulations is contained in the appendix at the end of this chapter.

5.3 THE TRIAL FUNCTIONS

The selection of trial functions in the F.E. methodology is based on interelement compatibility and completeness criteria to ensure convergence. The trial functions over the element must be chosen such that the expressions within the integrals in (5.3) are well behaved as the element size decreases. Also the validity of the summation in (5.3) must be preserved by requiring a specified level of continuity of the field variable and its derivatives between neighbouring elements.[10] These trial functions $N_i(x, y)$ are therefore typically polynomial interpolants between nodal values and span the domain of the problem on a mesh of finite elements with

$$\phi(x, y) = \sum N_i(x, y)\hat{\phi}_i = \mathbf{N}(x, y)\hat{\boldsymbol{\phi}} \tag{5.14}$$

To obtain trial functions from the B.I. formulations we first require a fundamental solution to the governing equations, that is a solution to the differential equation for a dirac delta 'forcing function'. This fundamental solution to (5.1) is the solution of

$$\nabla^2 G(p, q) = \delta(q - p) \tag{5.15}$$

where

$$\delta(q-p) = 0 \qquad p \neq q$$

and

$$\int_{\Omega_\infty} \delta(q-p)\, d\Omega = 1$$

where Ω_∞ denotes the infinite region in two or three dimensions. Typical solutions to (5.15) are the free space Green's functions with

$$G(p, q) = \frac{1}{r(p, q)} \qquad \text{in three dimensions}$$

and

$$G(p, q) = ln r(p, q)/r_1 \qquad \text{in two dimensions} \tag{5.16}$$

where $r(p, q)$ is the distance between the field point p and the singular point q, r_1 is a reference length.

These Green's functions for the Laplace equation apply to infinite regions and all boundaries must be imposed in the boundary solution process. They span the domain of the problem and we obtain the trial functions by specifying the variation of ϕ and $\partial\phi/\partial n$ on the surface elements.

For the indirect solution substituting from (5.13) into (5.12) and (5.11) we obtain

$$\phi(p) = \bar{\mathbf{N}}'(p)\hat{\boldsymbol{\mu}} \tag{5.17a}$$

from which the gradients follow as

$$\frac{\partial\phi}{\partial n}(p) = \bar{\mathbf{M}}'(p)\hat{\boldsymbol{\mu}} \tag{5.17b}$$

This defines ϕ in a similar way to the finite element method by trial functions and a set of variables $\hat{\boldsymbol{\mu}}$ defined on the boundary alone.

Alternatively, in the direct solution substituting from (5.8) into (5.7)

$$\phi(p) = \oint_\Gamma \frac{\partial G}{\partial n_\Gamma}(p, \Gamma)\mathbf{N}(\Gamma)\, \mathrm{d}\Gamma\, \hat{\boldsymbol{\phi}} - \oint_\Gamma G(p, \Gamma)\mathbf{M}(\Gamma)\, \mathrm{d}\Gamma\, \widehat{\frac{\partial \boldsymbol{\phi}}{\partial \mathbf{n}}} \tag{5.18}$$

which can be written

$$\phi(p) = \bar{\mathbf{N}}(p)\hat{\boldsymbol{\phi}} - \bar{\mathbf{M}}(p)\widehat{\frac{\partial \boldsymbol{\phi}}{\partial \mathbf{n}}}$$

Defining

$$\hat{\mathbf{a}} = \left[\hat{\boldsymbol{\phi}}, \widehat{\frac{\partial \boldsymbol{\phi}}{\partial \mathbf{n}}}\right]$$

gives

$$\phi(p) = \bar{\mathbf{N}}(p)\hat{\mathbf{a}} \tag{5.19}$$

again a typical trial function form but best suited to collocation procedures with the discrete parameters confined to the boundary of the region. However, we note that in above $\hat{\boldsymbol{\phi}}$ and $\widehat{\partial\boldsymbol{\phi}/\partial\mathbf{n}}$ cannot be independently specified as they are related by the identity when p corresponds to a point at which a discrete value of ϕ is defined.

Special fundamental solutions can be derived for prescribed problems to satisfy, for example, the free surface boundary condition on a half space *a priori*. The solution process then does not require discretization of this surface. Corresponding fundamental solutions can also be derived for the Navier equations for elasticity. Here Betti's theorem provides the reciprocal relationship for the B.I. formulation replacing (5.6).

5.3.1 Extension to Helmholtz equation

An equation frequently encountered in wave problems is one of the Helmholtz type which can be written as

$$\nabla^2\phi - k^2\phi = 0 \tag{5.20}$$

in which the variable ϕ is generally complex. The boundary conditions of type (5.2) are still generally specified and the solution of this equation in the finite element context presents no additional difficulties to that of the Laplace equation outlined.

The formulation of the boundary integral equation proceeds once more following precisely the steps leading from the identity (5.7) to the equations (5.9) which once again are valid as the terms involving ψ inside Ω vanish.[2] Indeed the only difference in the boundary integral treatment of this problem is the choice of the appropriate Green's function.

In the two-dimensional problem (5.16) is replaced by

$$\psi = H_0(kr)$$

where H_0 is a Hankel function. For three-dimensional problems the function given in (5.16) is replaced by

$$\psi = \frac{e^{-kr}}{r}$$

The remaining discretization follows precisely the pattern for the Laplace equation in this and following sections and indeed results in expressions which include the simple Laplacian solution as a special case obtainable by putting $k = 0$.

In the examples which follow both types of application will be indicated.

5.4 TREFFTZ ELEMENTS FOR MIXED SOLUTION PROCEDURES

Direct coupling of F.E. and boundary solution (B.S.) procedures is possible, at least in principle, by direct pairing of ϕ and $\partial\phi/\partial n$ values on common boundaries. Either the boundary distribution of $\partial\phi/\partial n$ in the B.I. formulation must be lumped to give nodal loads, or these loads on the boundary of the F.E. mesh must be replaced by an equivalent boundary distribution.

For example, we calculate the interface forces $\mathbf{F}$ of equation (5.5) from the values of the gradients $\partial\phi/\partial n$ on the element boundaries
Writing thus

$$\mathbf{F}_i = \int \mathbf{N}_i \frac{\overline{\partial\phi}}{\partial n}\, d\Gamma = \int \mathbf{N}_i \mathbf{M}\, d\Gamma \frac{\widehat{\boldsymbol{\partial\phi}}}{\boldsymbol{\partial}\mathbf{n}} \tag{5.21}$$

equation (5.5) then becomes

$$K\phi_{\mathrm{FE}} = \mathbf{C}\frac{\widehat{\partial\phi}}{\partial\mathbf{n}_{\mathrm{FE}}} \tag{5.22a}$$

In addition, from (5.9) we have

$$\mathbf{A}\hat{\boldsymbol{\phi}}_{\mathrm{BI}} = \mathbf{B}\frac{\widehat{\boldsymbol{\partial\phi}}}{\boldsymbol{\partial}\mathbf{n}_{\mathrm{BI}}} \tag{5.22b}$$

A direct coupling of the equations across a common interface can now follow by substituting

$$\hat{\boldsymbol{\phi}}_{\mathrm{FE}} = \hat{\boldsymbol{\phi}}_{\mathrm{BI}} \quad \text{and} \quad \frac{\widehat{\boldsymbol{\partial\phi}}}{\boldsymbol{\partial}\mathbf{n}_{\mathrm{FE}}} = -\frac{\widehat{\boldsymbol{\partial\phi}}}{\boldsymbol{\partial}\mathbf{n}_{\mathrm{BI}}} \tag{5.23}$$

where appropriate across common interfaces.

Alternatively, substituting for $\widehat{\boldsymbol{\partial\phi}/\boldsymbol{\partial}\mathbf{n}}$ from (5.9) into (5.21) we can write for nodes on the interface

$$\mathbf{F} = \left(\int \mathbf{N}^{\mathrm{T}}\mathbf{M}\, d\Gamma\right)\mathbf{B}^{-1}\mathbf{A}\boldsymbol{\phi} = \hat{\mathbf{K}}\boldsymbol{\phi}$$

where $\hat{\mathbf{K}}$ is the stiffness equivalent of the B.I. region.

We note immediately that the stiffness matrix $\hat{\mathbf{K}}$ is of a non-symmetric form

$$\hat{K}_{ij} = \int_{\Gamma} \mathbf{N}_i \mathbf{M}_j\, d\Gamma \mathbf{B}^{-1}\mathbf{A}$$

which is inconvenient. Indeed, in this form the equivalent stiffness matrix cannot be incorporated into 'standard' finite element software and much attention has been given to procedures which will result in symmetric matrices.[2,3] These involve the introduction of energy-variational concepts. If we now assume that ϕ on the interface between the B.I. and F.E. regions is continuous then the energy of the region can be written

$$U = U^{\mathrm{FE}} + U^{\mathrm{BI}} \tag{5.24}$$

with

$$U^{\mathrm{BI}}=\tfrac{1}{2}\int_{\Omega_{\mathrm{BI}}}\left[\left(\frac{\partial\phi}{\partial x}\right)^2+\left(\frac{\partial\phi}{\partial y}\right)^2\right]\mathrm{d}\Omega$$

for two dimensions. On using Green's theorem and the fact that ϕ satisfies the governing Laplacian equation, this can be written simply as

$$U^{\mathrm{BI}}=\tfrac{1}{2}\int_{\Gamma}\frac{\partial\phi}{\partial n}\phi\,\mathrm{d}\Gamma \tag{5.25}$$

where Γ is the boundary of the B.I. region. We require here that ϕ and $\partial\phi/\partial n$ constitute a solution satisfying the governing equations *a priori.* It should also be noted that strict adherence to (5.10) has been found necessary if successful formulations are to be achieved. A number of alternatives are available for deriving the Trefftz element and the discussion here will be divided into two distinct parts according as to whether the trial functions used are obtained from the direct or the indirect B.I. formulations.

5.4.1 Trial functions from the direct B.I. formulation

Here we concentrate on the use of the trial functions (5.19) to recover a more conventional finite element matrix which retains only the field variable ϕ in the energy functional. To recover this element we add an extra term to (5.25) to demand satisfaction of the boundary conditions (5.2b) and seek to minimize the functional

$$\pi=\tfrac{1}{2}\oint_{\Gamma}\frac{\partial\phi}{\partial n}\phi\,\mathrm{d}\Gamma-\int_{\Gamma_q}\phi\overline{\frac{\partial\phi}{\partial n}}\,\mathrm{d}\Gamma \tag{5.26}$$

On substituting interpolants of the form (5.8) this becomes

$$\pi=\tfrac{1}{2}\widehat{\frac{\partial\boldsymbol{\phi}}{\partial\mathbf{n}}}\oint_{\Gamma}\mathbf{M}^{\mathrm{T}}\mathbf{N}\,\mathrm{d}\Gamma\hat{\boldsymbol{\phi}}-\hat{\boldsymbol{\phi}}^{\mathrm{T}}\int_{\Gamma_q}\mathbf{N}^{\mathrm{T}}\overline{\frac{\partial\phi}{\partial n}}\,\mathrm{d}\Gamma \tag{5.27}$$

Now utilizing the trial function (5.19) we collocate at a number of points to recover the B.I. matrices (5.9)

$$\mathbf{B}\widehat{\frac{\partial\boldsymbol{\phi}}{\partial\mathbf{n}}}=\mathbf{A}\hat{\boldsymbol{\phi}} \tag{5.28}$$

Because numerical integration procedures are used for curved boundaries and the energy formulation is sensitive to the satisfaction of the equilibrium condition (5.10) we add this equation to the set (5.28) and get

$$\left[\begin{array}{c:c} \mathbf{B} & \mathbf{Q}^{\mathrm{T}} \\ \hdashline \mathbf{Q} & \mathbf{0} \end{array}\right] \left[\begin{array}{c} \widehat{\dfrac{\boldsymbol{\partial\phi}}{\mathbf{\partial n}}} \\ \lambda \end{array}\right] = \left[\begin{array}{c} \mathbf{A} \\ \hdashline \mathbf{0} \end{array}\right] [\hat{\boldsymbol{\phi}}] \tag{5.29}$$

Here $\mathbf{Q} = \oint \mathbf{M}\, \mathrm{d}\Gamma$ where from (5.8) $\partial\phi/\partial n = \mathbf{M}(\Gamma)\, \widehat{\boldsymbol{\partial\phi}/\mathbf{\partial n}}$
The addition of the extra equation ensures satisfaction of (5.10).

Substituting for $\widehat{\boldsymbol{\partial\phi}/\mathbf{\partial n}}$ from (5.29) into (5.27) and requiring $\partial\pi/\partial\hat{\boldsymbol{\phi}} = 0$ gives

$$\mathbf{K}'\hat{\boldsymbol{\phi}} + \mathbf{F}' = \mathbf{0} \tag{5.30}$$

with

$$\mathbf{K}' = \tfrac{1}{2}\mathbf{E}^{\mathrm{T}} \oint_{\mathrm{T}} \mathbf{M}^{\mathrm{T}}\mathbf{N}\, \mathrm{d}\Gamma + \tfrac{1}{2}\left[\mathbf{E}^{\mathrm{T}} \oint_{\mathrm{T}} \mathbf{M}^{\mathrm{T}}\mathbf{N}\, \mathrm{d}\Gamma\right]^{\mathrm{T}} \tag{5.31}$$

$$\mathbf{F}' = \int_{\Gamma_q} \mathbf{N}^{\mathrm{T}} \overline{\frac{\partial\phi}{\partial n}}\, \mathrm{d}\Gamma \tag{5.32}$$

and $\mathbf{E}$ defined by inversion and partitioning in (5.29). The matrix $\mathbf{K}'$ is symmetric and $\mathbf{K}'$ and $\mathbf{F}'$ can be assembled into (5.5) as contributions from a new element.

Compatibility with neighbouring finite elements will be ensured if the boundary interpolation functions $\mathbf{N}$ in (5.31) match those on the adjacent finite element boundary interpolation functions and positioning of the collocation point are discussed in the following section.

5.4.2 Trial functions from the indirect B.I. formulation

Here we aim to avoid the matrix inversion inherent in the formation of the Trefftz element stiffness in (5.31) and avoid the approximation introduced by discretizing the variation of both ϕ and $\partial\phi/\partial n$ on the Trefftz boundary.

Substituting from (5.13) into (5.11) and (5.12) gives

$$\begin{aligned} \phi(p) &= \int G(p, \Gamma)\mathbf{N}(\Gamma)\, \mathrm{d}\Gamma\, \hat{\boldsymbol{\mu}} + \sum Q_j b_j \\ &= \sum \bar{N}_i(p)\, \hat{\mu}_i + \sum Q_j b_j = \bar{\mathbf{N}}\hat{\boldsymbol{\mu}} + \mathbf{Q}\mathbf{b} \end{aligned} \tag{5.33}$$

and

$$\begin{aligned} \frac{\partial\phi}{\partial n}(p) &= \int \frac{\partial G}{\partial n}(p, \Gamma)\mathbf{M}(\Gamma)\, \mathrm{d}\Gamma\, \hat{\boldsymbol{\mu}} + \beta\mu(p) + \sum \frac{\partial Q_j}{\partial n} b_j \\ &= \sum \bar{M}_i(p)\hat{\mu}_i + \sum \frac{\partial Q_j}{\partial n} b_j = \bar{\mathbf{M}}\hat{\boldsymbol{\mu}} + \bar{\mathbf{P}}\mathbf{b} \end{aligned} \tag{5.34}$$

where Q_i is any supplementary solution of the governing equations such as the exact solution for a crack tip singularity or re-entrant corner. These expressions can now be substituted into (5.26) which is supplemented by a constraint that ϕ takes a compatible variation along the boundary it shares with finite elements. The functional now becomes[3]

$$\pi = \tfrac{1}{2}\int_{\Gamma} \frac{\partial \phi}{\partial n}\phi \,\mathrm{d}\Gamma - \int_{\Gamma_q} \phi \overline{\frac{\partial \phi}{\partial n}} \,\mathrm{d}\Gamma - \int_{\Gamma_{FE}} \left(\frac{\partial \phi}{\partial n}\right)^{\mathrm{BI}} (\phi^{\mathrm{BI}} - \phi^{\mathrm{FE}}) \,\mathrm{d}\Gamma \tag{5.35}$$

where the last term constrains the two sets of potentials to being equal along the boundary. If on the finite element we require $\phi = \mathbf{N}\hat{\boldsymbol{\phi}}$ the variation of (5.35) with respect to both $\hat{\boldsymbol{\phi}}$ and $\hat{\boldsymbol{\mu}}$ gives

$$\begin{bmatrix} \mathbf{0} & \mathbf{K}_{\mu\phi}^{\mathrm{T}} \\ \mathbf{K}_{\mu\phi} & \mathbf{K}_{\mu\mu} \end{bmatrix} \begin{bmatrix} \hat{\boldsymbol{\phi}} \\ \hat{\boldsymbol{\mu}} \end{bmatrix} = \begin{bmatrix} \mathbf{0} \\ \bar{\mathbf{F}}_e \end{bmatrix} \tag{5.36}$$

where

$$\mathbf{K}_{\mu\phi} = -\int \bar{\mathbf{M}}^{\mathrm{T}}\mathbf{N} \,\mathrm{d}\Gamma$$

$$\mathbf{K}_{\mu\mu} = \tfrac{1}{2}\int_{\Gamma_\phi} (\bar{\mathbf{N}}^{\mathrm{T}}\bar{\mathbf{M}} + \bar{\mathbf{M}}^{\mathrm{T}}\bar{\mathbf{N}}) \,\mathrm{d}\Gamma - \tfrac{1}{2}\int_{\Gamma_q} (\bar{\mathbf{N}}^{\mathrm{T}}\bar{\mathbf{M}} + \bar{\mathbf{M}}^{\mathrm{T}}\bar{\mathbf{N}}) \,\mathrm{d}\Gamma$$

and

$$\bar{\mathbf{F}} = \int_{\Gamma_q} \bar{\mathbf{N}}^{\mathrm{T}}\bar{\mathbf{q}} \,\mathbf{d}\Gamma.$$

Supplementary solutions are included directly by adding the extra parameters to $\hat{\boldsymbol{\mu}}$ from (5.33) and (5.34). Special care must be taken however, when supplementary solutions are added, to avoid singular or ill-conditional matrices as would occur if there was a linear dependence between trial solutions in a Ritz procedure. The presence of derivatives in each term in the energy functional also means that an extra equation must be added to define a constant (in terms of boundary values of ϕ).

To avoid ill-conditioning difficulties arising between the boundary integral and supplementary solution contributions in (5.34) and (5.35) a two-phase solution can be followed, first setting to zero the parameters $\boldsymbol{\mu}$ and solving for $\mathbf{b}$, then setting these values of $\mathbf{b}$ as boundary conditions and solving for a set of parameters $\boldsymbol{\mu}$ now dependent on $\mathbf{b}$. Alternative possibilities exist and the question of whether ill-conditioning will always occur as the boundary discretization for the boundary solution process is refined has not yet been resolved.

In (5.36) $\hat{\boldsymbol{\phi}}$ includes the nodal values of potential along the common interface between the finite elements and Trefftz element. Once again the matrix is symmetric but here the matrix inversion required in (5.31) has been removed and replaced by a double integration on the Trefftz element surface. The computational expenditure can here be alleviated somewhat for simple boundary interpolation as analytical expressions are available for the inner boundary integral and only one numerical boundary integration is then required.

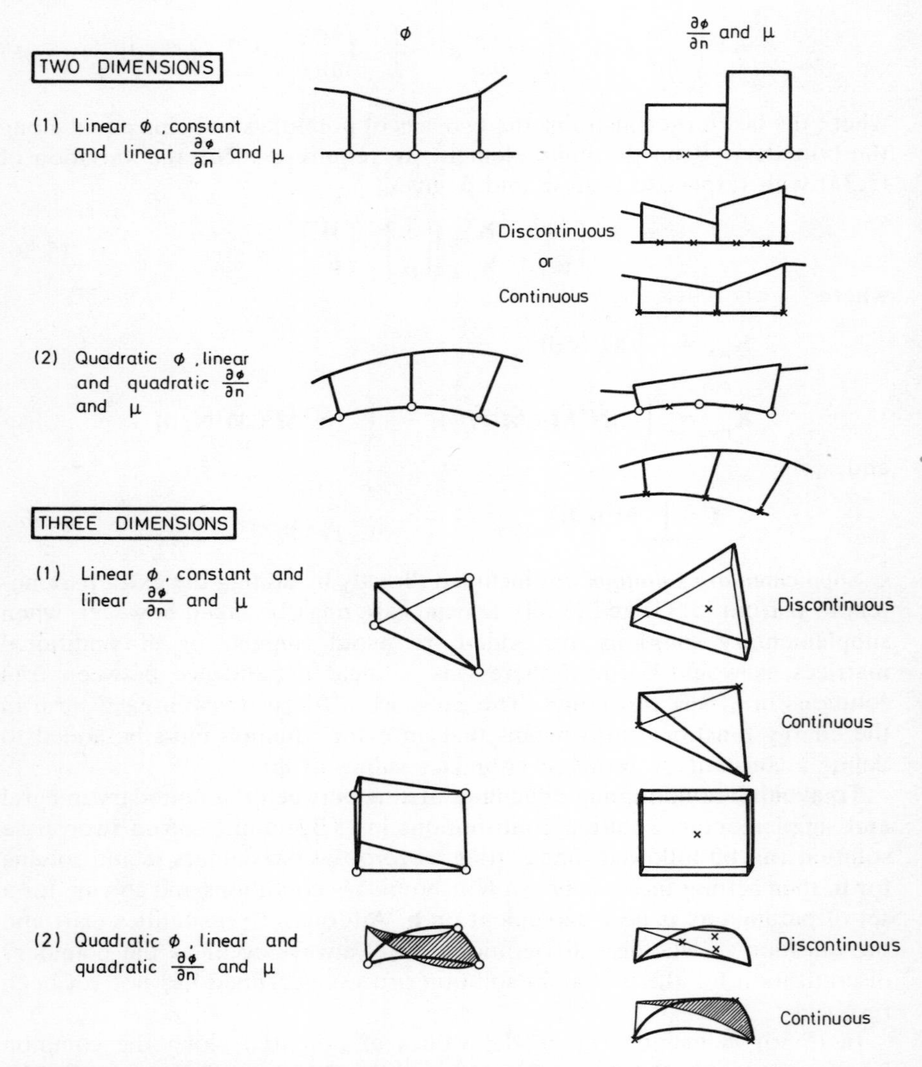

Figure 5.1 Surface interpolation for the direct and indirect B.I. formulations (× denotes position for collocation in the direct B.I. formulations)

5.5 SURFACE INTERPOLATION FOR THE TREFFTZ ELEMENTS

The trial functions for the finite elements (5.4) describe the variation of ϕ over the element and from them we can recover a functional variation of ϕ as, for example,

$$\phi(x, y) = ax + by + c \tag{5.37}$$

for the three-noded linear ϕ triangle. Once the problem has been solved the constants can be evaluated from the nodal values of ϕ and (5.37) can be used to define $\partial\phi/\partial n$ on the element.

On the boundary between Trefftz and finite elements compatibility requires the interpolation of ϕ to be the same. For the direct boundary integral formulation, however, interpolation between nodal values of both ϕ and $\partial\phi/\partial n$ is required in (5.8). In general we require ϕ continuous and $\partial\phi/\partial n$ discontinuous at nodes (especially at corners) and the lowest order of ϕ variation possible in coupling with finite elements is linear. As a result a family of surface interpolants is possible for two and three dimensions and these are shown in Figure 5.1. When a continuous $\partial\phi/\partial n$ variation is used on smooth boundaries a substitution for $\partial\phi/\partial n$ in terms of neighbouring ϕ values can overcome the difficulties caused by discontinuities at corners.

For the indirect boundary integral formulation only the distribution of the surface parameter μ is interpolated between nodal values. Again for regions requiring discontinuous $\partial\phi/\partial n$ the interpolant for μ is chosen discontinuous at nodes so that (5.12) will recover discontinuous values. A family of such interpolants would match those for $\partial\phi/\partial n$ in Figure 5.1.

5.6 THE PATCH TEST AND CONVERGENCE

The trial functions in the finite element method limit the infinite number of degrees of freedom of the system and the true minimum energy may never be reached, irrespective of the fineness of the subdivision. A sufficient though not necessary test to determine whether a finite element will converge to the correct solution is given by the patch test.[11] In the context of the Laplace equation we require that

(1) A uniform increment to ϕ should produce zero increment to $\partial\phi/\partial n$.
(2) The element should produce exact answers for a linear variation of the field variable ϕ since the distribution on the element will converge to this as the element mesh becomes finer.
(3) There should be no spurious contribution to the energy at the element interfaces.

Satisfaction of these criteria can be checked by taking a small patch of finite elements and applying boundary conditions consistent with a linear field solution and checking that the exact answers are recovered. In fact most finite elements, including those used in the examples here, will pass this test.

In the boundary integral formulations the refinement of the discretization to improve accuracy is confined to the region boundary or surface. Since the fundamental solutions satisfy the governing equations exactly, Green's identity (5.7) and the surface distributions (5.11) and (5.12) will provide the exact solution for continuously varying ϕ and $\partial\phi/\partial n$ in Green's identity, and μ in the surface distribution equations. Convergence will therefore occur in the limit as the surface discretization is made finer.

In a hybrid mesh therefore we would expect convergence to occur as the finite elements and surface segmentation of the B.I. are made finer with the B.I. domain retaining its original size. However, to show that accurate results can be achieved for coarse meshes and boundary segmentation in a hybrid discretization as shown in Figure 5.2, we can investigate the solution to problems governed by Laplace's equation for which the boundary interpolants of B.I. and the surface interpolants of F.E. can exactly reproduce the correct field variation.

In Figure 5.2(a) a direct linking of F.E. and B.I. as summarized in equations (5.22) and (5.23) is used to solve a simple one-dimensional problem. Linear ϕ and piecewise constant $\partial\phi/\partial n$ were assumed on the boundary of the B.I. region. The exact solution was obtained.

5.6.1 Inaccuracies in the discretization for the direct B.I. Trefftz elements

Inherent in the direct B.I. formulation is the interpolation of the surface variation of both ϕ and $\partial\phi/\partial n$. However, once the variation of one of these is defined, the variation of the other is uniquely specified so that the independent choice of interpolation functions for both ϕ and $\partial\phi/\partial n$ on the boundary introduces an error. This error does not affect the direct application of the procedure to the problem in Figure 5.2 where the assumed interpolation is correct. However, it appears that the derivatives of the energy functional taken to reproduce the Trefftz element matrices will not be correct.

In Figure 5.2(b) a simple rectangular domain is discretized using three-node triangular finite elements and a single Trefftz element. Boundary conditions consistent with a linear ϕ field are applied. When finite elements or the boundary integral formulation are applied independently to the problem (the boundary integral only discretizing the outer boundary) both produce the exact results shown. Here the surface interpolation in the B.I. formulation involves linear continuous ϕ and piecewise constant $\partial\phi/\partial n$. Collocation at the mid-point of each segment was used.

In Figure 5.2(b) results are given for application of the hybrid scheme with the Trefftz element based on the direct B.I. formulation and the procedures described above. A four-point Gauss rule was used in the numerical integration of (5.7).

5.6.2 Inaccuracy in the discretization for the indirect B.I. Trefftz elements

Similarly we note here that once the μ variation is specified, the variation of ϕ and $\partial\phi/\partial n$ on the boundary is uniquely defined by (5.11) and (5.12). Only in a limiting sense as the number of Trefftz element surface segments increases or a high order μ variation is taken will the constraint imposing compatibility of the boundary ϕ variation with neighbouring elements in (5.35) operate accurately. A solution for application of the indirect B.I. Trefftz element to a patch test problem is given as Figure 5.2(c).

We can here impose the linear variation of ϕ as a supplementary solution on the Trefftz zone by substituting

$$\sum Q_j b_j = b_1 + b_2 x + b_3 y$$

in (5.33) and (5.34) where b_1, b_2, and b_3 become additional auxiliary variables. The constraint in the variational formulation has now only to operate a force compatibility for the contributions from the surface distribution when the solution is not linear. The solution of the patch test problem now reproduced a more accurate result (Figure 5.2(c)).

When the Trefftz element is an exterior element it is apparent that the same supplementary solution cannot be imposed. However, this supplementary solution should only be necessary on domains with few boundary segments or for interior elements where inaccuracies at the interface boundary of even low order cannot be tolerated. For most exterior problems this will not apply.

It should also be noted here that there is no guarantee that the variation of μ will be well behaved. While this has not been ruled out, it has not caused difficulties in the problems considered especially when a linear supplementary solution is added and in general does not appear to hinder the use of the indirect B.I. formulation.

These aspects of the marriage between F.E. and B.I. are currently under investigation. One of the simplest though most convincing tests for the possibilities of achieving a complete union of the methods lies in the generation of the simplest finite element using the direct B.I. formulation. For the three-node triangle shown in Figure 5.3 ϕ is linear and $\partial\phi/\partial n$ piecewise constant on each side. The Trefftz element generated using this boundary interpolation and collocating at the mid-point of each side reproduces exactly the finite element matrix obtained by conventional F.E. methods. The accuracy of numerical integration is also indicated in the

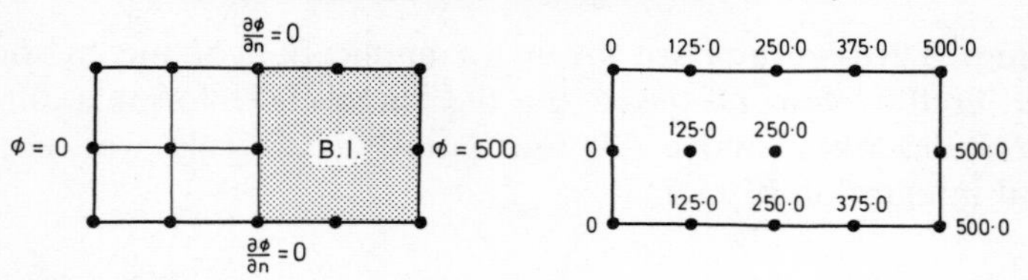

(a) Nonsymmetric linking of the direct boundary integral equation and finite elements

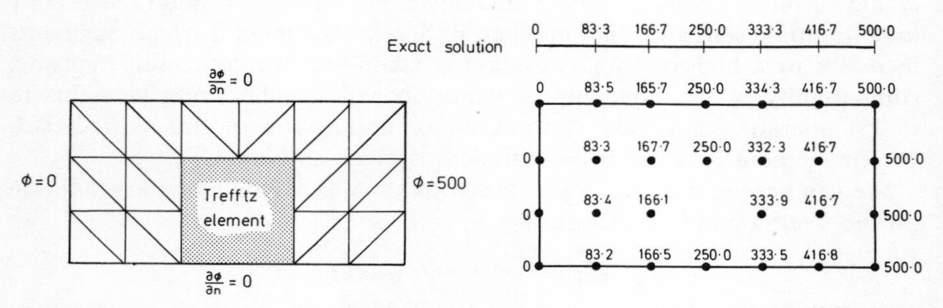

(b) Combined direct boundary integral symmetric Trefftz element and finite element solution

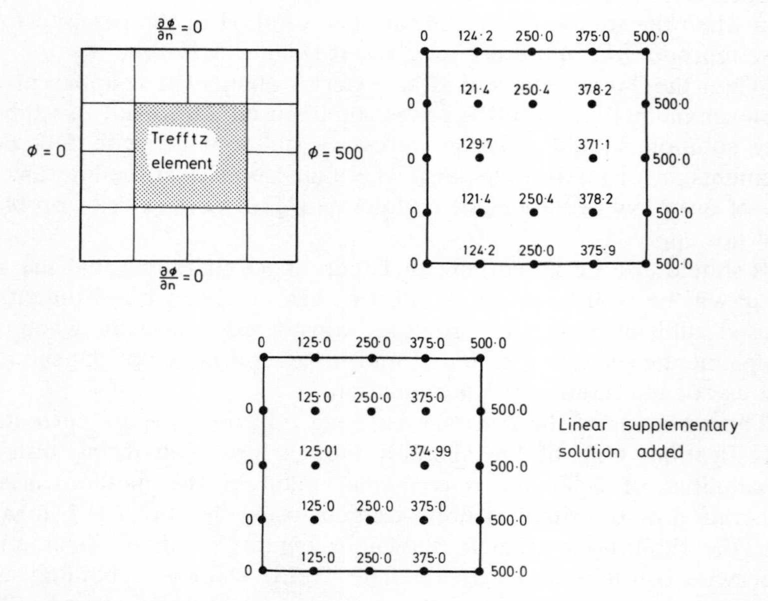

(c) Combined indirect boundary integral symmetric Trefftz element and finite element solution

Figure 5.2 Linear field examples

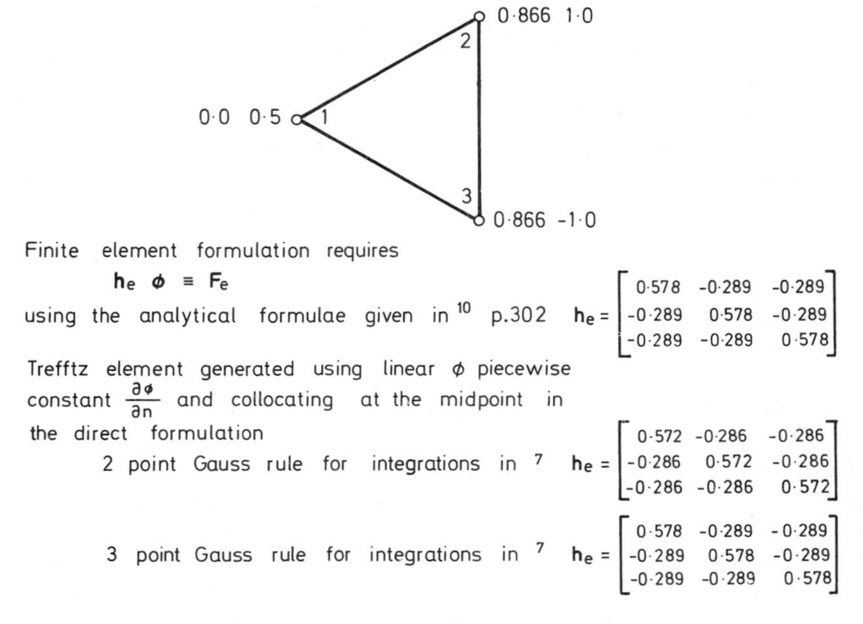

0·866 1·0
2
0·0 0·5
1
3
0·866 -1·0
Finite element formulation requires
he φ ≡ Fe
using the analytical formulae given in 10 p.302
he =
0·578 -0·289 -0·289
-0·289 0·578 -0·289
-0·289 -0·289 0·578
Trefftz element generated using linear φ piecewise
constant ∂φ/∂n and collocating at the midpoint in
the direct formulation
2 point Gauss rule for integrations in 7
he =
0·572 -0·286 -0·286
-0·286 0·572 -0·286
-0·286 -0·286 0·572
3 point Gauss rule for integrations in 7
he =
0·578 -0·289 -0·289
-0·289 0·578 -0·289
-0·289 -0·289 0·578

Figure 5.3 Element matrix for a three-node triangle

figure. Analytical evaluation of the integrals is possible for flat segments but numerical integration will, in general, be necessary for curved boundaries.

5.7 COMPUTATIONAL EXPENDITURE

In certain applications in three-dimensional analysis and exterior problems the B.I. procedures are more economical then F.E.[12] If the direct linking procedures are used these economies will be preserved. However, it is readily apparent that the non-symmetric matrices which result will be unacceptable if a small set of B.I. equations is linked to a large set of F.E. equations.

The B.I. procedures reduce the dimension of the problem by one and this leads to a reduction in the number of variables as internal nodes are eliminated from an equivalent finite element mesh and correspondingly reduced data preparation is required. In addition the approximation of truncating an exterior FE mesh is removed for exterior applications. In three dimensions where the volume to surface ratio is large the Trefftz elements may again be successful because large homogeneous domains could be modelled by a single macro-Trefftz element reducing the very rapid rise in the number of variables encountered in three-dimensional finite element meshes.

The viability of the symmetric Trefftz elements is therefore still to be proved and its evident that for simple interior two-dimensional problems they are several times more expensive than complete finite element models. The following examples however indicate the accuracy and flexibility of the hybrid models. The application to exterior problems in which the boundary conditions at infinity are automatically satisfied and the ease with which supplementary solutions can be added to the indirect B.I. Trefftz element have been demonstrated.

5.8 EXAMPLES

5.8.1 two-dimensional potential flow around an aerofoil

This example considers the two-dimensional flow around the symmetric aerofoil shown in Figure 5.4. The exterior field is modelled using exterior Trefftz elements based on both direct and indirect B.I. formulations as shown in the figures. Exterior potential flow problems have been solved using boundary solution procedures[14] but a model incorporating finite elements near the surface of the aerofoil should enable viscous and other boundary layer effects to be incorporated directly.

The analysis here is based on the stream function ψ which must satisfy Laplace's equation (5.1) in the exterior domain. Velocities can be recovered from the stream function according to

$$u = \frac{\partial \psi}{\partial y}$$

and

$$v = -\frac{\partial \psi}{\partial x}$$

Setting $\psi = \psi^{\infty} + \psi^1$ with $\psi^{\infty} = Uy$ where U is the free-stream velocity, leads to the problem of determining ψ^1 satisfying

$$\nabla^2 \psi^1 = 0 \quad \text{with} \quad \psi^1 \rightarrow 0 \text{ at } \infty \tag{5.38}$$

The surface of the cylinder or aerofoil must be a streamline with $\psi =$ constant giving on that surface

$$\psi^1 = c - Uy = -Uy \tag{5.39}$$

as the constant c is quite arbitrary.

The tangential velocity on the aerofoil surface is plotted in Figure 5.4. for both the direct and indirect B.I. Trefftz elements and compared with the distribution given in.[15] The number of boundary segments in the model was

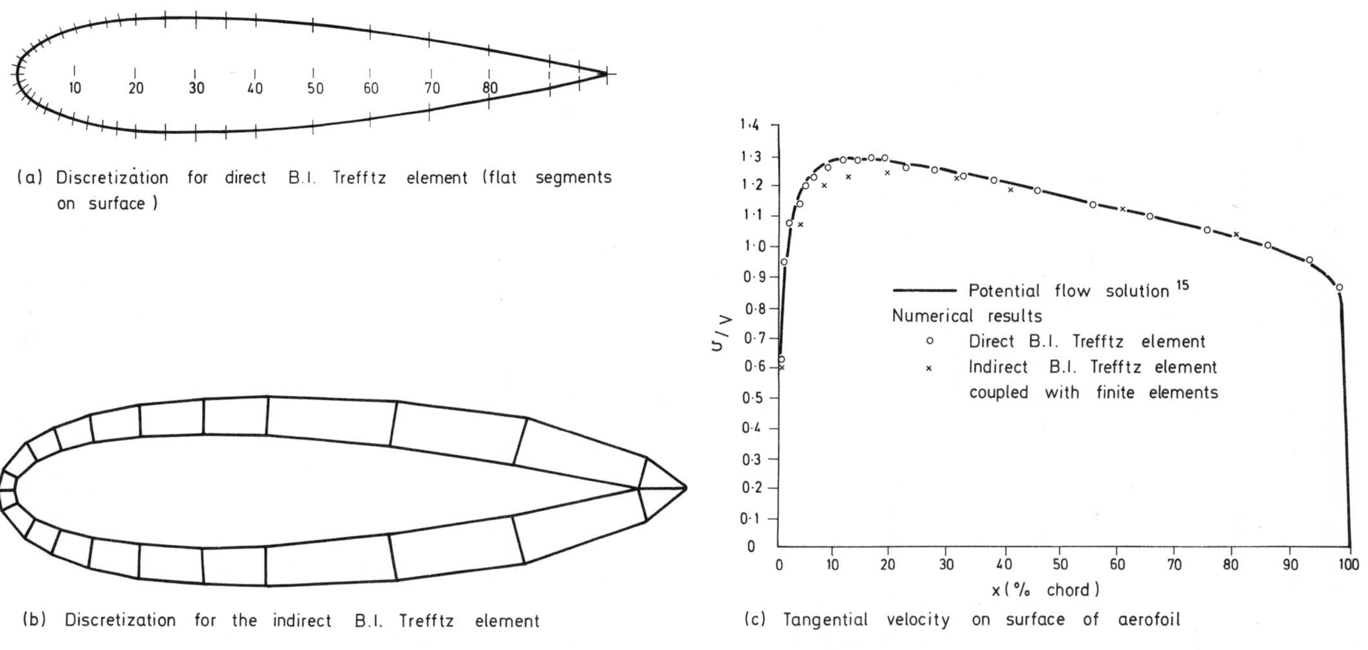

(a) Discretization for direct B.I. Trefftz element (flat segments on surface)

(b) Discretization for the indirect B.I. Trefftz element

(c) Tangential velocity on surface of aerofoil

Figure 5.4 Two-dimensional potential over a NACA 0018 wing section[15]

defined principally by the need to approximate the surface with piecewise flat segments. Improved accuracy and a reduction in the number of segments could be expected if a quadratic variation of ϕ and curved segments were used. It should be noted that a coarser discretization was used in the indirect form.

5.8.2 The scalar potential field for a C-shaped dipole magnet

A coupled finite element and boundary integral equation discretization was used in this example to determine the scalar potential field surrounding a C-magnet on a plane in which the problem can be adequately represented as two-dimensional. Finite elements were used to represent a symmetric inner region containing the iron magnet as shown in Figure 5.5. The homogeneous external field was modelled using a single boundary integral element based on the direct formulation.

The classical problem of magnetostatic field theory starts with the set of Maxwell's equations.

$$\boldsymbol{\nabla}^{\mathrm{T}} \times \mathbf{H} = -\mathbf{J}$$

$$\mathbf{B} = \mu \mathbf{H}$$

and

$$\boldsymbol{\nabla}^{\mathrm{T}} \cdot \mathbf{B} = 0 \tag{5.40}$$

where $\mathbf{H}$ is the magnetic field strength, $\mathbf{B}$ is the flux density, and $\mathbf{J}$ the current density and μ the magnetic permeability. All these quantities apart from μ are vectors.

It is shown in Reference 16, however, that taking

$$\mathbf{H} = \mathbf{H}_s + \mathbf{H}_m$$

where $\mathbf{H}_s$ is the field for a homogeneous region with $\mu = 1$ (which can be calculated by a direct integration), and substituting

$$\mathbf{H}_m = \boldsymbol{\nabla}\phi$$

the problem can be transformed to one requiring the solution for a scalar potential ϕ satisfying

$$\boldsymbol{\nabla}^{\mathrm{T}} \cdot \mu \boldsymbol{\nabla}\phi + \boldsymbol{\nabla}^{\mathrm{T}} \cdot \mu \mathbf{H}_s = 0 \tag{5.41}$$

where μ varies discontinuously across the air/iron interface.

The problem is now that of the general quasi-harmonic equation, the solution of which by the finite element process follows the standard pattern, now with a loading term applied. In Reference 16 it is shown that solutions satisfying this governing equation can be found by minimizing the functional

$$\pi = \int_\Omega \tfrac{1}{2}(\boldsymbol{\nabla}\phi)^{\mathrm{T}} \cdot \mu \boldsymbol{\nabla}\phi \, \mathrm{d}\Omega - \int_\Omega \phi \boldsymbol{\nabla}^{\mathrm{T}} \mu \mathbf{H}_s \, \mathrm{d}\Omega + \int_\Omega \phi H_n \, \mathrm{d}\Gamma$$

where H_n is the normal component of the field strength vector H. Setting

$$\phi = \mathbf{Na}$$

and requiring $\partial\pi/\partial\mathbf{a} = 0$ leads to $\mathbf{Ka} + \mathbf{f} = 0$ with

$$k_{ij} = \int_\Omega (\nabla N_i)^{\mathrm{T}} \mu \boldsymbol{\nabla} N_j \, \mathrm{d}\Omega$$

and

$$f_i = -\int_\Omega N_i \boldsymbol{\nabla}^{\mathrm{T}} \cdot \mu \mathbf{H}_s \, \mathrm{d}\Omega + \int_{\Gamma_q} N_i^{\mathrm{T}} H_n \, \mathrm{d}\Gamma$$

Integrating by parts gives

$$\begin{aligned} f_i &= \int_\Omega (\boldsymbol{\nabla}^{\mathrm{T}} N_i) \mu H_s \, \mathrm{d}\Omega - \int_\Gamma N_i \mu H_{sn} \, \mathrm{d}\Gamma + \int_{\Gamma_q} N_i^{\mathrm{T}} H_n \, \mathrm{d}\Gamma \\ &= f_i' + f_i'' \end{aligned}$$

where

$$f_i'' = \int_{\Gamma_q} N_i^{\mathrm{T}} H_n \, \mathrm{d}\Gamma = \int_{\Gamma_q} N_i^{\mathrm{T}} \frac{\partial \phi}{\partial n} \, \mathrm{d}\Gamma \tag{5.42}$$

In the exterior domain $\mu = 1$, so that the governing equation reduces to

$$\nabla^2 \phi = 0$$

and the standard procedures described earlier can be applied.

The field surrounding the magnet is given in Figure 5.5 and the homogeneity of the field between the jaws of the magnet is compared with similar results given by Trowbridge.[13] The relative permeability of the iron core of the magnet was here assumed constant at 1000. In practice, however, $\mu_I = \mu_I(H)$, a function of the field strength. Since the field strength varies in the iron core the problem becomes non-homogeneous over the iron as well as non-linear. The permeability of the exterior region, however, remains constant so that this region can be treated very efficiently using the exterior Trefftz element which need only be generated during the first of the iterations used to derive the non-linear solution.

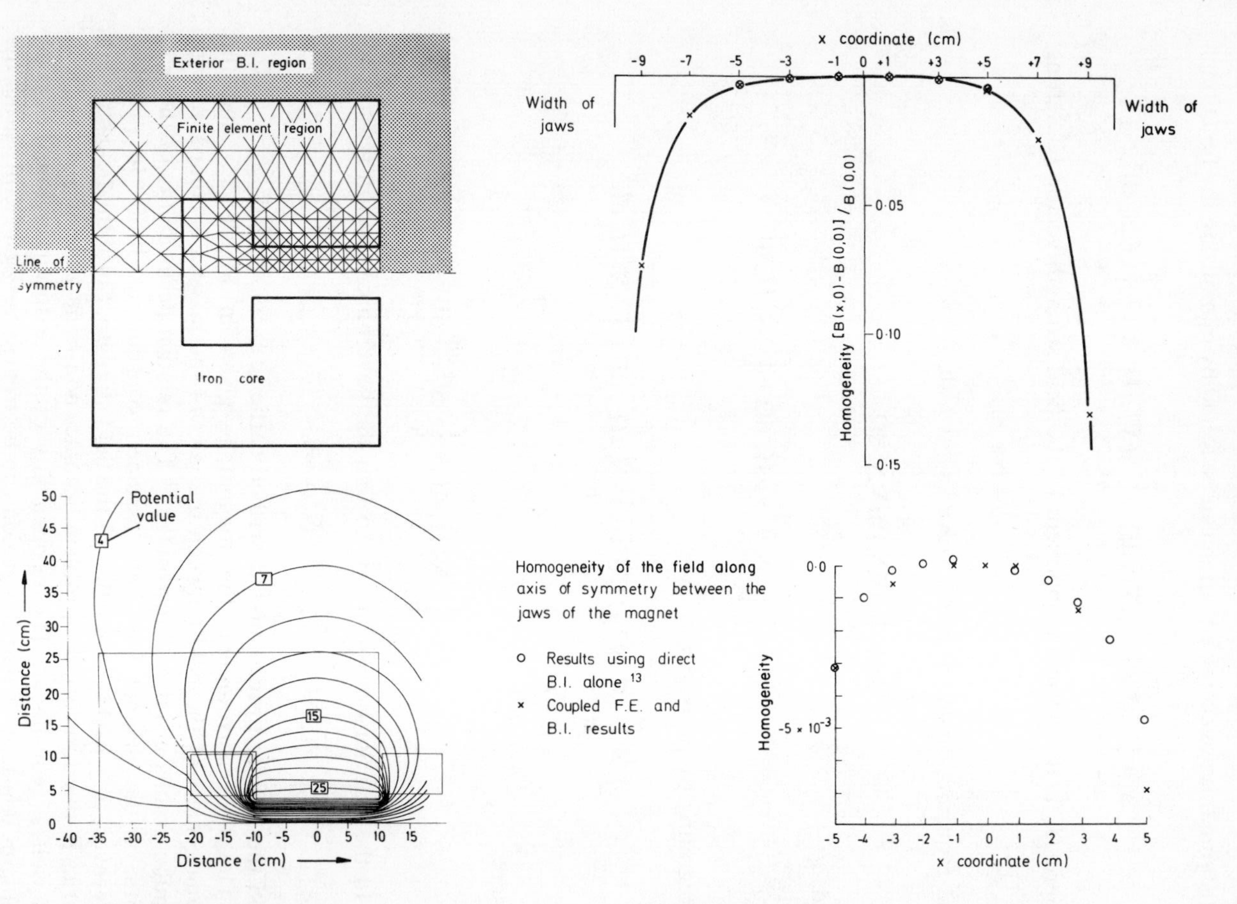

Figure 5.5 Discretization and numerical results for the scalar potential field of the C-shaped magnet

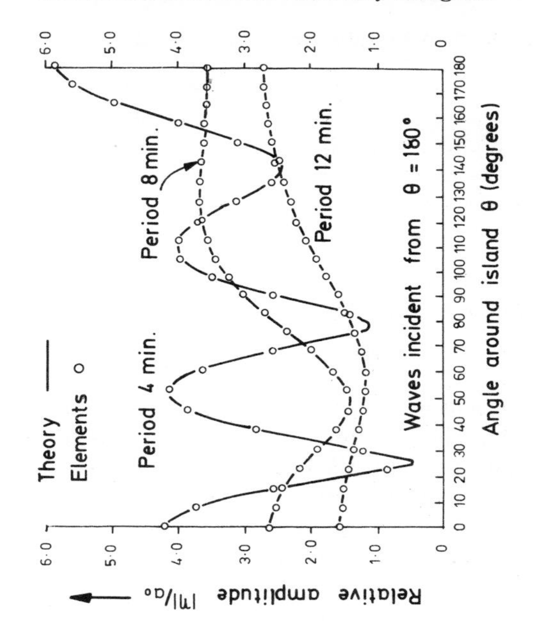

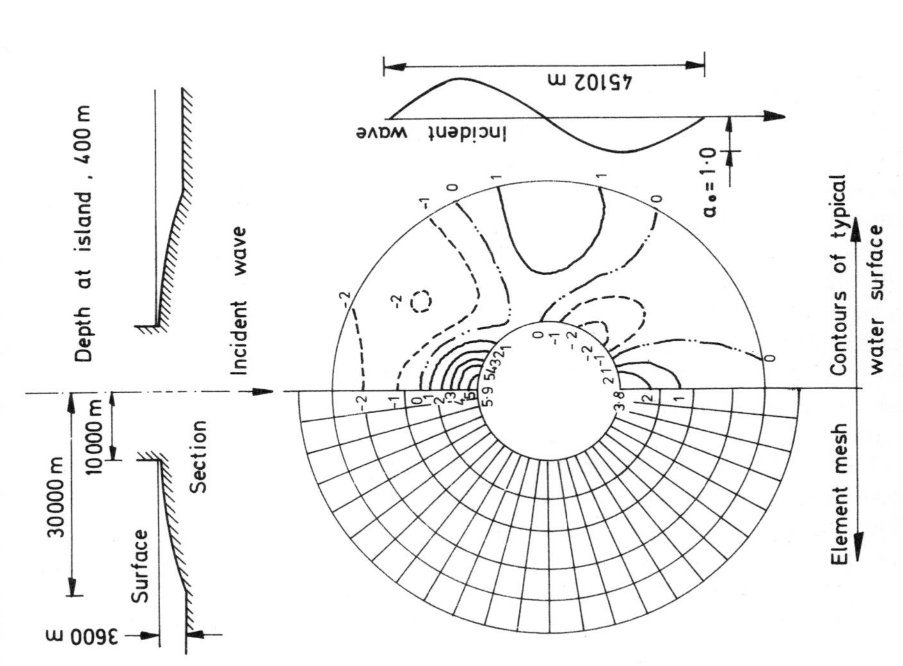

Figure 5.6 Coupled F.E. and B.I. solution of wave diffraction and refraction for the island on parabolic shoal problem

5.8.3 Extensions to non-homogeneous problems

In the preceding we have discussed the homogeneous problem such as given by equations (5.1) and (5.20) and assumed that this was to be solved in both the F.E. and B.I. regions. Clearly the restriction can be lifted for the F.E. domain where non-homogeneity presents no difficulties. Therefore in general we shall be concerned with the solution of equations of the type

$$\nabla^{\mathrm{T}} h \nabla \phi = 0 \tag{5.43}$$

in the place of equation (5.1) and

$$\nabla^{\mathrm{T}} h \nabla \phi - k^2 \phi = 0 \tag{5.44}$$

in the place of equation (5.20). These will be characterized by $h = 1$ pertaining in the domain for which the boundary integral method is used.

One such example has already been performed. The results have been published elsewhere[17] but are reproduced here in Figure 5.6. This problem involved the diffraction and refraction of waves by a parabolic shoal. The variation of the depth led to a governing equation of the form of (5.44). Finite elements were used in this region with the exterior constant depth solution being provided by a single Trefftz element.

5.9 CONCLUDING REMARKS

We have shown in this paper how trial functions which satisfy the governing equations and are defined entirely in terms of discrete parameters on the boundary can be defined and incorporated into standard finite element processes. The particular advantages of the B.I. processes which permit them to deal with infinite domains etc. can thus be made direct use of—indeed the extent to which B.I. or Trefftz 'elements' are used can be left to the choice of the user who will be able to determine their efficiency for this particular problem.

Direct coupling of the equations produced by standard B.I. processes and F.E. (with a non-standard form for the loading matrix) is also possible but leads to non-symmetric matrices. This form may be acceptable if the equation set is dominated by the B.I. equations but would be less attractive to a F.E. user who may only require the special properties of the Trefftz trial functions in a small subset of the otherwise symmetric equations.

The illustrative examples indicate the many possibilities for application of these procedures for several Laplace or Helmholtz type problems and extension to elastostatic situations will follow similar lines. We have also indicated two distinct possibilities of formulation of the symmetric Trefftz elements by direct or indirect boundary integral equation. Much work still remains to be done on streamlining the computational process of the latter but it may well prove more versatile.

ACKNOWLEDGEMENTS

The authors are indebted to the Science Research Council for financial support of part of this work. They would also like to thank Professor R. L. Taylor, Department of Civil Engineering, Berkeley, and Dr G. T. Symm and Dr G. Miller of the National Physical Laboratory, England, for their many helpful discussions.

APPENDIX: THE EQUIVALENCE OF THE DIRECT AND INDIRECT BOUNDARY INTEGRAL EQUATIONS

The direct boundary integral equation (5.7) is generally derived by a piece of formal mathematics—while the indirect, source distribution equation (5.11) has a physical connotation as a superposition of an infinite number of elementary solutions. While the first appears to give generality, incorporating within itself all possible solutions, the completeness of the latter is by no means obvious. If a mathematical link between the two types of equations can be established then proof of completeness is available. Below we quote a proof of such a link given by Lamb.[18]

Consider a region Ω giving for a set of singular functions ψ placed on the boundary the direct integral equation

$$\pi\phi = -\int_{\Gamma} \psi \frac{\partial \phi}{\partial n} \, d\Gamma + \int_{\Gamma} \frac{\partial \psi}{\partial n} \phi \, d\Gamma \tag{A.5.1}$$

Let us now write a similar equation for another function ϕ' but now placing the singularity poles just outside the boundary Γ. This gives on Γ

$$0 = -\int_{\Gamma} \psi \frac{\partial \phi'}{\partial n} \, d\Gamma + \int_{\Gamma} \frac{\partial \psi}{\partial n} \phi' \, d\Gamma \tag{A.5.2}$$

The new function ϕ' has now no special physical significance and indeed equation (A.5.2) will be satisfied by any function ϕ' which satisfies the governing equations.

Adding the two equations we can write

$$\pi\phi = -\int_{\Gamma} \psi \left(\frac{\partial \phi}{\partial n} + \frac{\partial \phi'}{\partial n} \right) d\Gamma + \int_{\Gamma} \frac{\partial \psi}{\partial n} (\phi + \phi') \, d\Gamma \tag{A.5.3}$$

Now we can choose ϕ' such that either one or the other bracketed terms disappear. If for instance we make

$$\phi + \phi' = 0 \tag{A.5.4}$$

and

$$\frac{\partial \phi}{\partial n} + \frac{\partial \phi'}{\partial n} \equiv -\pi\lambda$$

we can write

$$\phi = \int_\Gamma \psi\lambda \, d\Gamma \tag{A.5.5}$$

where λ is now interpreted as the source distribution and is equivalent to equation (5.11) where this quantity is not directly related to ϕ or its derivative. Similarly, if we choose

$$\frac{\partial \phi}{\partial n} + \frac{\partial \phi'}{\partial n} = 0 \quad \text{and} \quad \phi + \phi' \equiv \pi\lambda \tag{A.5.6}$$

we can write another indirect boundary integral equation corresponding to distributed doublets

$$\phi = \int_\Gamma \frac{\partial \phi}{\partial n} \lambda \, \mathrm{d}\Gamma \tag{A.5.7}$$

The link between the direct and indirect boundary integral equations has thus been established and clearly (A.5.5) and (A.5.7) are as general as (A.5.1).

REFERENCES

1. Trefftz, E., 'Gegenstuck Zum Ritz'schen Verfohren', *Proceedings of the Second International Congress on Applied Mechanics*, Zurich, 1926.
2. Zienkiewicz, O. C., 'The finite element method and boundary solution procedures as general approximation methods for field problems', *Proceedings of the World Congress on Finite element Methods in Structural Mechanics*. Bournemouth, October 1975.
3. Zienkiewicz, O. C., Kelly, D. W., and Bettess, P. 'The coupling of the finite element method and boundary solution procedures', *Int. Jnl. Num. Meth. Engng.*, **11,** No. 2, 355–375 (1977).
4. McDonald, B. H., and Wexler, A., 'Finite element solution of unbounded field problems', *IEEE Transactions on Microwave Theory and Techniques*, **MTT-20,** No. 12, December (1972).
5. Silvester, P., and Hseih, M. S., 'Finite element solution of two-dimensional exterior field problems', *Proc. IEE*, **118,** No. 12, December (1971).
6. Berkhoff, J. C. W., 'Linear wave propagation problems and the finite element method'. Ch. 13 in *Finite Element Methods in Fluid Mechanics*, Vol. 1 (ed. R. H. Gallagher, J. T. Oden, C. Taylor and O. C. Zienkiewicz), Wiley, London 1975.

7. Wood, W. L., 'On the finite element solution of an exterior boundary value problem', *Int. Jnl. Num. Meth. Engng.*, **10,** 885–891 (1976).
8. McDonald, B. H., Friedman, M., and Wexler, A., 'Variational solution of integral equations', *IEEE Transactions on Microwave Theory and Techniques,* **MTT-22,** No. 3 (1974).
9. Loof, H. W., 'The economic computation of stiffness matrices of large structural elements', *Proceedings of the International Symposium on the Use of Digital Computers in Structural Engineering,* University of Newcastle, 1966.
10. Zienkiewicz, O. C., *The Finite Element Method in Engineering Science,* McGraw-Hill, London 1971.
11. Bazeley, G. P., Cheung, Y. K., Irons, B. M., and Zienkiewicz, O. C., 'Triangular elements in bending—conforming and non-conforming solutions', *Proc. Conf. Matrix Methods in Struct. Mech.*, Air Force Inst. Techn. Wright Patterson A. F. Base, Ohio, 1965.
12. Cruse, T. A. 'Application of the boundary-integral equation method to three dimensional stress analysis', *Computers and Structures,* **3,** 509–527 (1973).
13. Trowbridge, G. W., 'Applications of integral equation methods, for the numerical solution of magnetostatic and eddy current problems', *International Conference on Numerical Methods in Electrical and Magnetic Field Problems,* Santa Margherita, Italy, June 1976.
14. Hess, J. L., 'Review of integral equation techniques for solving potential-flow problems with emphasis on the surface source method', *Comp. Meth. in Appl. Mech. and Eng.*, **5,** 145–196 (1975).
15. Abbott, I. H., and von Doenhoff, A., *Theory of Wing Sections,* Dover, New York, 1959.
16. Lyness, J., Owen, D. R. J., and Zienkiewicz, O. C., 'Three dimensional magnetic field determination using a scalar potential. A finite element solution', *Internal Civil Engineering Report.* C/R/275/76, University College of Swansea, 1976.
17. Zienkiewicz, O. C., Bettess, P., and Kelly, D. W. 'The finite element method for determining fluid loadings on rigid structures', *International Symposium on Numerical Methods in Offshore Engineering, Swansea,* January 1977.
18. Lamb, H., *Hydrodynamics,* 6th edn, Dover, New York, 1932.

Chapter 6

On Compatible and Equilibrium Models With Linear Stresses For Stretching of Elastic Plates

D. J. Allman

6.1 INTRODUCTION

The finite element method was conceived to provide a simple means of approximate analysis for complex engineering structures, the first recognized application[1] being motivated by the problem of evaluating the two-dimensional state of stress in the surface skin of an aircraft wing. The idealization involved a mesh of triangular elements, each with a constant stress distribution calculated from compatible linear displacements. This early development originated the conventional displacement method which is now identified as an application of the principle of minimum potential energy.[2] However, a constant stress idealization gives only average values of the stresses in each element, whereas the accurate calculation of stress concentrations in a stretched plate demands a more sophisticated approximation where point values of the stresses are available. Fortunately, the well-known compatible element[3,4] is easily derived using a quadratic displacement field. It provides a simple, yet accurate, method of analysis with linear stresses distributed over each element. The only unattractive feature is that stress discontinuities occur across element boundaries. Engineering applications of the displacement method therefore involve some kind of 'averaging' procedure at the nodes of a finite element mesh to give a piecewise-smooth stress distribution for design calculations.

Equilibrium finite element models for calculating the stress distribution in structures are based on the principle of minimum complementary energy.[2] They have a special appeal for practical design engineers, despite the current popularity of the conventional displacement method. The convenience of the equilibrium method for design purposes is a consequence of the exact transmission of stress resultants across boundaries between adjacent structural members or across discontinuities in plate thickness typified by material reinforcement at the edge of a hole. This avoids the need for 'averaging' procedures which are used to obtain unique nodal values of the stresses calculated from the displacement method. From a theoretical standpoint also, equilibrium finite element models provide a worthwhile analysis capability complementary to the displacement method. Indeed, the dual analysis of structures,[2,3] based on the alternate use of displacement and equilibrium models, is a traditional means of obtaining a direct estimate of convergence in energy. Equilibrium models are currently derived by two alternative techniques which use either stress functions[5] or stress components[2,3] as the fundamental variables. Yet it is apparent that each of these two techniques often gives equilibrium elements which are not entirely satisfactory, even for the basic plate stretching problem discussed in this paper. The elements can suffer from restrictions and deficiencies which render them unsuitable for general application.

The derivation of an equilibrium finite element model for plate stretching, using a continuous approximation for the Airy stress function and its first derivatives, is an example of a method with restrictive application. Appropriate forms for the stress function in an element are identical to the expressions already developed[6] for the lateral displacement field associated with compatible plate bending finite elements, but difficulties occur[5] in dealing with certain classes of stretching problem where the stress function is not single-valued. Notable examples are multiply-connected regions, as found in certain problems of plates pierced by holes, where correct solutions using a finite element approximation for the stress function can be obtained only if the continuity conditions are relaxed along fictitious 'cuts' in the plate. Other difficulties can also occur in assigning unique boundary values to the stress functions and their derivatives at plate edges where tractions are prescribed, cf. Merrifield[7] in the context of Southwell stress functions for plate bending problems. Furthermore, direct information on the associated displacement field is unavailable.

The alternative equilibrium method, pioneered by Fraeijs de Veubeke,[2,3] is the stiffness formulation based on a generalized complementary energy principle where continuity of stress resultants across element boundaries is enforced by Lagrange's method of multipliers.[8] This type of finite element analysis is unrestricted in range of application and provides values for the generalized displacements but, as found in the first derivation[2] of an equilibrium model with constant stresses, the approach can lead to zero stress states associated with kinematic deformations[2,3] other than the rigid body movements. Although the constant stress idealization has a well-behaved element stiffness matrix, the global stiffness matrix is singular even with the rigid body movements constrained and the final rank-deficient set of equations requires special numerical techniques for solution. Moreover, the associated set of generalized displacements is not unique, hence preventing any direct physical interpretation of the displacement values.

Unfortunately, an equilibrium element with a simple linear stress field also has insufficient stress parameters to ensure an elastic response to all admissible kinetic deformations. In this case the element stiffness matrix itself is rank-deficient. The present paper gives details of a triangular equilibrium element[9] for plate stretching with linear stresses derived in a way which avoids spurious kinematic modes. The linear stress field has nine stress parameters calculated from Airy stress functions, each defined over a triangular 'subregion' which is one-third the area of its parent element. Appropriate continuity conditions are enforced on the stress functions which ensure equilibrium of the tractions across the internal boundaries between subregions. Equilibrium of the tractions across the external boundaries between elements is enforced using Lagrange's method of multipliers. The complete technique is tantamount to assembling a number of smaller elements to form a satisfactory composite element, as described in the context of plate bending[10] and plate stretching[11] respectively. However, the present extension of this concept provides a convenient method with a minimum of algebraic complexity for calculating the stiffness matrix of an equilibrium element.

A useful guide to the relative accuracy of a compatible or an equilibrium analysis is provided by the number of unconstrained degrees of freedom involved in a finite element solution. The paper presents examples where the number of unconstrained degrees of freedom available to minimize the appropriate energy functional is determined. The examples comprise two basic problems, each with a small number of elements, and a generalized problem with a large number of elements. In the latter case it is clear that the compatible model provides more unconstrained degrees of freedom, with less global equations, than the equilibrium model; the computational efficiency of the compatible model for large problems is therefore confirmed. Numerical results are also presented for a compatible and an equilibrium analysis of the stresses in a pierced square plate under unidirectional tensile stress. They are compared to results selected from an approximate solution[12] obtained by a technique[13] different to the finite element method. It is expected that the compatible model gives a more accurate solution to the pierced plate problem, yet the equilibrium model provides a surprisingly accurate result for the maximum stress concentration.

6.2 THE COMPATIBLE DISPLACEMENT MODEL WITH LINEAR STRESSES

The compatible displacement model with linear stresses is presented by Fraeijs de Veubeke[3] and Argyris.[4] The formulation involves quadratic displacements u and v defined over the triangular element, shown in Figure 6.1,

$$\left.\begin{aligned} u &= u_O + \alpha_1 x + (\alpha_2 - \omega_O) y + \alpha_3 x^2 + \alpha_4 xy + \alpha_5 y^2 \\ v &= v_O + (\alpha_2 + \omega_O) x + \alpha_6 y + \alpha_7 x^2 + \alpha_8 xy + \alpha_9 y^2 \end{aligned}\right\} \tag{6.1}$$

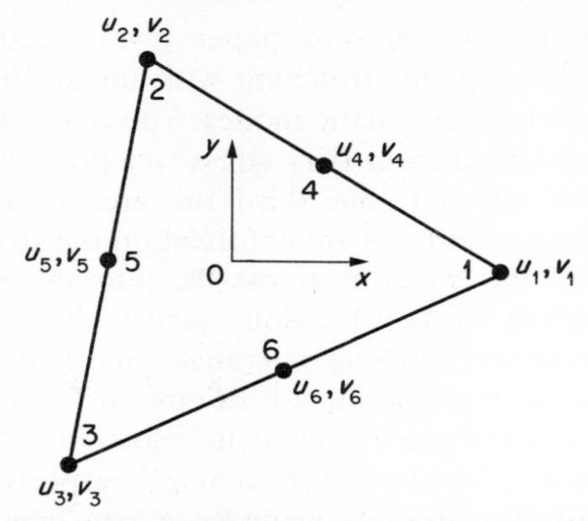

Figure 6.1 Triangular compatible element for plate stretching

where the rectangular Cartesian co-ordinates (x, y) have their origin at the centroid O. The three rigid body movements at the origin $x = y = 0$ are denoted u_O, v_O and ω_O respectively. The nine coefficients α_i, together with the three rigid body movements, are evaluated in terms of the vertex and mid-side nodal values of the displacements by inversion of equation (6.1). A mesh of elements connected by the nodal displacement values exhibits complete compatibility of the quadratic displacement field so that an associated finite element analysis is a valid application of the principle of minimum potential energy.

The components of direct stress σ_x, σ_y and shear stress τ_{xy} in an element are calculated from equation (6.1) using the strain–displacement relations

$$\varepsilon_x = \frac{\partial u}{\partial x}, \qquad \varepsilon_y = \frac{\partial v}{\partial y}, \qquad \gamma_{xy} = \frac{\partial u}{\partial y} + \frac{\partial v}{\partial x}$$

and the generalized Hooke's law[14] for plane stress. The linear stress field involves the nine coefficients α_i, viz.:

$$\left.\begin{aligned}
\sigma_x &= \frac{E}{1-\nu^2}[(\alpha_1 + \nu\alpha_6) + (2\alpha_3 + \nu\alpha_8)x + (\alpha_4 + 2\nu\alpha_9)y] \\
\sigma_y &= \frac{E}{1-\nu^2}[(\nu\alpha_1 + \alpha_6) + (2\nu\alpha_3 + \alpha_8)x + (\nu\alpha_4 + 2\alpha_9)y] \\
\tau_{xy} &= \frac{E}{2(1+\nu)}[2\alpha_2 + (\alpha_4 + 2\alpha_7)x + (2\alpha_5 + \alpha_8)y]
\end{aligned}\right\} \tag{6.2}$$

where E is Young's modulus and ν is Poisson's ratio relevant to an isotropic elastic material. The element stresses neither satisfy the equations of local stress equilibrium

$$\left.\begin{aligned}\frac{\partial \sigma_x}{\partial x}+\frac{\partial \tau_{xy}}{\partial y}=0\\ \frac{\partial \sigma_y}{\partial y}+\frac{\partial \tau_{xy}}{\partial x}=0\end{aligned}\right\} \tag{6.3}$$

nor give stress resultants in equilibrium across element boundaries. Instead, approximate satisfaction of the equations of equilibrium is achieved in accordance with the principle of minimum potential energy.

Calculation of the element stiffness matrix for the compatible displacement model with linear stresses follows the standard technique associated with the displacement method. A detailed description of the procedure is given by Fraeijs de Veubeke.[3]

6.3 A LINEAR EQUILIBRIUM STRESS FIELD WITH NINE STRESS PARAMETERS

The simple linear stress field which satisfies the equations (6.3) of local stress equilibrium is obtained from an Airy stress function $\phi(x, y)$ defined by a complete cubic polynomial

$$\phi = A + Bx + Cy + a_1x^2 + a_2xy + a_3y^2 + a_4x^3 + a_5x^2y + a_6xy^2 + a_7y^3 \tag{6.4}$$

where the rectangular Cartesian co-ordinates (x, y) have their origin at the centroid O of the triangular element shown in Figure 6.2. The stresses are calculated from the relations

$$\sigma_x = \frac{\partial^2 \phi}{\partial y^2}, \qquad \sigma_y = \frac{\partial^2 \phi}{\partial x^2}, \qquad \tau_{xy} = -\frac{\partial^2 \phi}{\partial x\, \partial y};$$

they involve seven stress parameters, namely the coefficients a_i of the stress function in equation (6.4), viz.:

$$\left.\begin{aligned}\sigma_x &= 2a_3 + 2a_6x + 6a_7y\\ \sigma_y &= 2a_1 + 6a_4x + 2a_5y\\ \tau_{xy} &= -a_2 - 2a_5x - 2a_6y\end{aligned}\right\} \tag{6.5}$$

The coefficients A, B, and C associated with the constant and linear terms in equation (6.4) can take any values without affecting the distribution of stresses.

The transmission of an equilibrium linear stress distribution across the sides of a triangular element requires twelve generalized forces for boundary stress connection, although nine only are independent because they are the boundary resultants of interior stresses satisfying three equations of equilibrium. Unfortunately, the simple linear stress field in equation (6.5) has only seven stress parameters; it is an insufficient number to provide an exact fit to all nine independent boundary stress modes. This has an important implication for an equilibrium element with twelve generalized displacements which is derived, using the stresses in equation (6.5), by the stiffness formulation due to Fraeijs de Veubeke.[2,3] The element stiffness matrix has a rank of seven and is therefore singular even if the three rigid body movements are constrained. Indeed, it can be shown that two independent distributions of quadratic displacement u and v, in addition to the three rigid body movements, are othogonal to the linear equilibrium stresses in the sense that no complementary work is obtained on the element boundary. In this case, the corresponding generalized displacements for the equilibrium model are zero and minimization of the complementary energy of an element produces zero values for the interior stresses. This type of behaviour is an undesirable feature in an equilibrium model because the global stiffness matrix is rank-deficient, unless suitable element patterns are used, and the final equations require special numerical techniques for solution. However, it is possible to derive an equilibrium element with linear stresses involving nine stress parameters so that spurious kinematic deformations are constrained. The approach described in this paper is tantamount to assembling a number of smaller equilibrium elements to form a satisfactory composite element,[11] but the technique employed here has the merit of a convenient and simple algebraic formulation.

The triangular finite element shown in Figure 6.2 is considered to be partitioned into three subregions by lines joining the vertices to the centroid O of the triangle. Equilibrium of the boundary stress resultants across the internal boundaries between subregions is assured if the stress functions and their normal derivatives are continuous.[5] This type of continuity across internal boundaries is achieved as follows: all three stress functions and their first derivatives are made zero at the centroid of the triangular element by choosing a special cubic form of equation (6.4) with $A=B=C=0$. In addition, at each vertex of the triangular element, the stress functions and their first derivatives are set equal to three new stress parameters to give a total of nine (unknown) stress parameters β_i for the entire element. A typical subregion therefore has six equations between the coefficients a_i of the stress functions and the stress parameters β_i shown in Figure 6.2, viz.:

$$Aa = T\beta \tag{6.6}$$

where the matrices A and T are defined below and the vectors a and β are

$$\left.\begin{aligned} a &= \{a_1, a_2, \ldots, a_7\}, \\ \beta &= \{\beta_1, \beta_2, \ldots, \beta_9\}, \end{aligned}\right\} \tag{6.7}$$

For the subregion I, the matrix A is

$$A = \begin{bmatrix} x_2^2 & x_2y_2 & y_2^2 & x_2^3 & x_2^2y_2 & x_2y_2^2 & y_2^3 \\ 2x_2 & y_2 & 0 & 3x_2^2 & 2x_2y_2 & y_2^2 & 0 \\ 0 & x_2 & 2y_2 & 0 & x_2^2 & 2x_2y_2 & 3y_2^2 \\ x_3^2 & x_3y_3 & y_3^2 & x_3^3 & x_3^2y_3 & x_3y_3^2 & y_3^3 \\ x_3^2 & x_3y_3 & y_3^2 & x_3^3 & x_3^2y_3 & x_3y_3^2 & y_3^3 \\ 2x_3 & y_3 & 0 & 3x_3^2 & 2x_3y_3 & y_3^2 & 0 \\ 0 & x_3 & 2y_3 & 0 & x_3^2 & 2x_3y_3 & 3y_3^2 \end{bmatrix} \tag{6.8}$$

and the Boolean matrix T, which selects the appropriate β_i, is

$$T = \begin{bmatrix} 0 & 0 & 0 & 1 & 0 & 0 & 0 & 0 & 0 \\ 0 & 0 & 0 & 0 & 1 & 0 & 0 & 0 & 0 \\ 0 & 0 & 0 & 0 & 0 & 1 & 0 & 0 & 0 \\ 0 & 0 & 0 & 0 & 0 & 0 & 1 & 0 & 0 \\ 0 & 0 & 0 & 0 & 0 & 0 & 0 & 1 & 0 \\ 0 & 0 & 0 & 0 & 0 & 0 & 0 & 0 & 1 \end{bmatrix} \tag{6.9}$$

Cyclic permutation of the indices 1, 2, 3 in the matrix A and an obvious rearrangement of the unit elements of the matrix T give the appropriate matrices for the other two subregions. Continuity of the stress functions and their first derivatives across internal boundaries is completed by matching the values of the normal derivatives at the mid-side point of an internal boundary. For the internal boundary O–3 between subregions I and II, this gives at the mid-side point L:

$$\left.\frac{\partial F_{\mathrm{I}}}{\partial \lambda}\right|_{\mathrm{L}} + \left.\frac{\partial F_{\mathrm{II}}}{\partial \lambda}\right|_{\mathrm{L}} = 0 \tag{6.10}$$

where the co-ordinate λ is the exterior normal to a subregion at an internal boundary. A total of three equations of this type is obtained by considering also the two other internal boundaries.

The stress functions now have the correct continuity properties to ensure that the linear stresses in equation (6.5) have resultants which are in equilibrium across the internal boundaries between subregions. Considering all three subregions, equations (6.6) and (6.10) together provide 21 equations which uniquely determine the coefficients a_i of the stress functions in terms of the nine stress parameters β_i. The vector of coefficients of the stress function in a typical subregion may be written

$$a = S\beta \tag{6.11}$$

where the (7×9) matrix $\mathcal{S}$ is a submatrix of the (21×9) matrix which is determined numerically by solution of the 21 equations.

Along an exterior boundary of a subregion, the normal and shear stress resultants respectively for a plate of thickness h are calculated from equations (6.5) as

$$\left.\begin{aligned} N_n &= h\sigma_x \cos^2\gamma + h\sigma_y \sin^2\gamma + 2h\tau_{xy}\sin\gamma\cos\gamma \\ N_{ns} &= h(\sigma_y - \sigma_x)\sin\gamma\cos\gamma + h\tau_{xy}(\cos^2\gamma - \sin^2\gamma) \end{aligned}\right\} \tag{6.12}$$

where the co-ordinate n is the exterior normal to the boundary and the co-ordinate s is measured from a vertex and along the boundary in an anti-clockwise sense for the parent element. The values of the stress resultants at the ends of the typical exterior side 1–2 of length l_{12} are denoted N_n^{12}, N_n^{21}, N_{ns}^{12}, and N_{ns}^{21} (see Figure 6.2) and they provide four generalized forces to allow a linear distribution of boundary traction to be transmitted between elements, viz.:

$$\left.\begin{aligned} N_n &= \left(1-\frac{s}{l_{12}}\right)N_n^{12} + \left(\frac{s}{l_{12}}\right)N_n^{21} \\ N_{ns} &= \left(1-\frac{s}{l_{12}}\right)N_{ns}^{12} + \left(\frac{s}{l_{12}}\right)N_{ns}^{21} \end{aligned}\right\} \tag{6.13}$$

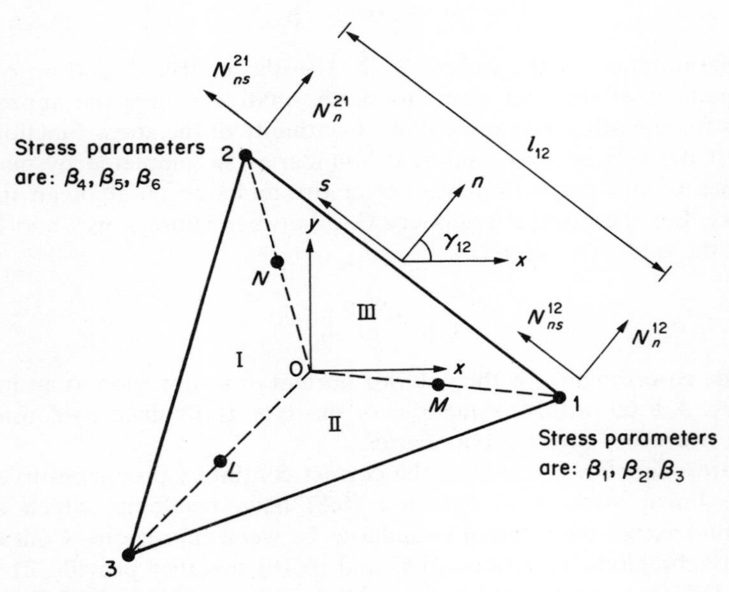

Figure 6.2 Triangular equilibrium element for plate stretching

6.4 CALCULATION OF THE EQUILIBRIUM ELEMENT STIFFNESS MATRIX

Assuming a state of plane stress in an isotropic plate of thickness h with Young's modulus E and Poisson's ratio ν, the contribution of a subregion to the total complementary strain energy of an element is

$$U=\frac{1}{2}\iint_{\text{subregion}}\frac{h}{E}[\sigma_x^2+\sigma_y^2-2\nu\sigma_x\sigma_y+2(1+\nu)\tau_{xy}^2]\,dx\,dy \qquad (6.14)$$

where it is noted that the area of a subregion is one-third of the area of the parent element. Substituting for the stresses from equations (6.5) and noting equation (6.11) it is found that the complementary strain energy is

$$U=\tfrac{1}{2}\beta^T f\beta \qquad (6.15)$$

where the subregion flexibility matrix is

$$f=S^T HS \qquad (6.16)$$

The positive definite matrix H is

$$H=\iint_{\text{subregion}}\frac{h}{E}\times\begin{bmatrix} 4 & 0 & -4\nu & 12x & 4y & -4\nu x & -12\nu y \\ & 2(1+\nu) & 0 & 0 & 4(1+\nu)x & 4(1+\nu)y & 0 \\ & & 4 & -12\nu x & -4\nu y & 4x & 12y \\ & & & 36x^2 & 12xy & -12\nu x^2 & -36\nu xy \\ & & & & 4[2(1+\nu)x^2+y^2] & 4(2+\nu)xy & -12\nu y^2 \\ & \text{symmetrical} & & & & 4[x^2+2(1+\nu)y^2] & 12xy \\ & & & & & & 36y^2 \end{bmatrix}dx\,dy \qquad (6.17)$$

where the integrations over a subregion can be written down immediately using the simple formulae given in the appendix at the end of this chapter.

Using equations (6.5), (6.11), and (6.12), the twelve generalized forces on an element boundary are found to be related to the nine stress parameters β_i by a (9×12) matrix C, viz.:

$$P=C^T\beta \qquad (6.18)$$

where the vector P is

$$P=\{N_n^{12}, N_n^{21}, N_{ns}^{12}, N_{ns}^{21}, N_n^{23}, N_n^{32}, N_{ns}^{23}, N_{ns}^{32}, N_n^{31}, N_n^{13}, N_{ns}^{31}, N_{ns}^{13}\} \qquad (6.19)$$

The twelve generalized displacements, corresponding to the generalized forces, which provide the Lagrange multipliers to enforce equilibrium of the stress resultants across element boundaries are the components of a vector p; it is written symbolically as

$$p = \{p_n^{12}, p_n^{21}, p_{ns}^{12}, p_{ns}^{21}, p_n^{23}, p_n^{32}, p_{ns}^{23}, p_{ns}^{32}, p_n^{31}, p_n^{13}, p_{ns}^{31}, p_{ns}^{13}\} \qquad (6.20)$$

For the typical side 1–2, of length l_{12}, the generalized displacements are defined by

$$\left.\begin{aligned} p_n^{12} &= \int_0^{l_{12}} \left(1 - \frac{s}{l_{12}}\right) u_n \,\mathrm{d}s, \qquad & p_n^{21} &= \int_0^{l_{12}} \left(\frac{s}{l_{12}}\right) u_n \,\mathrm{d}s \\ p_{ns}^{12} &= \int_0^{l_{12}} \left(1 - \frac{s}{l_{12}}\right) u_s \,\mathrm{d}s, \qquad & p_{ns}^{21} &= \int_0^{l_{12}} \left(\frac{s}{l_{12}}\right) u_s \,\mathrm{d}s \end{aligned}\right\} \qquad (6.21)$$

where u_n and u_s are the normal and tangential components of displacement respectively. Prescribed values of the displacements are included explicitly in the vector p using equations (6.21).

The complementary energy of an element[3] is

$$\Pi_c = \tfrac{1}{2}\beta^{\mathrm{T}} F \beta - P^{\mathrm{T}} p + p^{*\mathrm{T}} p \qquad (6.22)$$

where the (9×9) flexibility matrix F is given by the sum of the three subregion flexibility matrices in equation (6.16) and the vector P^* contains prescribed values of the twelve generalized forces of equation (6.19). The linear interior stresses with stress parameters β_i are local functions for each element, so that substitution from equation (6.18) and minimization of the complementary energy in equation (6.22) with respect to β provides

$$\beta = F^{-1} C p \qquad (6.23)$$

Equilibrium connection between elements is achieved by equating the values of the generalized displacements p_i. The total complementary energy of the plate is then obtained by summing the contribution, given in equation (6.22), from each of the elements. Substitution from equation (6.23) into the expression for the total complementary energy and minimization with respect to p_i then gives

$$\sum_{\text{elements}} [(C^{\mathrm{T}} F^{-1} C) p - P^*] = 0 \qquad (6.24)$$

where the positive semi-definite (12×12) element stiffness matrix is defined as

$$K = C^{\mathrm{T}} F^{-1} C \qquad (6.25)$$

The matrix of coefficients of the global equations (6.24) is positive definite when the rigid body displacements are constrained by specifying appropriate values for the generalized displacements p_i.

6.5 CALCULATION OF THE ELEMENT STRESSES

Local values of the stress components for the compatible finite element are defined uniquely over each element by equation (6.2), but stress discontinuities occur across element boundaries. In practical applications, where a piecewise-smooth stress distribution is required, it is customary to take the simple average of the nodal stress values. Nevertheless, it is useful to note that the severity of boundary stress discontinuity gives a useful ad hoc measure of the local accuracy of a finite element solution.

The linear stresses in each of the sub-regions of the triangular equilibrium element can be calculated from equations (6.5). However, this interpretation of the numerical results is inconvenient in a practical calculation because it gives two different values for each of the stress components σ_x, σ_y, and τ_{xy} corresponding to the two subregions at a vertex of the parent element. An alternative, yet strictly valid, procedure is to take the stress values directly from the interelement generalized forces given in equation (6.19); a stress component is obtained by dividing N_n^{12}, say, by the plate thickness h. This requires that the orientation of the mesh in a particular application is always such that either the typical generalized normal force N_n^{12} or the typical generalized shear force N_{ns}^{12} coincides with the desired stress component. In practice, there is little difficulty, in achieving this requirement and it has the merit, already discussed in the Section 6.1, of ensuring that quoted stress values are continuous across a common boundary between adjacent elements.

6.6 NUMERICAL APPLICATIONS AND COMPARISONS

The compatible finite element with linear stresses solves exactly simple plate stretching problems where the linear stress field is derived from quadratic displacements. The equilibrium finite element reproduces exact solutions to problems which involve either zero, constant, or linear stress states satisfying the homogeneous equations of equilibrium. A useful guide to the relative accuracy of the two alternative methods of analysis is provided by the number of unconstrained degrees of freedom involved in a finite element solution. As examples, the number of unconstrained degrees of freedom available to minimize the appropriate energy functional is determined for both a compatible and an equilibrium formulation of two basic problems, each with a small number of elements, and for a generalized problem with a large number of elements. However, it is important to investigate the numerical accuracy which can be expected from more complicated practical applications. Results from a compatible and an equilibrium finite element analysis are presented for the stresses in a pierced square plate under unidirectional tensile stress. They are compared to results selected from an approximate solution[12] obtained by a technique[13] different to the finite element method.

Consider the two basic problems shown in Figures 6.3 and 6.4 where a square plate, idealized by eight elements, is stretched first by prescribed tractions and then be prescribed displacements; both types of prescribed condition are applied around the entire external boundary. In the first problem, shown in Figure 6.3, the compatible model involves 47 unconstrained displacement degrees of freedom in the global numerical solution,

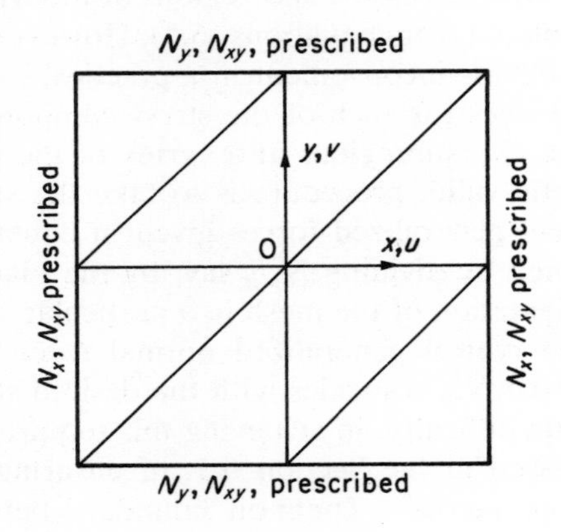

	Compatible element	Equilibrium element	
		Stresses	Displacements
Total dof	50	72	64
Kinematic constraints	3		3
Equilibrium constraints		61	
Unconstrained dof	47	11	61

N.B.: dof = degrees of freedom.

Figure 6.3 Basic problem with prescribed boundary tractions

all of which are available to minimize the potential energy. On the other hand, although the numerical solution of this problem using the equilibrium model involves 61 unconstrained displacement degrees of freedom, these are the Lagrange multipliers which enforce the equilibrium constraints at the element boundaries. In fact, only eleven unconstrained stress parameter degrees of freedom from a total of 72 are available to minimize the complementary energy. It is expected, therefore, that the compatible model gives more accurate numerical results in this case. In the second problem, shown in Figure 6.4 the previous situation is completely reversed. Now, the compatible model has 18 unconstrained displacement degrees of freedom only, while the equilibrium model has 40 unconstrained stress parameter degrees of freedom, so that the latter is expected to be more accurate. Clearly, the number of unconstrained degrees of freedom available to

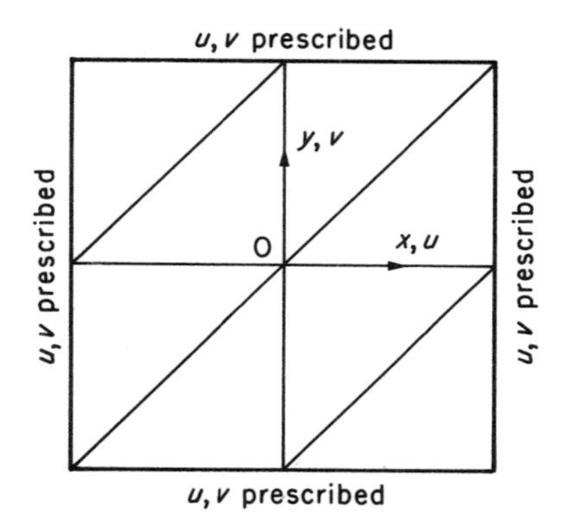

	Compatible element	Equilibrium element	
		Stresses	Displacements
Total dof	50	72	64
Kinematic constraints	32		32
Equilibrium constraints		32	
Unconstrained dof	18	40	32

N.B.: dof = degrees of freedom.

Figure 6.4 Basic problem with prescribed boundary displacements

minimize the energy functional for a compatible or an equilibrium solution is problem-dependent. A specific way of examining this dependence is to consider the generalized problem of a square plate subdivided into an $(n \times n)$ mech with $2n^2$ triangular elements. Tables 6.1 and 6.2 show the number of degrees of freedom corresponding to the cases of prescribed boundary tractions and prescribed boundary displacements respectively. The last row of each table gives a close approximation to the number of unconstrained degrees of freedom for large n, which is obtained by neglecting terms not involving n^2. For a fine mesh, where $n \gg 1$, it is clear that the compatible model provides more unconstrained degrees of freedom, with less global equations, than the equilibrium model.The computational efficiency of the compatible model for large problems is therefore confirmed.

Table 6.1 Degrees of freedom for $(n \times n)$ model of square plate under prescribed boundary tractions

	Compatible element	Equilibrium element	
		Stresses	Displacements
Total dof	$2(2n+1)^2$	$18n^2$	$4n(3n+2)$
Kinematic constraints	3		3
Equilibrium constraints		$4n(3n+2)-3$	
Unconstrained dof	$2(2n+1)^2-3$	$2n(3n-4)+3$	$4n(3n+2)-3$
Unconstrained dof with $n \gg 1$	$8n^2$	$6n^2$	$12n^2$

N.B.: dof = degrees of freedom.

Table 6.2 Degrees of freedom for $(n \times n)$ model of square plate under prescribed boundary displacements

	Compatible element	Equilibrium element	
		Stresses	Displacements
Total dof	$2(2n+1)^2$	$18n^2$	$4n(3n+2)$
Kinematic constraints	$16n$		$16n$
Equilibrium constraints		$4n(3n-2)$	
Unconstrained dof	$2(2n-1)^2$	$2n(3n+4)$	$4n(3n-2)$
Unconstrained dof with $n \gg 1$	$8n^2$	$6n^2$	$12n^2$

N.B.: dof = degrees of freedom.

As a practical application, the problem of a homogeneous square plate pierced by a circular hole of diameter equal to half its side length is considered. The plate is loaded by a constant unidirectional tensile stress σ^* applied along opposite sides. Because of symmetry, it is possible to perform the analysis with a mesh of 48 elements covering a quarter of the plate only, as shown in Figure 6.5, with appropriate kinematic and static constraints imposed along the symmetry lines AB and CD. Numerical results for the normal stresses along the lines AB and CD are presented in Table 6.3, both for the equilibrium stress model and for the compatible displacement model. The normal stresses at the points A and C are also the circumferential stresses at the edge of the hole and these are compared with the values obtained by Hengst[12] using a technique due to Trefftz.[13] The number of unconstrained displacement degrees of freedom involved in the compatible analysis is found to be very nearly twice the number of unconstrained stress

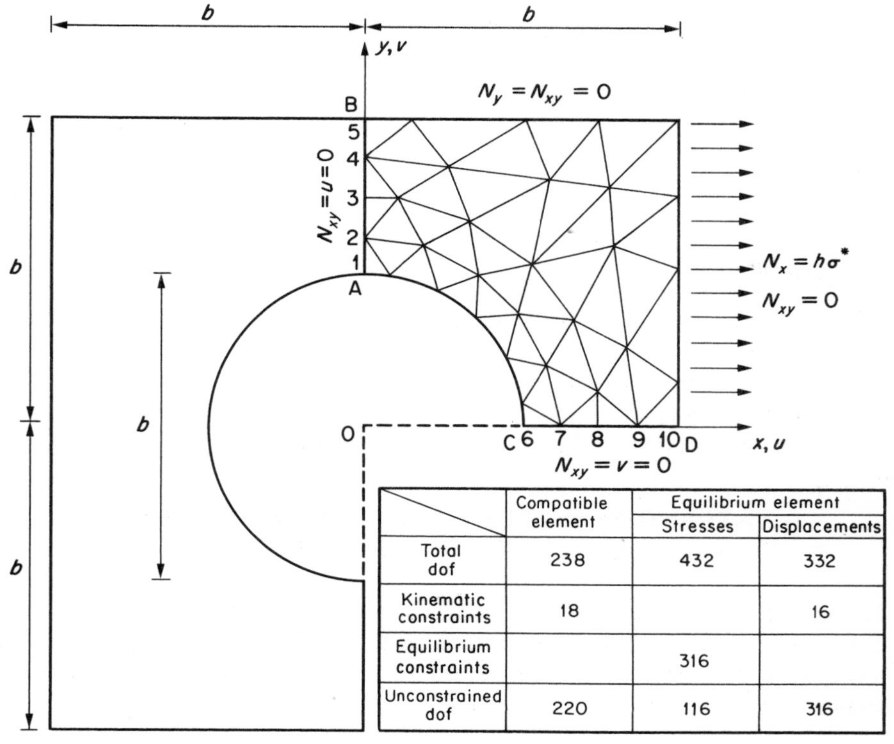

	Compatible element	Equilibrium element	
		Stresses	Displacements
Total dof	238	432	332
Kinematic constraints	18		16
Equilibrium constraints		316	
Unconstrained dof	220	116	316

N.B.: dof = degrees of freedom.

Figure 6.5 Finite element mesh for a pierced square plate of thickness h

parameter degrees of freedom involved in the equilibrium analysis (see Figure 6.5) and, for the same reasons put forward in the previous discussion of basic problems, it is expected that the former solution is more accurate. Nevertheless, an examination of Table 6.3 reveals that the stresses from the equilibrium analysis compare favourably with those both from the compatible analysis and from Hengst's solution. The maximum stress concentration

Table 6.3 Distribution of normal stresses along symmetry lines of pierced square plate

σ_x along AB			σ_y along CD		
Boundary node	Compatible element	Equilibrium element	Boundary node	Compatible element	Equilibrium element
1	6.55 (6.328)[a]	6.06 (6.328)[a]	6	−4.03 (−3.912)[a]	−3.52 (−3.912)[a]
2	2.41	3.22	7	−0.44	−1.21
2	3.00	3.09	7	−0.97	−1.01
3	1.70	1.67	8	0.19	0.18
3	1.70	1.62	8	0.19	0.26
4	0.54	0.61	9	1.27	1.20
4	0.58	0.40	9	1.24	1.49
5	−0.82	−0.67	10	2.82	2.69
	Multiplier $= \sigma^*$			Multiplier $= \sigma^*$	

[a] Values in parentheses are quoted from Hengst's paper.[12]

for the plate, which occurs in the circumferential stress at point A, takes the value 6.328 according to Hengst's paper. This value for the stress concentration is overestimated by 3.5 per cent using the compatible method and underestimated by 4.3 per cent using the equilibrium method, the latter result being surprisingly accurate despite the significantly smaller number of unconstrained degrees of freedom involved in the equilibrium solution.

APPENDIX: INTEGRATION FORMULAE FOR AN ELEMENT SUBREGION

The calculation of the flexibility matrix for a typical subregion requires the evaluation of area integrals expressed in terms of local co-ordinates (x, y) located at the centroid of the parent element. The area of the element shown in Figure 6.2 is

$$\Delta = \tfrac{1}{2}[x_1(y_2 - y_3) + x_2(y_3 - y_1) + x_3(y_1 - y_2)]$$

where the co-ordinates (x_1, y_1), (x_2, y_2) and (x_3, y_3) refer to the element vertices. The values of the integrals over the typical subregion I are

$$\iint\limits_{\mathrm{I}} \mathrm{d}x\,\mathrm{d}y = \frac{\Delta}{3}$$

$$\iint\limits_{\mathrm{I}} x\,\mathrm{d}x\,\mathrm{d}y = -\left(\frac{\Delta}{9}\right)x_1$$

$$\iint\limits_{\mathrm{I}} y\,\mathrm{d}x\,\mathrm{d}y = -\left(\frac{\Delta}{9}\right)y_1$$

$$\iint\limits_{\mathrm{I}} x^2\,\mathrm{d}x\,\mathrm{d}y = \frac{\Delta}{18}(x_1^2 - x_2x_3)$$

$$\iint\limits_{\mathrm{I}} xy\,\mathrm{d}x\,\mathrm{d}y = \frac{\Delta}{36}[2x_1y_1 - (x_2y_3 + x_3y_2)]$$

$$\iint\limits_{\mathrm{I}} y^2\,\mathrm{d}_x\,\mathrm{d}y = \frac{\Delta}{18}(y_1^2 - y_2y_3)$$

The values of these integrals over the subregions II and III are obtained by a cyclic permutation of the integers 1, 2, and 3.

REFERENCES

1. M. J. Turner, R. W. Clough, H. C. Martin, and L. J. Topp, 'Stiffness and deflection analysis of complex structures', *J. Aeron. Sci.*, **23,** No. 9 (1956).
2. B. Fraeijs de Veubeke, 'Upper and lower bounds in matrix structural analysis', *Matrix Methods of Structural Mechanics* (ed. by B. Fraeijs de Veubeke), AGARDograph, 72, Pergamon Press, Oxford, 1964, pp. 165–201.
3. B. Fraeijs de Veubeke, 'Displacement and equilibrium models in the finite element method', in *Stress Analysis* (ed. by O. C. Zienkiewicz and G. S. Holister), Wiley, London, 1965, Ch. 9, pp. 145–197.
4. J. H. Argyris, 'Triangular elements with linearly varying strain for the matrix displacement method', *J. Royal Aero. Soc.*, **69,** No. 658, 711–713 (1965).
5. B. Fraeijs de Veubeke and O. C. Zienkiewicz, 'Strain energy bounds in finite element analysis by slab analogy', *J. Strain Analysis*, **2,** No. 4, 265–271 (1967).
6. K. Bell, 'Triangular plate bending elements', in *Finite Element Methods in Stress Analysis* (ed. by I. Holand and K. Bell), TAPIR, Trondheim, 1969, pp. 213–252.
7. B. C. Merrifield, 'A computer program for the finite element analysis of a class of plate bending problems', *RAE Technical Report* 69270, 1969, p.7.
8. R. Courant and D. Hilbert, *Methods of mathematical physics*, Vol. 1, 5th edn, Interscience, New York, 1965, p. 165.

9. D. J. Allman, 'An equilibrium finite element for plate stretching,' *RAE Technical Report* 77049, 1977.
10. R. W. Clough and J. L. Tocher, 'Finite element stiffness matrices for analysis of plate bending', in *Matrix Methods in Structural Mechanics* (ed. by J. S. Przemieniecki *et al.*, AFFDL-TR-66-80, 1966, pp. 515–546.
11. G. Sander, 'Application of the dual analysis principle'. *High-speed Computing of Elastic Structures*, Vol. 1, University of Liège, 1970, pp. 167–207.
12. H. Hengst, 'Beitrag zur Beurteilung des Spannungszustandes einer Gelochten Scheibe', *Ztschr. f. angew. Math. und Mech.*, **18,** 44–48, (1938)
13. E. Trefftz, *Mathematische Annalen,* **100,** 503–521 (1928).
14. I. S. Sokolnikoff, *Mathematical Theory of Elasticity*, 2nd edn, McGraw-Hill, New York, 1956.

Chapter 7

Note on a Finite Element Stiffness Formulation in Shell Theory

L. S. D. Morley

7.1 INTRODUCTION

The inception, development, and understanding of the stiffness formulation in the finite element method owes much to the pioneering work accomplished by Fraeijs de Veubeke. In a stiffness formulation, the element connectors may be generalized displacements and generalized slopes rather than the actual physical quantities. It is the present purpose to examine the details of a formulation of this kind for the solution of the mathematical problem which is posed by first approximation linear theory of thin-walled shells.

The motivation arises out of awareness that the current position in the finite element analysis of shells is, to say the least, paradoxical. On the one hand we know that the derivation and comprehension of a satisfactory first approximation theory has taken a long time; resolution of its subtleties has attracted the attention of many distinguished contributors, e.g. Love[1,2] (1890 and 1927), Reissner[3] (1941), Novozhilov[4] (1959), Koiter[5] (1960), Gol'denveizer[6] (1961), Naghdi[7] (1963). On the other hand, we have evidence that numerical analysis of shell behaviour by a stiffness formulation of the finite element method known as Semiloof, developed on heuristic arguments and due largely to Irons[8] (1976), can produce adequate and sometimes remarkably accurate results. Irons is the first to admit that Semiloof rests upon a contrived shell theory which owes nothing to the modern and consistent version; it rests upon deliberate use of the inaccuracies of approximate integration, upon incompatibility between and inside the elements, upon a (curvilinear) facet representation of the shell geometry, upon a rank-deficient stiffness matrix, etc. This finite element is, nonetheless, outstanding and has received widespread acclaim in engineering technology although it is understood that every new application involves some speculation. The security of the foundations of modern shell theory and the encouraging results from Semiloof are manifestations that the ultimate objective—the practical calculation of authentic numerical results—

can be achieved by the finite element method. The present formulation is distinct from that of Semiloof, it is a contribution towards the objective from within the confines of modern shell theory.

The viewpoint is adopted that Koiter's[5] (1960) first approximation shell theory provides the most suitable foundations upon which to develop a consistent finite element formulation. In this theory, the neglected effects of transverse direct stress and of transverse shear stress provide error norms which can be invoked whenever appropriate. Here, a surface co-ordinate system is used to examine convergence criteria in the finite element method specifically for shells with developable middle surface. This leads to the conclusion that it is generally advisable to discretize the elastic behaviour by mutually supplementing displacement and equilibrium sets of trial functions. The concepts developed by Fraeijs de Veubeke guide the selection and topology of element connectors appropriate to a stiffness formulation and also the relation of the connectors with the supplementing interior fields through consideration of modified complementary energy. The combination of surface co-ordinates on a developable surface leads to certain difficulties in interpolating the connectors around the element boundary. This is briefly examined for the special case of circular cylindrical shells—as is also the phenomenon of edge effect.

There is much that remains to be achieved, although it is already of interest to note the correspondences between the present proposals and with Irons's Semiloof finite element. Both make use of a two-field discretization for the interior with identical topology for the connectors.

7.2 BASIC TRIAL FUNCTIONS IN FIRST APPROXIMATION SHELL THEORY

Koiter[5] (1960) derives his complete theory of thin elastic shells in a consistent first approximation from Love's approximate strain energy expression. This strain energy expression is derived on the basis of three assumptions:

(i) The shell is thin so that h/R is small in comparison with unity where h is the shell thickness and R is the smallest principal radius of curvature of the middle surface.
(ii) The strains are small everywhere and the strain energy is represented by the quadratic function of the strain components for an isotropic solid.
(iii) The state of stress is approximately plane so that the effects of transverse shear stress and of transverse direct stress are neglected.

The theory proceeds from the point of view that the neglected stresses imply errors in the strain energy. In particular, the transverse direct stress is of order h/R times the membrane stresses while the transverse shear stress is of order h/L times the bending stresses, where L is the wavelength of distortion. Thus, the strain energy is calculated at best only to within a relative error h^2/R^2 or h^2/L^2 whichever of these is critical.

The strain energy density (SED) of the middle surface, which according to Love's approximate result is the sum of the membrane and bending strain energies, is given by

$$\text{SED}=\frac{1}{2}\frac{Eh}{1-\nu^2}\{(\varepsilon_{11}+\varepsilon_{22})^2-2(1-\nu)(\varepsilon_{11}\varepsilon_{22}-\varepsilon_{12}^2)\}$$

$$+\frac{1}{2}\frac{Eh^3}{12(1-\nu^2)}\{(\kappa_{11}+\kappa_{22})^2-2(1-\nu)(\kappa_{11}\kappa_{22}-\bar{\kappa}_{12}^2)\} \quad (7.1)$$

where ε_{11}, κ_{11}, ... are the strains and curvature changes along orthogonal curves on the middle surface with co-ordinates ξ_1, ξ_2. The quantity E is Young's modulus and ν denotes Poisson's ratio. In view of the foregoing remark that the neglected transverse direct stress is of order h/R times the membrane stresses it is clearly not worth trying to determine any of the strains ε_{11}, ε_{22}, ε_{12} to within a better accuracy than that which is provided by $\varepsilon(1\pm h/R)$ where ε is any physical strain. In like vein, since the outer fibre strain from the physical curvature change κ is $\kappa h/2$ it follows that:

in solving a problem according to first approximation shell theory it is pointless to seek a solution for a curvature change κ to within a better accuracy than is provided by $\kappa\pm\varepsilon/R$ (7.2)

Moreover, in the above expression for the SED we may replace any κ by $\kappa\pm\varepsilon/R$.

Naturally, this has ramifications with respect to the finite element displacement method where the stiffness matrix is calculated directly from the SED of equation (7.1). There are well established criteria for convergence of this displacement method for problems in the plane. The criteria require that each finite element recovers, in some global co-ordinate system, constant independent strains ε_{11}, ε_{22}, ε_{12} for problems of plane stress and, similarly, constant independent curvature changes κ_{11}, κ_{22}, $\bar{\kappa}_{12}$ for plate bending problems. Let us now examine these criteria in relation to the shell problem. For this purpose it is convenient to specialize attention to developable surfaces where both co-ordinates ξ_1, ξ_2 measure arc length and form an orthogonal system of geodesic lines on the middle surface. The conclusions from this examination remain essentially valid outside of the specialization. The strain and curvature change relations for this co-ordinate system, see, for example, Morley[9] (1976), are

$$
\left.
\begin{aligned}
\varepsilon_{11} &= \frac{\partial U_1}{\partial \xi_1} + \frac{W}{R_1} \\
\varepsilon_{22} &= \frac{\partial U_2}{\partial \xi_2} + \frac{W}{R_2} \\
\varepsilon_{12} &= \frac{1}{2}\left(\frac{\partial U_1}{\partial \xi_2} + \frac{\partial U_2}{\partial \xi_1} - \frac{2W}{R_{12}}\right) \\
\kappa_{11} &= -\frac{\partial^2 W}{\partial \xi_1^2} + \frac{\partial}{\partial \xi_1}\left\{\frac{U_1}{R_1} - \frac{U_2}{R_{12}}\right) - \frac{1}{2R_{12}}\left(\frac{\partial U_2}{\partial \xi_1} - \frac{\partial U_1}{\partial \xi_2}\right) \\
\kappa_{22} &= -\frac{\partial^2 W}{\partial \xi_2^2} + \frac{\partial}{\partial \xi_2}\left(\frac{U_2}{R_2} - \frac{U_1}{R_{12}}\right) + \frac{1}{2R_{12}}\left(\frac{\partial U_2}{\partial \xi_1} - \frac{\partial U_1}{\partial \xi_2}\right) \\
\bar{\kappa}_{12} &= -\frac{\partial^2 W}{\partial \xi_1 \, \partial \xi_2} + \frac{1}{2}\frac{\partial}{\partial \xi_1}\left(\frac{U_2}{R_2} - \frac{U_1}{R_{12}}\right) + \frac{1}{2}\frac{\partial}{\partial \xi_2}\left(\frac{U_1}{R_1} - \frac{U_2}{R_{12}}\right) \\
&\quad - \frac{1}{4}\left(\frac{1}{R_1} - \frac{1}{R_2}\right)\left(\frac{\partial U_2}{\partial \xi_1} - \frac{\partial U_1}{\partial \xi_2}\right)
\end{aligned}
\right\} \tag{7.3}
$$

where U_1, U_2 are insurface displacements along the co-ordinate directions ξ_1, ξ_2 and W is the transverse displacement with R_1, R_2, R_{12} the radii of curvature and of torsion appropriate to the co-ordinate system.

Consider a basic system of displacement trial functions

$$
\left.
\begin{aligned}
U_{1\alpha} &= \alpha_1 \xi_1 + \alpha_3 \xi_2 \\
U_{2\alpha} &= \alpha_2 \xi_2 + \alpha_3 \xi_1 \\
W &= 0
\end{aligned}
\right\} \tag{7.4}
$$

where α_1, α_2, α_3 are constants. These trial functions, when substituted into equations (7.3), give constant strains

$$
\varepsilon_{11\alpha} = \alpha_1, \qquad \varepsilon_{22\alpha} = \alpha_2, \qquad \varepsilon_{12\alpha} = \alpha_3 \tag{7.5}
$$

and the following curvature changes

$$
\left.
\begin{aligned}
\kappa_{11\alpha} &= \underline{\frac{\alpha_1}{R_1} - \frac{\alpha_3}{R_{12}}} + U_{1\alpha}\frac{\partial}{\partial \xi_1}\left(\frac{1}{R_1}\right) - U_{2\alpha}\frac{\partial}{\partial \xi_1}\left(\frac{1}{R_{12}}\right) \\
\kappa_{22\alpha} &= \underline{\frac{\alpha_2}{R_2} - \frac{\alpha_3}{R_{12}}} + U_{2\alpha}\frac{\partial}{\partial \xi_2}\left(\frac{1}{R_2}\right) - U_{1\alpha}\frac{\partial}{\partial \xi_2}\left(\frac{1}{R_{12}}\right) \\
\bar{\kappa}_{12\alpha} &= \frac{1}{2}\left[\underline{-\frac{\alpha_1 + \alpha_2}{R_{12}} + \frac{\alpha_3}{2}\left(\frac{1}{R_1} + \frac{1}{R_2}\right)}\right. \\
&\quad + U_{1\alpha}\left\{\frac{\partial}{\partial \xi_2}\left(\frac{1}{R_1}\right) - \frac{\partial}{\partial \xi_1}\left(\frac{1}{R_{12}}\right)\right\} \\
&\quad \left. + U_{2\alpha}\left\{\frac{\partial}{\partial \xi_1}\left(\frac{1}{R_2}\right) - \frac{\partial}{\partial \xi_2}\left(\frac{1}{R_{12}}\right)\right\}\right]
\end{aligned}
\right\} \tag{7.6}
$$

The underlined terms represent curvature changes κ which are of order $\pm\varepsilon/R$ and an appeal to the statement in equation (7.2) shows that they are comparatively insignificant. The remaining terms in $\bar{\kappa}_{12\alpha}$ equate to zero because of the Mainardi–Codazzi relations

$$\left.\begin{aligned}\frac{\partial}{\partial \xi_2}\left(\frac{1}{R_1}\right)=\frac{\partial}{\partial \xi_1}\left(\frac{1}{R_{12}}\right)\\ \frac{\partial}{\partial \xi_1}\left(\frac{1}{R_2}\right)=\frac{\partial}{\partial \xi_2}\left(\frac{1}{R_{12}}\right)\end{aligned}\right\} \tag{7.7}$$

as expressed for developable surfaces. The remaining terms in $\kappa_{11\alpha}$, $\kappa_{22\alpha}$ vanish at the co-ordinate origin ($U_{1\alpha}=U_{2\alpha}=0$), they vanish everywhere if the radii R_1, R_2, R_{12} are constant (i.e. for the circular cylinder); in other circumstances these terms may be very large because there is no restriction apart from equations (7.7), upon the rates of change of the radii of curvature. We conclude from this that the trial functions of equations (7.4) provide three constant independent strains $\varepsilon_{11\alpha}$, $\varepsilon_{22\alpha}$, $\varepsilon_{12\alpha}$ only for sufficiently small finite elements at the origin of the global co-ordinate system ξ_1, ξ_2. In other words it seems that the constant independent strain criterion, as established for convergence of the finite element displacement method in plane stress problems, can at best be contrived only in a piecewise context for shell problems.

Let us consider other displacement trial functions, this time with the intention of providing three constant and independent curvature changes κ_{11}, κ_{22}, $\bar{\kappa}_{12}$. The trial functions

$$\left.\begin{aligned}U_{1\alpha}, U_{2\alpha}&=0\\ W_\alpha&=-\tfrac{1}{2}\alpha_4\xi_1^2-\tfrac{1}{2}\alpha_5\xi_2^2-\alpha_6\xi_1\xi_2\end{aligned}\right\} \tag{7.8}$$

where α_4, α_5, α_6 are constants, give constant curvature changes

$$\kappa_{11\alpha}=\alpha_4, \qquad \kappa_{22\alpha}=\alpha_5, \qquad \bar{\kappa}_{12\alpha}=\alpha_6 \tag{7.9}$$

but are accompanied by strains

$$\varepsilon_{11\alpha}=W_\alpha/R_1, \qquad \varepsilon_{22\alpha}=W_\alpha/R_2, \qquad \varepsilon_{12\alpha}=-W_\alpha/R_{12} \tag{7.10}$$

Now, the statement in equation (7.2) shows that curvature changes κ may be calculated to within a tolerance $\kappa\pm\varepsilon/R$ and, in the present instance, we have

$$\kappa_\alpha\pm\varepsilon_\alpha/R\simeq\alpha(1\pm\xi^2/R^2) \tag{7.11}$$

Thus, in a solution to a problem in first approximation shell theory, the curvature changes of equations (7.9) are suitable candidates for a constant and independent curvature change criterion only when the finite element, which is at the origin of co-ordinates, has dimensions of the same order of magnitude as the shell thickness, i.e. when

$$\xi_1, \xi_2 \leqslant h \tag{7.12}$$

(note that in engineering situations we are certainly prepared to accept an accuracy to within one per cent and this provides a relaxed criterion ξ_1, $\xi_2 \leqslant 0.1\ (h/R)^{1/2}$).

It is, furthermore, generally impracticable to secure inextensional bending solutions, i.e. where $\varepsilon_{11\alpha}$, $\varepsilon_{22\alpha}$, $\varepsilon_{12\alpha} \equiv 0$, other than

$$\left.\begin{aligned} U_{1\alpha} &= \alpha_7 + \alpha_9 \xi_2 \\ U_{2\alpha} &= \alpha_8 - \alpha_9 \xi_1 \end{aligned}\right\} \tag{7.13}$$

where α_7, α_8, α_9 are constants. None of these trial functions leads to constant curvature change in the general developable shell and, moreover, the terms α_7 and α_8 lead to curvature changes which are wholly dependent upon derivatives of the radii of curvature of the shell middle surface. Inextensional curvature changes must satisfy the compatibility condition

$$\frac{\kappa_{11}}{R_2} + \frac{\kappa_{22}}{R_1} - \frac{2\bar{\kappa}_{12}}{R_{12}} = 0 \tag{7.14}$$

so that it is in any case fruitless to seek displacement trial functions from inextensional bending solutions with the intention of providing constant and independent curvature changes κ_{11}, κ_{12}, $\bar{\kappa}_{12}$.

From all this, we conclude that:

an element stiffness matrix which is formulated solely from trial functions in the displacements $U_{1\alpha}$, $U_{2\alpha}$, W_α is likely to exhibit overstiffness in response to bending actions. It cannot, in any case, reflect a criterion for constant and independent curvature changes $\kappa_{11\alpha}$, $\kappa_{22\alpha}$, $\bar{\kappa}_{12\alpha}$ unless the dimensions of the finite element are of the same order of magnitude as the thickness of the shell (7.15)

Koiter[5] (1960), in deriving his shell theory, may just as well have started from the standpoint that the strain energy is represented by the quadratic function of the stress components for an isotropic solid, i.e. rather than of the strain components. Then, instead of equation (7.1), the strain energy density is given by

$$\begin{aligned} \text{SED} = {} & \frac{1}{2}\frac{1}{Eh}\{(N_{11}+N_{22})^2 - 2(1+\nu)(N_{11}N_{22} - \bar{N}_{12}^2)\} \\ & + \frac{1}{2}\frac{12}{Eh^3}\{(M_{11}+M_{22})^2 - 2(1+\nu)(M_{11}M_{22} - M_{12}^2)\} \end{aligned} \tag{7.16}$$

where N_{11}, M_{11}, ... are stress resultants and stress couples along orthogonal curves on the middle surface in the co-ordinate system ξ_1, ξ_2. This time, the remark that the neglected transverse stress is of order h/R times the membrane stresses means that in the above expression for the SED we may replace any physical stress resultant N by $N \pm M/R$ where M is any physical stress couple; in other words:

in solving a problem according to first approximation shell theory it is pointless to seek a solution for a stress resultant N to within a better accuracy than is provided by $N \pm M/R$ (2.17)

This statement has its own ramifications with respect to the finite element equilibrium method where the flexibility matrix is calculated directly from the SED of equation (7.16). The criteria here for convergence of the finite element method for problems in the plane require a recovery, in some global co-ordinate system, of constant independent stress resultants N_{11}, N_{22}, $\bar{N}_{12}$ for plane stress and of constant independent stress couples M_{11}, M_{22}, M_{12} for plate bending. As before, in examining these criteria in relation to the shell problem, we specialize attention to developable surfaces with orthogonal geodesic co-ordinates ξ_1, ξ_2. An equilibrium field of stress resultants and stress couples, i.e. where the partial differential equations of equilibrium are satisfied with zero surface forces, can then be expressed in terms of stress functions χ_1, χ_2, ψ by the expressions

$$\left.\begin{aligned}
N_{11} &= \frac{\partial^2 \psi}{\partial \xi_2^2} - \frac{\partial}{\partial \xi_2}\left(\frac{\chi_2}{R_2} - \frac{\chi_1}{R_{12}}\right) - \frac{1}{2R_{12}}\left(\frac{\partial \chi_2}{\partial \xi_1} - \frac{\partial \chi_1}{\partial \xi_2}\right) \\
N_{22} &= \frac{\partial^2 \psi}{\partial \xi_1^2} - \frac{\partial}{\partial \xi_1}\left(\frac{\chi_1}{R_1} - \frac{\chi_2}{R_{12}}\right) + \frac{1}{2R_{12}}\left(\frac{\partial \chi_2}{\partial \xi_1} - \frac{\partial \chi_1}{\partial \xi_2}\right) \\
\bar{N}_{12} &= -\frac{\partial^2 \psi}{\partial \xi_1 \partial \xi_2} + \frac{1}{2}\frac{\partial}{\partial \xi_1}\left(\frac{\chi_2}{R_2} - \frac{\chi_1}{R_{12}}\right) + \frac{1}{2}\frac{\partial}{\partial \xi_2}\left(\frac{\chi_1}{R_1} - \frac{\chi_2}{R_{12}}\right) \\
&\quad - \frac{1}{4}\left(\frac{1}{R_1} - \frac{1}{R_2}\right)\left(\frac{\partial \chi_2}{\partial \xi_1} - \frac{\partial \chi_1}{\partial \xi_2}\right) \\
M_{11} &= \frac{\partial \chi_2}{\partial \xi_2} + \frac{\psi}{R_2} \\
M_{22} &= \frac{\partial \chi_1}{\partial \xi_1} + \frac{\psi}{R_1} \\
M_{12} &= \frac{1}{2}\left(-\frac{\partial \chi_1}{\partial \xi_2} - \frac{\partial \chi_2}{\partial \xi_1} + \frac{2\psi}{R_{12}}\right)
\end{aligned}\right\} \quad (7.18)$$

Basic trial functions

$$\left.\begin{aligned}\chi_{1\beta} &= \beta_1\xi_1 + \beta_3\xi_2 \\ \chi_{2\beta} &= \beta_2\xi_2 + \beta_3\xi_1 \\ \psi_\beta &= 0\end{aligned}\right\} \tag{7.19}$$

where β_1, β_2, β_3 are constants, lead to constant stress couples

$$M_{11\beta} = \beta_2, \qquad M_{22\beta} = \beta_1, \qquad M_{12\beta} = -\beta_3 \tag{7.20}$$

and to the following stress resultants

$$\left.\begin{aligned}N_{11\beta} &= \underline{-\frac{\beta_2}{R_2} + \frac{\beta_3}{2R_{12}}} - \chi_{2\beta}\frac{\partial}{\partial\xi_2}\left(\frac{1}{R_2}\right) + \chi_{1\beta}\frac{\partial}{\partial\xi_2}\left(\frac{1}{R_{12}}\right) \\ N_{22\beta} &= \underline{-\frac{\beta_1}{R_1} + \frac{\beta_3}{2R_{12}}} - \chi_{1\beta}\frac{\partial}{\partial\xi_1}\left(\frac{1}{R_1}\right) + \chi_{2\beta}\frac{\partial}{\partial\xi_1}\left(\frac{1}{R_{12}}\right) \\ \bar{N}_{12\beta} &= \underline{\frac{1}{2}\left[-\frac{\beta_1+\beta_2}{R_{12}} + \frac{\beta_3}{2}\left(\frac{1}{R_1} + \frac{1}{R_2}\right)\right]}\end{aligned}\right\} \tag{7.21}$$

Now, equations (7.18) to (7.21) are seen to agree in form with equations (7.3) to (7.6). This is in accordance with Gol'denweizer's[6] (1961) static geometric analogy which has the following correspondence of variables

$$\left.\begin{aligned}&\varepsilon_{11} \to -M_{22}, &&\kappa_{11} \to N_{22} \\ &\varepsilon_{22} \to -M_{11}, &&\kappa_{22} \to N_{11} \\ &\varepsilon_{12} \to M_{12}, &&\bar{\kappa}_{12} \to -\bar{N}_{12} \\ &U_1 \to -\chi_1, &&\alpha_1 = -\beta_1 \\ &U_2 \to -\chi_2, &&\alpha_2 = -\beta_2 \\ &W \to -\psi, &&\alpha_3 = -\beta_3\end{aligned}\right\} \tag{7.22}$$

and makes it unnecessary to repeat the arguments which were made about displacement trial functions. We may say, right away, that the basic trial functions of equations (7.19) provide three constant independent stress couples M_{11}, M_{22}, M_{12} for the finite element(s) at the origin of co-ordinates and that a constant stress couple criterion for convergence can be contrived, at best, only in a piecewise context. We conclude also that:

a flexibility matrix which is formulated solely from stress functions $\chi_{1\beta}$, $\chi_{2\beta}$, ψ_β (so that the partial differential equations of equilibrium are satisfied with zero surface forces) is likely to exhibit overflexibility in response to membrane actions. It cannot, in any case, reflect a criterion for constant independent stress resultants $N_{11\beta}$, $N_{22\beta}$, $\bar{N}_{12\beta}$ unless the dimensions of the finite element are of the same order of magnitude as the thickness of the shell (7.23)

7.3 CHOICE OF CONNECTORS FOR A HYBRID FINITE ELEMENT MODEL

A discretization which takes cognizance of the foregoing arguments violates both equilibrium and compatibility conditions within the finite element and consequently excludes application of the straightforward limitation principles which provide direct information on convergence. It is appropriate to mention that while Fraeijs de Veubeke[10] (1965) correctly pointed out the limited usefulness of variational principles which allow simultaneous approximations to both displacements and stresses, the present approach relates to mutually supplementing approximations in a hybrid model; his analysis of transmission characteristics of interior fields (Fraeijs de Veubeke[11] (1974), Fraeijs de Veubeke and Millard[12] (1976)) across element boundaries is of special interest in this context.

Now, the assumed stress hybrid model discretizes also the displacement and slope along the element boundary and, as introduced by Pian[13] (1964), provides compatibility both along the boundary and at the vertex points; it is categorized by Fraeijs de Veubeke[11] (1974) as a 'statically admissible but weakly diffusive' finite element. However, hybrid models are versatile so that this categorization is not universal, it is dependent upon the topology of the connectors, i.e. whether vertex or side. Although the underlying complementary energy theorem demands compatibility along the element boundary it does not demand, with the exception of W displacements, compatibility at the element vertex points. Indeed, each side connector in this model really acts as an undetermined Lagrange multiplier in achieving equilibrium across the element boundary of its generalized load, i.e. the integral along the boundary of the associated traction weighted by the shape function belonging to the side connector. Thus, the topology of the connectors can be arranged to transform completely the assumed stress hybrid model into the category of 'statically admissible and strongly diffusive', i.e. the equilibrium model. As a specific example, the equilibrium model for plate bending due to Fraeijs de Veubeke and Sander[14] (1968) can be readily derived as a 'statically admissible and strongly diffusive' assumed stress hybrid model where application of the complementary energy principle does nothing to restore compatibility in the interior of an individual finite element under given generalized boundary loads.

In any hybrid model, the objective is to achieve a relationship between the interior and boundary discretizations which does not impose unintended restraint on the interior discretization. This is difficult to achieve in the present model because the mutually supplementing non-equilibrium α field and non-compatible β field makes it impossible to assign each connector to an exclusive role. We may require only, for example, that the U_1, U_2 vertex connectors provide *a path* for unrestrained transmission of the basic α

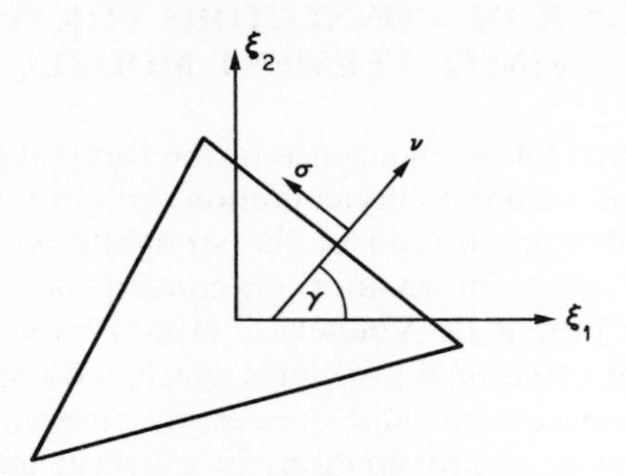

Figure 7.1. Triangular finite element

displacement field while side connectors, in conjunction with W vertex connectors, provide *a path* to achieve unrestrained transmission of the basic β stress field.

Let us examine these aspects in more detail by continuing with the specialization to developable surfaces. Consider a triangular finite element constructed from geodesic lines on the shell middle surface. One side of this element has its insurface normal ν inclined at the angle γ to the $O\xi_1$ co-ordinate axis as shown in Figure 7.1, while arc length along the side is measured by the co-ordinate σ; note that the angle γ is generally a function of σ. The components of the traction, displacement, and slope continuity conditions at this element side interface are

$$\left.\begin{array}{rcl} N_{\nu\nu}-\dfrac{M_{\nu\sigma}}{R_{\nu\sigma}} & \text{or} & U_\nu \\ N'_{\nu\sigma}+\dfrac{M_{\nu\sigma}}{R_\sigma} & \text{or} & U_\sigma \\ M_{\nu\nu} & \text{or} & \phi_\nu \\ V_\nu & \text{or} & W \end{array}\right\} \tag{7.24}$$

and at the vertices

$$M_{\nu\sigma} \quad \text{or} \quad W \tag{7.25}$$

where the stress resultants and stress couples are expressed by

$$\left.\begin{aligned} N_{\nu\nu} &= N_{11}\cos^2\gamma + N_{22}\sin^2\gamma + 2\bar{N}_{12}\sin\gamma\cos\gamma \\ N'_{\nu\sigma} &= N_{\nu\sigma} - \frac{1}{2R_{12}}(M_{11}-M_{22}) - \frac{1}{2}\left(\frac{1}{R_1}-\frac{1}{R_2}\right)M_{12} \\ N_{\nu\sigma} &= -(N_{11}-N_{22})\sin\gamma\cos\gamma + \bar{N}_{12}(\cos^2\gamma - \sin^2\gamma) \\ M_{\nu\nu} &= M_{11}\cos^2\gamma + M_{22}\sin^2\gamma + 2M_{12}\sin\gamma\cos\gamma \\ M_{\nu\sigma} &= -(M_{11}-M_{22})\sin\gamma\cos\gamma + M_{12}(\cos^2\gamma - \sin^2\gamma) \end{aligned}\right\} \tag{7.26}$$

and the Kirchhoff force is given by

$$V_\nu = Q_\nu + \frac{\partial M_{\nu\sigma}}{\partial \sigma} \tag{7.27}$$

with

$$\left.\begin{aligned} Q_\nu &= Q_1 \cos\gamma + Q_2 \sin\gamma \\ Q_1 &= \frac{\partial M_{11}}{\partial \xi_1} + \frac{\partial M_{12}}{\partial \xi_2} \\ Q_2 &= \frac{\partial M_{12}}{\partial \xi_1} + \frac{\partial M_{22}}{\partial \xi_2} \\ \frac{\partial}{\partial \sigma} &= -\sin\gamma \frac{\partial}{\partial \xi_1} + \cos\gamma \frac{\partial}{\partial \xi_2} \end{aligned}\right\} \tag{7.28}$$

The displacements and slopes are given by

$$\left.\begin{aligned} U_\nu &= U_1 \cos\gamma + U_2 \sin\gamma \\ U_\sigma &= -U_1 \sin\gamma + U_2 \cos\gamma \\ \phi_\nu &= \phi_1 \cos\gamma + \phi_2 \sin\gamma \\ \phi_\sigma &= -\phi_1 \sin\gamma + \phi_2 \cos\gamma \\ \phi_1 &= \frac{U_1}{R_1} - \frac{U_2}{R_{12}} - \frac{\partial W}{\partial \xi_1} \\ \phi_2 &= \frac{U_2}{R_2} - \frac{U_1}{R_{12}} - \frac{\partial W}{\partial \xi_2} \end{aligned}\right\} \tag{7.29}$$

and the curvatures by

$$\left.\begin{aligned} \frac{1}{R_\sigma} &= \frac{\sin^2\gamma}{R_1} + \frac{\cos^2\gamma}{R_2} + \frac{2\sin\gamma\cos\gamma}{R_{12}} \\ \frac{1}{R_{\nu\sigma}} &= -\left(\frac{1}{R_1} - \frac{1}{R_2}\right)\sin\gamma\cos\gamma + \frac{1}{R_{12}}(\cos^2\gamma - \sin^2\gamma) \end{aligned}\right\} \tag{7.30}$$

As a first choice, consider the familiar arrangement of connectors U_1, U_2, W, ϕ_1, ϕ_2 at the vertices and U_1, U_2, ϕ_ν at the mid-side positions of the element as shown in Figure 7.2. Basic interpolation of these 13 connectors

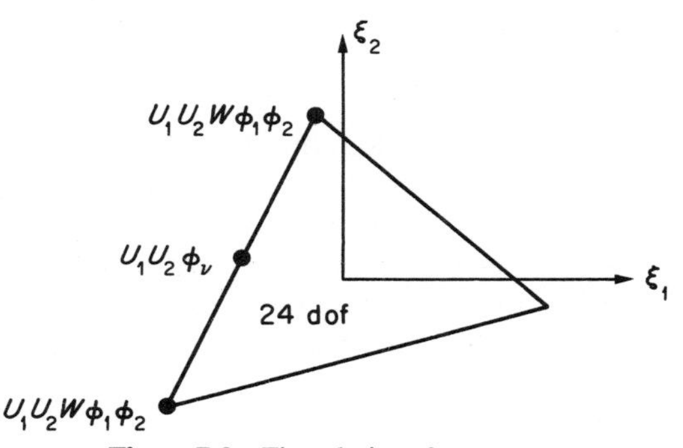

Figure 7.2 First choice of connectors

along each side must include an arbitrary six-degree-of-freedom rigid body movement together with freedom to recover the displacement and slopes from the basic α field. Thus, along each side

$$\left.\begin{aligned} U_1(\sigma) &= a_1 + a_2\xi_1(\sigma) + a_3\xi_2(\sigma) + \text{rigid body terms} \\ U_2(\sigma) &= a_4 + a_3\xi_1(\sigma) + a_5\xi_2(\sigma) + \text{rigid body terms} \\ W(\sigma) &= \text{rigid body terms} \\ \phi_\nu(\sigma) &= \{a_1 + a_2\xi_1(\sigma) + a_3\xi_2(\sigma)\}\left(\frac{\cos\gamma}{R_1} - \frac{\sin\gamma}{R_{12}}\right) \\ &\quad + \{a_4 + a_3\xi_1(\sigma) + a_5\xi_2(\sigma)\}\left(\frac{\sin\gamma}{R_2} - \frac{\cos\gamma}{R_{12}}\right) \\ &\quad + \text{rigid body terms} \end{aligned}\right\} \tag{7.31}$$

where $a_1, \ldots, a_{11}$ are constants with $a_6, \ldots, a_{11}$ referring to rigid body terms. The required interpolation of the 13 connectors is therefore always feasible and a path for unrestrained transmission of the basic α field of equations (7.4) is secured by equating vertex values of $U_{1\alpha}$, $U_{2\alpha}$ with those from equations (7.31). Turning now to the basic β field, equations (7.19), the side connectors U_1, U_2. ϕ_ν are available to serve as undetermined Lagrange multipliers to secure a path for equilibrium across the element boundary of the generalized loads

$$\left.\begin{aligned} &\int_{\text{side}} \left\{N_{\nu\nu\beta}(\sigma) - \frac{M_{\nu\sigma\beta}(\sigma)}{R_{\nu\sigma}(\sigma)}\right\}\hat{U}_\nu(\sigma)\,d\sigma \\ &\int_{\text{side}} \left\{N'_{\nu\sigma\beta}(\sigma) + \frac{M_{\nu\sigma\beta}(\sigma)}{R_\sigma(\sigma)}\right\}\hat{U}_\sigma(\sigma)\,d\sigma \\ &\int_{\text{side}} M_{\nu\nu\beta}(\sigma)\hat{\phi}_\nu(\sigma)\,d\sigma \end{aligned}\right\} \tag{7.32}$$

where the circumflex denotes shape function of the side connector. The stress resultants $N_{\nu\nu\beta}$, $N'_{\nu\sigma\beta}$ and the stress couples $M_{\nu\nu\beta}$, $M_{\nu\sigma\beta}$ are calculated from equations (7.20) and (7.21) with the aid of equations like (7.26). Vertex connectors W are also available to equilibriate the summed

$$M^-_{\nu\sigma\beta} - M^+_{\nu\sigma\beta} \tag{7.33}$$

at the conjunction of vertices. The Kirchhoff force $V_{\nu\beta}$, see equations (7.24) and (7.27), is identically zero for the basic β field and hence does not require a connector. Note for problems of the plane ($R_{\nu\sigma}$, $R_\sigma = \infty$) that the shape functions $\hat{U}_\nu$, $\hat{U}_\sigma$, $\hat{\phi}_\nu$ are even in the co-ordinate σ so that for linearly varying $N_{\nu\nu\beta}$, $N'_{\nu\sigma\beta}$, $M_{\nu\nu\beta}$ the equations (7.32) enforce transmission of their point values at the mid-side in this assumed stress hybrid model.

The ten basic α field trial functions of equations (7.4) when augumented with six rigid body terms, plus the four basic β field trial functions of equations (7.19) are inadequate to provide a rank sufficient stiffness matrix for the 24-degree-of-freedom triangular element shown in Figure 7.2. Let us therefore extend these fields so that

$$\left.\begin{aligned} U_{1\alpha} &= \alpha_1 + \alpha_2\xi_1 + \alpha_3\xi_2 + \alpha_4\xi_1^2 + \alpha_5\xi_1\xi_2 + \alpha_6\xi_2^2 \\ &\quad + \text{rigid body terms} \\ U_{2\alpha} &= \alpha_7 + \alpha_8\xi_1 + \alpha_9\xi_2 + \alpha_{10}\xi_1^2 + \alpha_{11}\xi_1\xi_2 + \alpha_{12}\xi_2^2 \\ &\quad + \text{rigid body terms} \\ W_\alpha &= \alpha_{13} + \alpha_{14}\xi_1 + \alpha_{15}\xi_2 + \text{rigid body terms} \end{aligned}\right\} \tag{7.34}$$

and

$$\left.\begin{aligned} \chi_{1\beta} &= \beta_1 + \beta_2\xi_1 + \beta_3\xi_2 + \beta_4\xi_1^2 + \beta_5\xi_1\xi_2 + \beta_6\xi_2^2 \\ \chi_{2\beta} &= \beta_7 + \beta_8\xi_1 + \beta_9\xi_2 + \beta_{10}\xi_1^2 + \beta_{11}\xi_1\xi_2 + \beta_{12}\xi_2^2 \\ \psi_\beta &= \beta_{13} + \beta_{14}\xi_1 + \beta_{15}\xi_2 \end{aligned}\right\} \tag{7.35}$$

where $\alpha_1, \ldots, \alpha_{21}$ and $\beta_1, \ldots, \beta_{15}$ are new constants. The constants $\alpha_{16}, \ldots, \alpha_{21}$ refer to rigid body terms the expressions for which may be derived from the vector equation describing the geometry of the middle surface of the given shell. These fields provide trial functions which form complete polynomials. They include terms which are superfluous to the underlying role of the respective fields (e.g. the terms α_{13}, α_{14}, α_{15} in W_α and the inextensional curvature changes provided by the terms α_1, α_7; similarly, the corresponding terms β_{13}, β_{14}, β_{15} in ψ_β and the purely membrane stresses provided by β_1, β_7). Note must now be taken of experience with flat plate bending (Allman[15] (1976)) and with circular cylindrical shells (Morley[9] (1976)) which shows that the hybrid model of Figure 7.2 does not provide a rank-sufficient element stiffness matrix even in conjunction with the extended α and β fields of equations (7.34) and (7.35). However, there is an alternative choice of finite element model which has U_1, U_2, W connectors at the vertices and mid-side positions and two connectors ϕ_ν on each side as shown in Figure 7.3. Interpolation according to equations (7.31) is now just feasible for the eleven connectors along each side. This choice of connectors is known to lead to rank-sufficiency for plane problems because the topology is identical with that of the Fraeijs de Veubeke[10] (1965) quadratically varying displacement plane stress element when augumented with the Fraeijs de Veubeke and Sander[14] (1968) equilibrium element for plate bending.

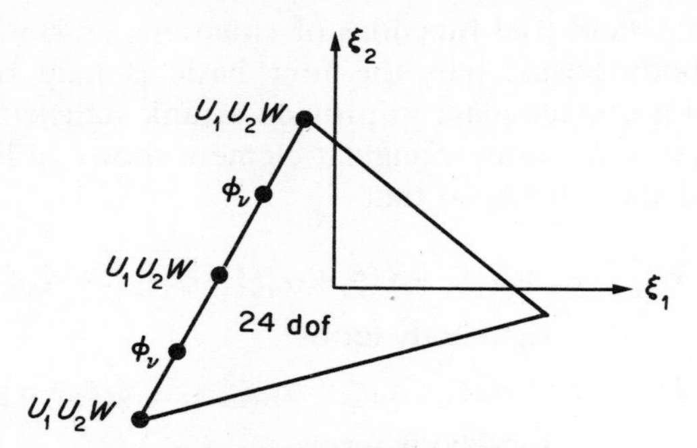

Figure 7.3 Second choice of connectors

7.4 ELEMENT STIFFNESS MATRIX

The use of mutually supplementing fields to discretize the elastic behaviour of the hybrid finite element requires special techniques with which to relate the α and β fields to the connectors and thence to the derivation of the element stiffness matrix. For this purpose, it is convenient to introduce further notation so that displacements and slopes, strains and curvature changes, stress resultants and stress couples, together with surface forces may be expressed as column matrices.

Let

$$\mathbf{U}'_C = (U_\nu U_\sigma \phi_\nu \phi_\sigma W)^{\mathrm{T}} \tag{7.36}$$

represent interpolation of the connectors around the boundary C of the finite element; correspondingly let

$$\mathbf{U}'_{\alpha C} = (U_{\nu\alpha} U_{\sigma\alpha} \phi_{\nu\alpha} \phi_{\sigma\alpha} W_\alpha)^{\mathrm{T}} \tag{7.37}$$

represent the displacements and slopes around C as calculated from the extended α field of equations (7.34) with the aid of equations like (7.29). Note that the slopes ϕ_ν, ϕ_σ at the geodesic side may be expressed by

$$\left.\begin{aligned} \phi_\nu &= \frac{U_\nu}{R_\nu} - \frac{U_\sigma}{R_{\nu\sigma}} - \frac{\partial W}{\partial \nu} \\ \phi_\sigma &= \frac{U_\sigma}{R_\sigma} - \frac{U_\sigma}{R_{\nu\sigma}} - \frac{\partial W}{\partial \sigma} \end{aligned}\right\} \tag{7.38}$$

For the interior A of the finite element, denote

$$\left.\begin{aligned} \mathbf{U} &= (U_1 U_2 W)^{\mathrm{T}} \\ \bar{\boldsymbol{\varepsilon}} &= (\varepsilon_{11}\varepsilon_{22} 2\varepsilon_{12}\kappa_{11}\kappa_{22}\bar{\kappa}_{12})^{\mathrm{T}} \\ \bar{\mathbf{N}} &= (N_{11}N_{22}\bar{N}_{12}M_{11}M_{22}2M_{12})^{\mathrm{T}} \\ \mathbf{p} &= (p_1 p_2 p) \end{aligned}\right\} \tag{7.39}$$

where p_1, p_2, p are surface forces which act in the same directions respectively as U_1, U_2, W and are calculated from the partial differential equations of equilibrium. The notation of equation (7.39) is freely extended by the addition of appropriate suffixes so that, for example,

$$\bar{\mathbf{N}}_\beta = (N_{11\beta} N_{22\beta} \bar{N}_{12\beta} M_{11\beta} M_{22\beta} 2M_{12\beta})^{\mathrm{T}} \tag{7.40}$$

but note that

$$\mathbf{p}_\beta = (0 \quad 0 \quad 0)^{\mathrm{T}} \tag{7.41}$$

because the β field is calculated from stress functions so that the equations of equilibrium are satisfied with zero surface forces. The surface forces $\mathbf{p}_\alpha$ satisfy Green's formula

$$\iint\limits_A (\bar{\mathbf{N}}_\alpha^{\mathrm{T}} \bar{\boldsymbol{\varepsilon}}_\alpha - \mathbf{p}_\alpha^{\mathrm{T}} \mathbf{U}_\alpha)\, \mathrm{d}\xi_1\, \mathrm{d}\xi_2 - \int\limits_C \mathbf{N}_{\alpha C}^{\prime \mathrm{T}} \mathbf{U}_{\alpha C}^{\prime}\, \mathrm{d}\sigma = 0 \tag{7.42}$$

The stress resultants and stress couples $\bar{\mathbf{N}}_\alpha$ of the α field are calculated from $\bar{\boldsymbol{\varepsilon}}_\alpha$ using constitutive relations derived from the strain energy density of equation (7.1), namely

$$\left.\begin{aligned} N_{11} &= \frac{Eh}{1-\nu^2}(\varepsilon_{11} + \nu\varepsilon_{22}), & M_{11} &= \frac{Eh^3}{12(1-\nu^2)}(\kappa_{11} + \nu\kappa_{22}) \\ N_{22} &= \frac{Eh}{1-\nu^2}(\varepsilon_{22} + \nu\varepsilon_{11}), & M_{22} &= \frac{Eh^3}{12(1-\nu^2)}(\kappa_{22} + \nu\kappa_{11}) \\ \bar{N}_{12} &= \frac{Eh}{1+\nu}\varepsilon_{12}, & M_{12} &= \frac{Eh^3}{12(1+\nu)}\bar{\kappa}_{12} \end{aligned}\right\} \tag{7.43}$$

The strains and curvature changes $\bar{\boldsymbol{\varepsilon}}_\beta$ of the β field are calculated from $\bar{\mathbf{N}}_\beta$ by using the inverse of these equations. Stress resultants, stress couples and shearing forces at the boundary C are denoted by $\mathbf{N}_C^{\prime}$ in equation (7.42) where

$$\mathbf{N}_C^{\prime} = (N_{\nu\nu} N_{\nu\sigma}^{\prime} M_{\nu\nu} M_{\nu\sigma} Q_\nu)^{\mathrm{T}} \tag{7.44}$$

and $N_{\nu\nu}, \ldots, Q_\nu$ are calculated with the help of equations (7.26) and (7.28).

Let us decide to obtain first a fit of the α displacement field of equations (7.33) to the interpolation $\mathbf{U}_C^{\prime}$ of the connectors and accordingly consider the application of a modified complementary energy principle

$$\delta\left\{\frac{1}{2}\iint\limits_A \bar{\mathbf{N}}_\alpha^{\mathrm{T}} \bar{\boldsymbol{\varepsilon}}_\alpha\, \mathrm{d}\xi_1\, \mathrm{d}\xi_2 - \iint\limits_A \mathbf{p}_\alpha^{\mathrm{T}} \mathbf{U}_\alpha\, \mathrm{d}\xi_1\, \mathrm{d}\xi_2 - \int\limits_C \mathbf{N}_{\alpha C}^{\prime \mathrm{T}} \mathbf{U}_C^{\prime}\, d\sigma\right\} = 0 \tag{7.45}$$

where $\mathbf{U}_\alpha$ acts as an undetermined Lagrange multiplier in restoring equilibrium with zero applied surface forces, i.e.

$$\mathbf{p}_\alpha \to 0 \tag{7.46}$$

A substitution from the Green's formula of equation (7.42) transforms this into a modified potential energy principle

$$\delta\left\{\frac{1}{2}\iint_A \bar{\mathbf{N}}_\alpha^{\mathrm{T}}\bar{\boldsymbol{\varepsilon}}_\alpha \,\mathrm{d}\xi_1\,\mathrm{d}\xi_2 + \int_C \mathbf{N}_{\alpha C}^{\prime\mathrm{T}}(\mathbf{U}_C^{\prime} - \mathbf{U}_{\alpha C}^{\prime})\,\mathrm{d}\sigma\right\} = 0 \tag{7.47}$$

which is simpler to use. Each coefficient $\alpha_1, \ldots, \alpha_{21}$ is available for arbitrary variation but the modified principles provide a kinematic model where the interior displacement field is restrained in transmission across C; moreover, the generalized boundary loads like those of equations (7.31) are no longer derived from an equilibrium field and are therefore also restrained in transmission. Consequently, and recalling the earlier discussion, it is advisable to constrain the variations so that

$$U_{1\alpha} - U_1 = 0, \qquad U_{2\alpha} - U_2 = 0 \tag{7.48}$$

at each vertex of the finite element so as to provide a path for unrestrained transmission of the basic α field of equations (7.4).

Meanwhile, the extended β equilibrium stress field of equations (7.35) awaits its formal role to correct the mismatch $\mathbf{U}_C^{\prime} - \mathbf{U}_{\alpha C}^{\prime}$ and for this purpose we make use of the complementary energy principle

$$\delta\left\{\frac{1}{2}\iint_A \bar{\mathbf{N}}_\beta^{\mathrm{T}}\bar{\boldsymbol{\varepsilon}}_\beta \,\mathrm{d}\xi_1\,\mathrm{d}\xi_2 - \int_C \mathbf{N}_{\beta C}^{\prime\mathrm{T}}(\mathbf{U}_C^{\prime} - \mathbf{U}_{\alpha C}^{\prime})\,\mathrm{d}\sigma\right\} = 0 \tag{7.49}$$

where the coefficients $\beta_1, \ldots, \beta_{15}$ are the only quantities subject to arbitrary variation with no constraints like those of equations (7.48).

Use in succession of the two principles, equations (7.45) and (7.49), provides one means with which to relate the α and β fields with the connectors. This discretization of the elastic behaviour in the element is simple and illustrative but it is not wholly satisfactory. Its qualification rests upon a disregard of some of the cross-coupling between the α and β fields. This amounts to an assumption which neglects h/R in comparison with unity in the expressions (7.1), (7.16) for the strain energy density and is reminiscent of the approximation which is made in shallow shell theory. It is certainly unwise to repeat the assumption in the final derivation of the stiffness matrix from the total strain energy

$$\frac{1}{2}\iint_A (\bar{\mathbf{N}}_\alpha^{\mathrm{T}} + \bar{\boldsymbol{N}}_\beta^{\mathrm{T}})(\bar{\boldsymbol{\varepsilon}}_\alpha + \bar{\boldsymbol{\varepsilon}}_\beta)\,\mathrm{d}\xi_1\,\mathrm{d}\xi_2 \tag{7.50}$$

Another discretization process, more formalistic and thus perhaps more satisfactory, starts from a generalization of the modified complementary energy principle of equation (7.45)

$$\delta\left\{\frac{1}{2}\iint\limits_A(\bar{\mathbf{N}}_\alpha^{\mathrm{T}}+\bar{\mathbf{N}}_\beta^{\mathrm{T}})(\bar{\boldsymbol{\varepsilon}}_\alpha+\bar{\boldsymbol{\varepsilon}}_\beta)\,\mathrm{d}\xi_1\,\mathrm{d}\xi_2-\iint\limits_A\mathbf{p}_\alpha^{\mathrm{T}}\mathbf{U}_\lambda\,\mathrm{d}\xi_1\,\mathrm{d}\xi_2-\int\limits_C(\mathbf{N}_{\alpha C}'^{\mathrm{T}}+\mathbf{N}_{\beta C}'^{\mathrm{T}})\mathbf{U}_C'\right\}\mathrm{d}\sigma=0 \quad (7.51)$$

where the Lagrange multiplier $\mathbf{U}_\lambda$ is an additional displacement field; all three fields α, β, λ are subject to arbitrary variation. (The option is available to set $\mathbf{U}_\lambda=\mathbf{U}_\alpha$ in this principle.) It remains necessary to constrain the variations of the α field according to equations (7.48). The conjugate principle is obtained by making use of Green's formula

$$-\iint\limits_A\mathbf{p}_\alpha^{\mathrm{T}}\mathbf{U}_\lambda\,\mathrm{d}\xi_1\,\mathrm{d}\xi_2=-\iint\limits_A\bar{\mathbf{N}}_\alpha^{\mathrm{T}}\bar{\boldsymbol{\varepsilon}}_\lambda\,\mathrm{d}\xi_1\,\mathrm{d}\xi_2+\int\limits_C\mathbf{N}_{\alpha C}'^{\mathrm{T}}\mathbf{U}_{\lambda C}'\,\mathrm{d}\sigma \quad (7.52)$$

7.5 INTERPOLATION OF CONNECTORS

It was convenient for the purposes of the foregoing analysis to specialize attention to developable surfaces but a penalty is incurred when a side of the finite element coincides with a generator. The interpolation formulae for the connectors then suffer degeneracy and this feature is now briefly illustrated, along with the remedy, for a geodesic line drawn on the middle surface of a circular cylindrical shell.

The coordinate ξ_1 measures arc length along a generator, with ξ_2 measuring the circumferential arc. There is no loss in generality by assuming that the mid-side position coincides with the origin of co-ordinates as shown in Figure 7.4. Let the diameter of the middle surface of the circular cylinder be $2R$ and replace the co-ordinates ξ_1, ξ_2 by

$$\xi_1=Rx, \qquad \xi_2=R\theta \quad (7.53)$$

and arc length σ along the geodesic side by

$$\sigma=Rs \quad (7.54)$$

Consequently,

$$x=-s\sin\gamma, \qquad \theta=s\cos\gamma \quad (7.55)$$

where the angle γ is constant for cylinders and

$$\frac{1}{R_1}=\frac{1}{R_{12}}=0, \qquad \frac{1}{R_2}=\frac{1}{R} \quad (7.56)$$

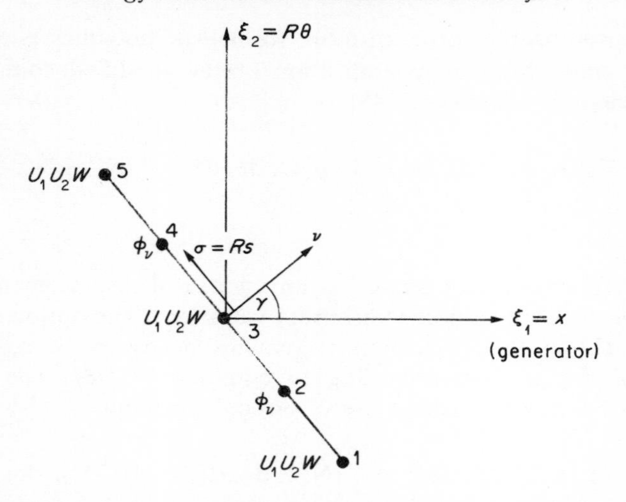

Figure 7.4. Notation for geodesic side on circular cylinder with second choice of connectors

A path for the independent and unrestrained transmission of the basic α field of displacement trial functions is provided if we substitute equations (7.53) to (7.56) into (7.31) to obtain interpolation formulae

$$\left.\begin{aligned} U_1(s) &= a_1 + a_2'a + \underline{a_3's^2} + a_{10}\sin\theta + a_{11}\cos\theta \\ U_2(s) &= a_4 + a_5's + \underline{a_6's^2} - a_8\sin\theta + a_9\cos\theta \\ &\quad + x(-a_{10}\cos\theta + a_{11}\sin\theta) \\ W(s) &= \underline{a_7'} + a_8\cos\theta + a_9\sin\theta - x(a_{10}\sin\theta + a_{11}\cos\theta) \\ R\phi_\nu(s) &= (a_4 + a_5's)\sin\gamma + (a_{10}\sin\theta + a_{11}\cos\theta)\cos\gamma \end{aligned}\right\} \quad (7.57)$$

where

$$\left.\begin{aligned} a_2' &= R(-a_2\sin\gamma + a_3\cos\gamma) \\ a_5' &= R(-a_3\sin\gamma + a_5\cos\gamma) \end{aligned}\right\} \quad (7.58)$$

The constants a_1, a_4, $a_8, \ldots, a_{11}$ refer to rigid body terms, while the underlined terms are additional and replace those which are lost because of special geometrical characteristics of the circular cylinder. The primes are disregarded in the sequel.

Along a generator, where $\gamma = -\pi/2$, $s = x$, $\theta = 0$, the interpolation of equations (7.57) degenerates to

$$\left.\begin{aligned} U_1(x) &= a_1 + a_2x + a_3x^2 + a_{11} \\ U_2(x) &= a_4 + a_5x + a_6x^2 + a_9 - a_{10}x \\ W(x) &= a_7 + a_8 - a_{11}x \\ R\phi_\nu(x) &= -a_4 - a_5x \end{aligned}\right\} \quad (7.59)$$

so that the constants a_8, a_9, and a_{10} now refer to terms which play no useful role in the interpolation process. The degeneracy is best resolved by introducing new transcendental functions $F_j(s, \gamma)$ and $G_j(s, \gamma)$, see Morley[9] (1976), where

$$\left.\begin{aligned}
F_0(s,\gamma) &= \cos\theta, \qquad F_1(s,\gamma)\cos\gamma = \sin\theta \\
F_j(s,\gamma) &= s^j - \frac{F_{j+2}\cos^2\gamma}{(j+2)(j+1)} \\
&= \frac{j!}{\cos^j\gamma}\left\{\frac{\theta^j}{j!} - \frac{\theta^{j+2}}{(j+2)!} + \frac{\theta^{j+4}}{(j+4)!} - \dots\right\} \\
G_j(s,\gamma) &= \tfrac{1}{2}j(sF_{j-1} - F_j) + F_j
\end{aligned}\right\} \tag{7.60}$$

which have the very useful properties

$$\left.\begin{aligned}
&F_j(s,\gamma) \to G_j(s,\gamma) \to s^j \quad \text{as} \quad s \to 0 \\
&F_j(s,\pi/2) = G_j(s,\pi/2) = s^j \quad \text{for all} \quad s \\
&F_j(s,\gamma+\pi) = F_j(s,\gamma) \qquad G_j(s,\gamma+\pi) = G_j(s,\gamma) \\
&\frac{\partial F_j}{\partial s} = jF_{j-1} \qquad \frac{\partial G_j}{\partial s} = jG_{j-1}
\end{aligned}\right\} \tag{7.61}$$

The interpolation formulae of equations (7.57) may now be rewritten in the entirely equivalent form

$$\left.\begin{aligned}
U_1(s) &= A_1 + A_2 s + A_3 s^2 - \Phi(s,\gamma)\cos^2\gamma \\
U_2(s) &= B_1 + B_2 s + B_3 s^2 - \Phi(s,\gamma)\sin\gamma\cos\gamma - \Psi(s,\gamma)\cos\gamma \\
&\quad + \Gamma(s,\gamma)\sin\gamma\cos^3\gamma \\
W(s) &= C_1 + C_2 s + C_3 F_2(s,\gamma) + C_4 F_3(s,\gamma) - \Gamma'(s,\gamma)\sin\gamma\cos^2\gamma \\
R\phi_\nu(s) &= D_1 + D_2 s + D_3 s^2 - \Phi(s,\gamma)\cos^3\gamma
\end{aligned}\right\} \tag{7.62}$$

where

$$\left.\begin{aligned}
\Phi(s,\gamma) &= \tfrac{1}{6}\Delta_1 F_3 + \tfrac{1}{12}\Delta_2 F_4 \\
\Gamma(s,\gamma) &= \tfrac{1}{60}\Delta_1 G_5 + \tfrac{1}{180}\Delta_2 G_6 \\
\Psi(s,\gamma) &= \tfrac{1}{3}C_3 F_3 + \tfrac{1}{4}C_4 F_4 \\
\Gamma'(s,\gamma) &= \frac{\partial\Gamma}{\partial s} = \tfrac{1}{12}\Delta_1 G_4 + \tfrac{1}{30}\Delta_2 G_5 \\
\Delta_1 &= D_2\cos\gamma - B_2\sin\gamma\cos\gamma + 2C_3\sin\gamma \\
\Delta_2 &= D_3\cos\gamma - B_3\sin\gamma\cos\gamma + 3C_4\sin\gamma
\end{aligned}\right\} \tag{7.63}$$

The $A_1, \ldots, D_3$ are new constants and for the interpolation of the second choice of connectors, see Figures 7.3 and 7.4, the following relationships hold for C_4 and D_3,

$$\left.\begin{aligned} C_4 &= \tfrac{1}{6}\cos^2\gamma\{2B_1\sin^2\gamma\cos\gamma - C_2(1+2\sin^2\gamma) - 2D_1\sin\gamma\cos\gamma\} \\ D_3 &= \tfrac{1}{2}\{(B_1\sin\gamma - D_1)(1-2\sin^2\gamma)\cos^2\gamma + 2B_3\sin\gamma + 2C_2\sin^3\gamma\cos\gamma\} \end{aligned}\right\} \tag{7.64}$$

The interpolation formulae of equations (7.62) are robust and completely free from numerical hazard; they retain all the properties of equations (7.57) without the degeneracy exhibited by equations (7.59). The constants $A_1, \ldots, D_3$ are related to $a_1, \ldots, a_{11}$ by

$$\left.\begin{aligned} A_1 &= a_1 + a_{11}, & B_1 &= a_4 + a_9 \\ A_2 &= a_2 + a_{10}\cos\gamma, & B_2 &= a_5 - a_8\cos\gamma + a_{10}\sin\gamma \\ A_3 &= a_3 - \tfrac{1}{2}a_{11}\cos^2\gamma, & B_3 &= a_6 - \tfrac{1}{2}a_9\cos^2\gamma - a_{11}\sin\gamma\cos\gamma \\ C_1 &= a_7 + a_8, & D_1 &= a_4\sin\gamma + a_{11}\cos\gamma \\ C_2 &= a_9\cos\gamma + a_{11}\sin\gamma, & D_2 &= a_5\sin\gamma + a_{10}\cos^2\gamma \\ C_3 &= -\tfrac{1}{2}a_8\cos^2\gamma + a_{10}\sin\gamma\cos\gamma, & D_3 &= a_6\sin\gamma - \tfrac{1}{2}a_{11}\cos^3\gamma \\ C_4 &= -\tfrac{1}{6}a_9\cos^3\gamma - \tfrac{1}{2}a_{11}\sin\gamma\cos^2\gamma, & & \\ \Delta_1 &= a_{10}\cos\gamma & & \\ \Delta_2 &= -\tfrac{1}{2}a_{11}\cos^2\gamma & & \end{aligned}\right\} \tag{7.65}$$

7.6 EDGE EFFECT

Edge effect is a highly damped state of stress and strain and is a phenomenon peculiar to thin-walled shells and the presence of curvature. It is natural, therefore, to want to enquire into the sensibility of its finite element representation when the discretization seeks to rest merely upon constant strains ε_{11}, ε_{22}, ε_{12} and constant stress couples M_{11}, M_{22}, M_{12}.

Accordingly, let us examine the phenomenon of edge effect along a geodesic edge of the same circular cylindrical shell which was considered in Section 7.5. Arc length along this edge is measured by the co-ordinate σ, while arc length along the geodesic normal co-ordinate is measured by ν, see Figure 7.5.

For simple edge effect the solution is independent of the co-ordinate σ; it

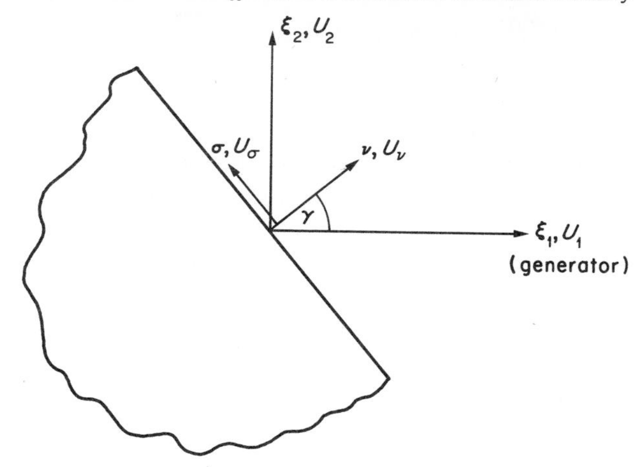

Figure 7.5 Notation for geodesic edge on circular cylinder

may be written

$$\left.\begin{aligned} W &= e^{\lambda\nu/R}(a_1 \cos \lambda\nu/R + a_2 \sin \lambda\nu/R) \\ U_\nu &= -(\sin^2 \gamma + \nu_p \cos^2 \gamma)\int W \, d(\nu/R) \\ U_\sigma &= -2 \sin \gamma \cos \gamma \int W \, d(\nu/R) \end{aligned}\right\} \tag{7.66}$$

where the displacements U_1 and U_2 may be derived with the aid of equations (7.29). In equations (7.66) the a_1, a_2 are arbitrary constants which are determined by boundary conditions at the edge, $\nu=0$; the constant λ is given by

$$\lambda = \{3(1-\nu_p^2)R^2/h^2\}^{\frac{1}{4}} \cos \gamma \tag{7.67}$$

and ν_p is used here to denote the Poisson's ratio. The co-ordinate ν takes on only negative values, see Figure 7.5.

The following considerations allow the solution to be derived strictly from Koiter's shell theory, The wavelength L of distortion is

$$L = 2\pi R/\lambda \tag{7.68}$$

and so the strain energy of edge effect is calculated at best only to within the relative error h^2/L^2 where

$$\frac{h^2}{L^2} \simeq 0.04 \frac{h}{R} \cos^2 \gamma \tag{7.69}$$

This is more critical than h^2/R^2 provided that

$$\left||\gamma|-\frac{\pi}{2}\right|>5\left(\frac{h}{R}\right)^{\frac{1}{2}}, \qquad -\pi\leqslant\gamma\leqslant\pi \tag{7.70}$$

i.e. unless the edge is sufficiently closely aligned with a generator. It follows from equation (7.69) that it is reasonable to neglect the quantity

$$\frac{h^2\lambda^2}{12R^2}\simeq 3\frac{h^2}{L^2} \qquad \text{cf. unity} \tag{7.71}$$

in the calculation of edge effect for angles γ which satisfy equation (7.70); in other words it is reasonable to ignore derivatives like

$$\frac{h^2}{12}\frac{\partial^2}{\partial\nu^2} \qquad \text{cf. unity} \tag{7.72}$$

This affords considerable simplification in the equations of equilibrium when they are expressed in terms of the displacements U_1, U_2, W and leads directly to the solution expressed by equations (7.66).

All physical quantities have the same wavelength in the solution for simple edge effect. Thus, the absolute depth l_Z of the zone of penetration from the shell edge is determined by

$$e^{-\lambda l_Z/R}=\frac{h}{L}$$

which is the same as

$$l_Z=-\frac{L}{2\pi}\ln\left(\frac{h}{L}\right) \tag{7.73}$$

For $\gamma=0$ this gives

$$\left.\begin{aligned} l_Z&\simeq 0.40R\simeq 20h \quad \text{for} \quad h/R=0.02\\ l_Z&\simeq 0.17R\simeq 84h \quad \text{for} \quad h/R=0.002\end{aligned}\right\} \tag{7.74}$$

The length l_{ZH} of half decay is given by

$$l_{ZH}=\frac{L}{2\pi}\ln 2 \tag{7.75}$$

and for $\gamma=0$ this gives

$$\left.\begin{aligned} l_{ZH}&\simeq 0.078R\simeq 4h \quad \text{for} \quad h/R=0.02\\ l_{ZH}&\simeq 0.025R\simeq 12h \quad \text{for} \quad h/R=0.002\end{aligned}\right\} \tag{7.76}$$

Thus, the peak intensity is confined to a very narrow region at the edge of the shell. To achieve a fully converged solution with a constant strain finite element discretization it is required that the element adjacent to the edge has depth l no greater than that given by

$$1-e^{-\lambda l/R} \leqslant \frac{h}{l}$$

where substitution from equation (7.68) provides

$$l \leqslant \frac{h}{2\pi} \tag{7.77}$$

Since this depth l is less than the wall thickness, it is advantageous to employ special quadrilateral finite elements in analysing edge effect. The α and β fields of trial functions are then more readily augmented with edge effect solutions, as well as with the customary additional polynomial terms.

It is usually too difficult a matter to obtain edge effect solutions like those of equation (7.66) for other than circular cylindrical shells. Thus, we may wish to use these equations for a wider range of shell shapes and for this purpose the radii of curvature R_ν, R_σ, $R_{\nu\sigma}$ are introduced by

$$\frac{1}{R_\nu}=\frac{\sin^2\gamma}{R}, \qquad \frac{1}{R_\sigma}=\frac{\cos^2\gamma}{R}, \qquad \frac{1}{R_{\nu\sigma}}=-\frac{\sin\gamma\cos\gamma}{R}$$

and it is assumed that these radii are constant within the finite element and equal to their values obtaining at a point on the shell edge. This is often satisfactory because the peak intensity of edge effect is generally confined to a very narrow region. Thus, in extending further the α field of trial functions we may employ

$$\left.\begin{aligned} W_\alpha &= e^{\lambda'\nu/R_\sigma}(a_1\cos\lambda'\nu/R_\sigma + a_2\sin\lambda'\nu/R_\sigma) \\ U_{\nu\alpha} &= -\left(\frac{1}{R_\nu}+\frac{\nu_p}{R_\sigma}\right)\int W\,d\nu \\ U_{\sigma\alpha} &= \frac{2}{R_{\nu\sigma}}\int W\,d\nu \end{aligned}\right\} \tag{7.78}$$

where

$$\lambda' = \{3(1-\nu_p^2)R_\sigma^2/h^2\}^{\frac{1}{4}} \tag{7.79}$$

Similarly, in extending further the β field of trial functions we may make use of

$$\left.\begin{aligned} \psi_\beta &= e^{\lambda'\nu/R_\sigma}(b_1\cos\lambda'\nu/R_\sigma + b_2\sin\lambda'\nu/R_\sigma) \\ \chi_{\nu\beta} &= -\left(\frac{1}{R_\nu}-\frac{\nu_p}{R_\sigma}\right)\int W\,d\nu \\ \chi_{\sigma\beta} &= \frac{2}{R_{\nu\sigma}}\int W\,d\nu \end{aligned}\right\} \tag{7.80}$$

where note the change in sign of Poisson's ratio ν_p. The stress functions $\chi_{1\beta}$, $\chi_{2\beta}$ are determined from

$$\left.\begin{aligned}\chi_{1\beta} &= \chi_{\nu\beta}\cos\gamma - \chi_{\sigma\beta}\sin\gamma\\ \chi_{2\beta} &= \chi_{\nu\beta}\sin\gamma + \chi_{\sigma\beta}\cos\gamma\end{aligned}\right\} \tag{7.81}$$

Equations (7.78) and (7.80) are equivalent to each other in the special case of circular cylindrical shells.

7.7 CONCLUSIONS

Koiter's first approximation shell theory is used to determine details of a consistent finite element stiffness formulation. It is concluded that it is advisable to use a mutually supplementing two field discretization, i.e. a displacement and an equilibrium field, in the interior of the finite element. The displacements and slopes at the element boundary are then separately discretized and related to the interior fields by a modified principle of complementary energy with constraints. The best topology of the connectors for this model is identical with the topology of the Semiloof finite element and is an amalgam of the Fraeijs de Veubeke quadratically varying displacement plane stress element and the Fraeijs de Veubeke and Sander equilibrium element for plate bending.

The problems of edge effect are briefly examined and it found advisable to employ special trial functions for accurate investigation of this phenomenon.

REFERENCES

1. A. E. H. Love, 'The small free vibrations and deformation of a thin elastic shell', *Phil. Trans. R. Soc., Lond.*, **A179,** 491 (1890).
2. A. E. H. Love, *Mathematical Theory of Elasticity*, 4th edn, Cambridge University Press, London, 1927.
3. E. Reissner, 'A new derivation of the equations for the deformation of elastic shells', *Am. J. Math.*, **63,** 177 (1941).
4. V. V. Novozhilov, *The Theory of Thin Shells*, Noordhoff, Groningen, 1959.
5. W. T. Koiter, 'A consistent first approximation in the general theory of thin elastic shells', in *Theory of Thin Elastic Shells*, (ed. W. T. Koiter), North-Holland Publishing Co., Amsterdam, 1960, p. 12.
6. A. L. Gol'denveizer, *Theory of Elastic Thin Shells*, Pergamon, Oxford, 1961.
7. P. M. Naghdi, 'Foundations of elastic shell theory', in *Progress in Solid Mechanics*, Vol. 4 (ed. I. N. Sneddon and R. Hill), North-Holland Publishing Co., Amsterdam, 1963.
8. B. M. Irons, 'The Semiloof shell element', in *Finite Elements for Thin Shells and Curved Members*, Wiley, London, 1976, p. 197.
9. L. S. D. Morley, 'Analysis of developable shells with special reference to the finite element method and circular cylinders', *Phil. Trans. R. Soc., Lond.*, **A281,** 113 (1976).

10. B. Fraeijs de Veubeke, 'Displacement and equilibrium models in the finite element method', in *Stress Analysis* (ed. O. C. Zienkiewicz and G. S. Hollister), Wiley, London, 1965, p. 145.
11. B. Fraeijs de Veubeke, 'Variational principle and the patch test', *Int. J. Num. Methods Engng.*, **8,** 783 (1974).
12. B. Fraeijs de Veubeke and A. Millard, 'Discretization of stress fields in the finite element method', *J. Franklin Inst.*, **302,** 389 (1976).
13. T. H. H. Pian, 'Derivation of element stiffness matrices by assumed stress distributions', *Am. Inst. Aviation Astronaut, J.*, **2,** 1333 (1964).
14. B. Fraeijs de Veubeke and G. Sander, 'An equilibrium model for plate bending', *Int. J. Solids Structs.*, **4,** 447 (1968).
15. D. J. Allman, 'A simple cubic displacement element for plate bending', *Int. J. Num. Methods Engng.*, **10,** 263 (1976).

Chapter 8

Complementary Energy with Penalty Functions in Finite Element Analysis

R. L. Taylor and O. C. Zienkiewicz

8.1 INTRODUCTION

A solution of finite element problems using stress methods was proposed by Fraeijs de Veubeke in References 1 and 2. In his original work only very simple types of elements were considered and it was not evident how to systematically generalize his work to more complicated elements. Since that time a number of authors have performed stress method analyses using stress functions (see, for example, References 3, 4, and 5). The introduction of stress functions usually leads to further complications in performing finite element analyses. Many stress functions lead to high order differential equations requiring higher order continuity requirements than corresponding displacement analyses. For example, if the Airy stress function is introduced a finite element model must have C^1 continuity, whereas, a displacement analysis requires only C^0 continuity. In addition the stresses are no longer the primary dependent variable and are deduced by differentiating the stress function which implies they are less accurate than the stress function. Use of stress functions also make it difficult to recover displacements if they are desired. Finally, use of stress functions usually means that special conditions must be employed when multiply-connected regions are considered.

Alternatives to pure stress methods are hybrid[6] or mixed methods[7] (including least squares analyses[8]) where both displacements and stresses are retained in the analysis. If both stresses and displacements are retained in the global equation quite large numbers of unknowns exists at each node (e.g. in two-dimensional elasticity five degrees of freedom per node exist—two displacements and three stresses) leading to very expensive computation in large problems. Furthermore, the global equations for mixed methods are often indefinite and special care must be exercised in performing a direction solution (often pivoting is necessary or special numbering schemes for the nodes so that boundary points do not cause the solution process to fail).

In hybrid methods the stress variables are eliminated at the element level resulting in considerable savings compared with the global elimination method. Nevertheless, the element matrices include all the variables and are larger than those for either a stress or a displacement method. Also the cost for elimination at the element level is not negligible. Finally, in hybrid methods it is necessary to construct particular solutions for situations where body forces exist.

In the present study we propose a stress method which appears to overcome most of the difficulties cited above. The primary variables are stresses and the global equations include only stress degrees of freedom. We use complementary energy (or the principal of virtual stresses) as a starting point and impose the requirement that stresses satisfy equilibrium using a penalty function. The method was originally proposed in Reference 9; however, success is only now possible with a better understanding of how penalty functions behave.[10] The relationship between penalty function and mixed methods allows us to compute displacements in elements by a process analogous to computing stresses in displacement methods. We employ a projection method to obtain a continuous representation of displacements in terms of their nodal values. Finally, we include numerical examples to indicate the type of results which can be obtained.

8.2 GOVERNING EQUATIONS

In this section we state the field equations for a body Ω with boundary Γ. We will assume, for convenience, that the body is defined by Cartesian co-ordinates

$$\mathbf{x} = [x_1, x_2, x_3]^{\mathrm{T}}$$

and undergoes small deformations so that linearized strain–displacement relations are appropriate. The generalization to finite deformation is straightforward and involves using any set of consistent deformation and stress measures.

The displacement field is denoted by

$$\mathbf{u}(\mathbf{x}) = [u_1(\mathbf{x}), u_2(\mathbf{x}), u_3(\mathbf{x})]^{\mathrm{T}} \tag{8.1}$$

and the strain–displacement equations are written as

$$\boldsymbol{\varepsilon} = \mathbf{L}\mathbf{u} \tag{8.2}$$

where

$$\boldsymbol{\varepsilon} = [\varepsilon_{11}, \varepsilon_{22}, \varepsilon_{33}, 2\varepsilon_{12}, 2\varepsilon_{23}, 2\varepsilon_{31}]^{\mathrm{T}} \tag{8.3}$$

are the components of strain written, for convenience later, as a vector and

$$\mathbf{L}=\begin{bmatrix} \frac{\partial}{\partial x_1} & 0 & 0 & \frac{\partial}{\partial x_2} & 0 & \frac{\partial}{\partial x_3} \\ 0 & \frac{\partial}{\partial x_2} & 0 & \frac{\partial}{\partial x_1} & \frac{\partial}{\partial x_3} & 0 \\ 0 & 0 & \frac{\partial}{\partial x_3} & 0 & \frac{\partial}{\partial x_2} & \frac{\partial}{\partial x_1} \end{bmatrix}^{\mathrm{T}} \tag{8.4}$$

The equilibrium equations for quasi-static conditions are given by

$$\mathbf{L}^{\mathrm{T}}\boldsymbol{\sigma}+\mathbf{b}=0 \tag{8.5}$$

where

$$\sigma=[\sigma_{11}, \sigma_{22}, \sigma_{33}, \sigma_{12}, \sigma_{23}, \sigma_{31}]^{\mathrm{T}} \tag{8.6}$$

are the components of stress written, again for convenience later, as a vector; $\mathbf{b}$ is the body force vector; and $\mathbf{0}$ is a null vector.

The boundary conditions for a problem consist of specifying either displacements or tractions on the boundary Γ. Accordingly on that part of Γ where displacements are known we set

$$\mathbf{u}=\hat{\mathbf{u}}(\mathbf{x}); \qquad \mathbf{x} \text{ on } \Gamma_1 \tag{8.7}$$

and on that part where tractions are specified we set

$$\mathbf{t}=\hat{\mathbf{t}}(\mathbf{x}); \qquad \mathbf{x} \text{ on } \Gamma_2 \tag{8.8}$$

where $\mathbf{t}$ is the traction, ( ) indicates a specified quality, Γ_1 and Γ_2 are parts of Γ such that $\Gamma_1 \cup \Gamma_2=\Gamma$ and $\Gamma_1 \cap \Gamma_2=0$. The traction vector is determined from the stresses by

$$\mathbf{t}=\mathbf{H}\boldsymbol{\sigma}=[t_1, t_2, t_3]^{\mathrm{T}} \tag{8.9}$$

where $\mathbf{H}$ is a matrix of direction cosines for an outward normal to the boundary, $\mathbf{n}$.

In subsequent developments it is convenient to transform the boundary stresses from $\mathbf{x}$ components to $\mathbf{n}$ components in order to impose boundary conditions. In this case we will write a transformation

$$\boldsymbol{\sigma}=\mathbf{T}\boldsymbol{\sigma}_n \tag{8.10}$$

where

$$\boldsymbol{\sigma}_n=[\sigma_{nn}, \sigma_{ss}, \sigma_{tt}, \sigma_{ns}, \sigma_{st}, \sigma_{tn}]^{\mathrm{T}} \tag{8.11}$$

and the transformed co-ordinate axes are now

$$\mathbf{n}=[n, s, t]^{\mathrm{T}} \tag{8.12}$$

The specification of the governing equation for a problem is complete once appropriate constitutive relations are introduced. In the sequel we will write the constitutive equations as

$$\boldsymbol{\varepsilon} = \boldsymbol{\varepsilon}(\boldsymbol{\sigma}, \dot{\boldsymbol{\sigma}}) \tag{8.13}$$

to indicate that strain may be recovered from stress and stress rate. In cases where the constitutive equations are in terms of strain rate (e.g. viscoplasticity) it will be necessary to solve the rate equation before performing the solution. In the sections on implementation in two dimensions and numerical examples we consider only linear elastic constitutive equations which we write in the form

$$\boldsymbol{\varepsilon} = \mathbf{C}(\boldsymbol{\sigma} - \boldsymbol{\sigma}^0) \tag{8.14}$$

where $\boldsymbol{\sigma}^0$ is an initial stress state (e.g. tectonic stresses, thermal expansion effects $\mathbf{D}\boldsymbol{\alpha}\,\Delta T$ where $\mathbf{D} = \mathbf{C}^{-1}$). The arrays $\mathbf{D}$ and $\mathbf{C}$ are the elastic moduli and compliances, respectively, $\boldsymbol{\alpha}$ is a vector of thermal expansion coefficients, and ΔT is a change in temperature from a stress free state.

In the next section we will consider a formulation which casts the governing equation in a weak form using stress as the primary dependent variable.

8.3 STRESS FORMULATION

In the stress formulation we consider the static case only and use stress as the primary dependent variable (instead of deriving stress from strain via the constitutive relation and strain–displacement relations, which introduces loss of accuracy over that of the displacements). In linear elasticity the use of stresses as primary variable leads eventually to the overdetermined set of Beltrami–Michell compatability equations (see, for example, Reference 11), and there are often difficulties associated with finding the linearly independent equations. One approach of overcoming this difficulty is to introduce a stress function (e.g. those of Maxwell, Morera, Airy, etc.) but this immediately implies that stresses are no longer the primary dependent variables and once more must be derived by differentiation. In this section we indicate a method which appears to be very promising for solving problems with stresses as the primary dependent variables.

Since we are interested in obtaining a formulation suitable for use in a finite element model we begin from the principle of virtual stress which we write as

$$\int_{\Omega} \delta\boldsymbol{\sigma}^{\mathrm{T}}\boldsymbol{\varepsilon}(\boldsymbol{\sigma}, \dot{\boldsymbol{\sigma}})\,\mathrm{d}\Omega - \int_{\Gamma_1} \delta\mathbf{t}^{\mathrm{T}}\hat{\mathbf{u}}\,\mathrm{d}\Gamma = 0 \tag{8.15}$$

Equation (8.15) is known to lead to appropriate compatibility equations provided the stresses satisfy equilibrium and the traction boundary conditions. It is at this stage that the stress function is introduced and the usual compatibility equation in terms of the stress function deduced, see, for example, Reference 11. Instead we will approximate stresses in each element directly by the relation

$$\boldsymbol{\sigma} \approx \sum_{j=1}^{m} N_j \mathbf{a}_j \tag{8.16}$$

where now $\mathbf{a}_j$ are nodal stresses and $N_j(\mathbf{x})$ are the usual C^0 shape functions.

Substitution of (8.16) into (8.15) leads to the approximation

$$\mathbf{Q}_i(\mathbf{a}_j, \dot{\mathbf{a}}_j) = \mathbf{G}_i \tag{8.17}$$

where

$$\mathbf{Q}_i = \sum_e \int_{\Omega^e} N_i \boldsymbol{\varepsilon}(\mathbf{a}_j, \dot{\mathbf{a}}_j)\, \mathrm{d}\Omega$$

and

$$\mathbf{G}_i = \sum_e \int_{\Gamma_1^e} N_i \mathbf{H}^{\mathrm{T}} \hat{\mathbf{u}}\, \mathrm{d}\Gamma \tag{8.18}$$

To ensure that the stress are admissible we will use a penalty function approach to constrain the stresses to satisfy the equilibrium equations.[9]

The term added to (8.17) takes the form

$$\alpha \int_{\Omega} \delta[\mathbf{L}^{\mathrm{T}}\boldsymbol{\sigma} + \mathbf{b}]^{\mathrm{T}}[\mathbf{L}^{\mathrm{T}}\boldsymbol{\sigma} + \mathbf{b}]\, \mathrm{d}\Omega \tag{8.19}$$

which upon using (8.16) and noting that $\delta\mathbf{b}$ is zero becomes

$$\alpha \sum_e \int_{\Omega} \delta\mathbf{a}_i^{\mathrm{T}} \mathbf{B}_i (\mathbf{B}_j^{\mathrm{T}} \mathbf{a}_j + \mathbf{b})\, \mathrm{d}\Omega = \alpha\, \delta\mathbf{a}_i \mathbf{K}_{S_{ij}} \mathbf{a}_j + \alpha\, \delta\mathbf{a}_i \mathbf{F}_{S_i} \tag{8.20}$$

where α is a parameter and we note that the same $\mathbf{B}_i$ as in a displacement formulation occurs since the equilibrium equations involve the transpose of the operator used to define strains. The above term is the one which requires the traction vector to be C^0 continuous across element boundaries. The actual continuity condition to be imposed at each element edge is for traction. For elements with nodes at corners it is not possible to impose continuity on traction only, consequently, we make stress continuous throughout. This is acceptable except at material interfaces where only traction is continuous. Along material interfaces transformation to stresses normal and tangent to the interface can be performed using (8.10) and then allowing σ_{ss}, σ_{tt}, and σ_{st} to be discontinuous. Again difficulties will be encountered if more than one interface exists at any node. In this context we note that the use of a Loof type element[12] which contains only midside nodes, may be helpful. Unfortunately Loof elements are incompatible and this leads to further complications which can be assessed only by considerable computations and patch tests.

As pointed out in References 10 and 13 extreme care must be exercised in evaluating $\mathbf{K}_s$. If we compute the integral exactly we will introduce too many constraints and cause the solution to be highly inaccurate. The most effective procedure to construct $\mathbf{K}_s$ is to evaluate the integral numerically and use as few integration points as possible. The requirements are that $\mathbf{K}_s$ must be singular (there must be degrees of freedom available to satisfy compatibility) and that mechanisms must not exist in the total matrix. In a finite element context a mechanism is a zero energy mode which is not sampled by the quadrature formula.

With the penalty function imposed we now solve the problem

$$\mathbf{Q}_i(\mathbf{a}_j, \dot{\mathbf{a}}_j) + \alpha \mathbf{K}_{sij}\mathbf{a}_j = \mathbf{G}_i \tag{8.21}$$

for a large value of α. For the special case of the linear elastic constitutive equation (8.14) $\mathbf{Q}_i$ becomes

$$\mathbf{Q}_i = -\mathbf{Q}_{0i} + \mathbf{K}_{cij}\mathbf{a}_j \tag{8.22}$$

where

$$\mathbf{Q}_{0i} = \sum_e \int_{\Omega^e} N_i \mathbf{C}\boldsymbol{\sigma}_0 \, \mathrm{d}\Omega \tag{8.23}$$

and

$$\mathbf{K}_{cij} = \sum_e \int_{\Omega^e} N_i \mathbf{C} N_j \, \mathrm{d}\Omega \tag{8.24}$$

Thus for the linear elastic problem we solve

$$(\mathbf{K}_{cij} + \alpha \mathbf{K}_{sij})\mathbf{a}_j = \mathbf{G}_i + \mathbf{Q}_{0i} \tag{8.25}$$

In solving (8.21) or (8.25) where boundaries have specified stresses we transform each boundary node using (8.10). This involves a premultiplication of $\mathbf{K}_{cij} + \alpha\mathbf{K}_{sij}$ and $\mathbf{G}_i + \mathbf{Q}_{0i}$ by $\mathbf{T}^{\mathrm{T}}$ and a postmultiplication of $\mathbf{K}_{cij} + \alpha\mathbf{K}_{sij}$ by $\mathbf{T}$ for each boundary node. As a result the computed values at the boundary nodes produce answers according to (8.11). The non-specified parts of (8.11) are usually values which prescribe locations of maximum stress.

In the stress method we can compute strains in the material from the constitutive equations; the recovery of the displacements requires further computation.

8.4 RELATIONSHIP OF PENALTY FUNCTION FORMULATIONS WITH MIXED VARIATIONAL PRINCIPLES—RECOVERY OF DISPLACEMENTS

A penalty function formulation can always be related to a mixed variational principle and, thus, often allow the recovery of additional information. In the present case of the complementary energy formulation with penalty function equilibrium constraint the associated mixed variational principle is Reissner's principle in the form (for elastic problems)

$$I=\frac{1}{2}\int_{\Omega}[W(\boldsymbol{\sigma})+\mathbf{u}^{\mathrm{T}}(\mathbf{L}^{\mathrm{T}}\boldsymbol{\sigma}+\mathbf{b})]\,\mathrm{d}\Omega-\int_{\Gamma_2}\mathbf{u}^{\mathrm{T}}(\mathbf{t}-\hat{\mathbf{t}})\,\mathrm{d}\Gamma-\int_{\Gamma_1}\mathbf{t}^{\mathrm{T}}\hat{\mathbf{u}}\,\mathrm{d}\Gamma=\text{stationary} \tag{8.26}$$

In the implementation of the penalty function formulation we have assumed that the traction boundary conditions are satisfied identically; consequently, in the sequel, we shall drop this term. The penalty function formulation is equivalent to (8.26) only in the limit as α tends to infinity. For any finite value of α we are in fact solving the modified principle.

$$I_m=\frac{1}{2}\int_{\Omega}\left[W(\boldsymbol{\sigma})+\mathbf{u}^{\mathrm{T}}(\mathbf{L}^{\mathrm{T}}\boldsymbol{\sigma}+\mathbf{b})-\frac{1}{2\alpha}\,\mathbf{u}^{\mathrm{T}}\mathbf{u}\right]\mathrm{d}\Omega-\int_{\Gamma_1}\mathbf{t}^{\mathrm{T}}\hat{\mathbf{u}}\,\mathrm{d}\Gamma \tag{8.27}$$

where we can immediately observe that the variations with respect to $\mathbf{u}$ gives

$$\delta I_u=\frac{1}{2}\int_{\Omega}\delta\mathbf{u}^{\mathrm{T}}\left[\mathbf{L}^{\mathrm{T}}\boldsymbol{\sigma}+\mathbf{b}-\frac{1}{\alpha}\,\mathbf{u}\right]\mathrm{d}\Omega=0 \tag{8.28}$$

so that the displacement field is given by

$$\mathbf{u}=\alpha(\mathbf{L}^{\mathrm{T}}\boldsymbol{\sigma}+\mathbf{b}) \tag{8.29}$$

Upon substitution of (8.29) into (8.27) we recover the penalty function formulation

$$I_m=\int_{\Omega}\left[W(\boldsymbol{\sigma})+\frac{\alpha}{2}(\mathbf{L}^{\mathrm{T}}\boldsymbol{\sigma}+\mathbf{b})^{\mathrm{T}}(\mathbf{L}^{\mathrm{T}}\boldsymbol{\sigma}+\mathbf{b})\right]\mathrm{d}l-\int_{\Gamma_1}\mathbf{t}^{\mathrm{T}}\hat{\mathbf{u}}\,\mathrm{d}\Gamma \tag{8.30}$$

The variation of I_m with respect to $\boldsymbol{\sigma}$ gives

$$\delta I_m=\int_{\Omega}\left[\delta\boldsymbol{\sigma}\frac{\partial W}{\partial\boldsymbol{\sigma}}+\alpha\delta(\mathbf{L}^{\mathrm{T}}\boldsymbol{\sigma})^{\mathrm{T}}(\mathbf{L}^{\mathrm{T}}\boldsymbol{\sigma}+\mathbf{b})\right]\mathrm{d}\Omega-\int_{\Gamma_1}\delta\mathbf{t}^{\mathrm{T}}\hat{\mathbf{u}}\,\mathrm{d}\Gamma=0 \tag{8.31}$$

which upon substituting

$$\frac{\partial W}{\partial\boldsymbol{\sigma}}=\boldsymbol{\varepsilon}(\boldsymbol{\sigma}), \tag{8.32}$$

generalizing for constitutive equations which depend on stress rates, and using (8.16) gives (8.20). This relationship of the penalty function form with the mixed variational theorem allows us to compute the displacement field in an analogous manner to that of computing stresses and strains in a displacement formulation. From (8.27) we also can observe how the equilibrium constraint is being satisfied by the penalty function formulation.

8.5 FINITE ELEMENT IMPLEMENTATION FOR TWO-DIMENSIONAL PLANE LINEAR ELASTIC ANALYSIS

To demonstrate the applicability of the procedures given in the previous sections we will consider two-dimensional plane linear elastic problems. In this case the domain Ω can be taken as the x_1, x_2 plane as shown in Figure 8.1. The boundary Γ together with the co-ordinates associated with an outward normal are also shown in Figure 8.1. The relevant arrays for the two-dimensional problem become:

Displacements

$$\mathbf{u}=[u_1, u_2]^{\mathrm{T}} \tag{8.33}$$

Strains

$$\boldsymbol{\varepsilon}=[\varepsilon_{11}, \varepsilon_{22}, 2\varepsilon_{12}]^{\mathrm{T}} \tag{8.34}$$

Strain–displacement operator

$$L=\begin{bmatrix} \dfrac{\partial}{\partial x_1} & 0 & \dfrac{\partial}{\partial x_2} \\ 0 & \dfrac{\partial}{\partial x_2} & \dfrac{\partial}{\partial x_1} \end{bmatrix}^{\mathrm{T}} \tag{8.35}$$

Stress and traction

$$\begin{aligned} \boldsymbol{\sigma} &= [\sigma_{11}\sigma_{22}\sigma_{12}]^{\mathrm{T}} \\ \boldsymbol{\sigma}_n &= [\sigma_{nn}\sigma_{ss}\sigma_{ns}]^{\mathrm{T}} \\ \mathbf{t} &= [t_1, t_2]^{\mathrm{T}} \end{aligned} \tag{8.36}$$

Traction transformation matrix

$$H=\begin{bmatrix} \cos\theta & 0 & \sin\theta \\ 0 & \sin\theta & \cos\theta \end{bmatrix} \tag{8.37}$$

and finally stress transformation matrix

$$T=\begin{bmatrix} \cos^2\theta & \sin^2\theta & -\sin\theta\cos\theta \\ \sin^2 & \cos^2\theta & 2\sin\theta\cos\theta \\ \sin\theta\cos\theta & -\sin\theta\cos\theta & \cos^2\theta-\sin^2\theta \end{bmatrix} \tag{8.38}$$

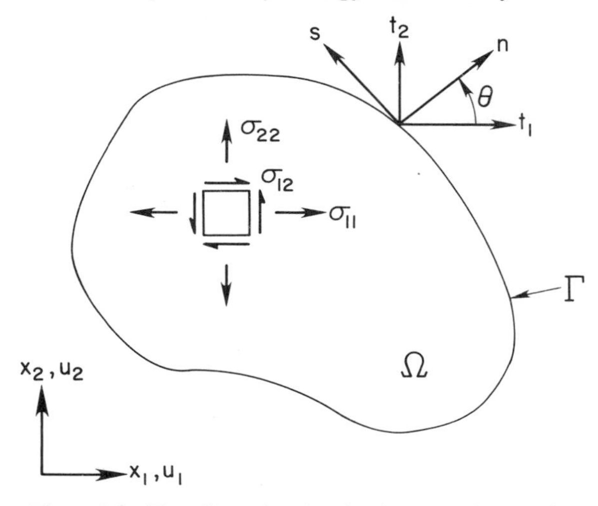

Figure 8.1 Two-dimensional region for stress formulation

In the subsequent numerical experiments we will use an isotropic linear elastic constitutive equation which for two dimensions becomes

$$\begin{Bmatrix} \varepsilon_{11} \\ \varepsilon_{22} \\ 2\varepsilon_{12} \end{Bmatrix} = \begin{bmatrix} C_{11} & C_{12} & 0 \\ C_{12} & C_{11} & 0 \\ 0 & 0 & 2(C_{11}-C_{12}) \end{bmatrix} \begin{Bmatrix} \sigma_{11}-\sigma_{11}^0 \\ \sigma_{22}-\sigma_{22}^0 \\ \sigma_{12}-\sigma_{12}^0 \end{Bmatrix} \tag{8.39}$$

For generalized plane stress

$$C_{11} = \frac{1}{E} \qquad C_{12} = -\frac{\nu}{E} \tag{8.40}$$

whereas for plane strain

$$C_{11} = \frac{(1-\nu^2)}{E} \qquad C_{12} = \frac{-\nu(1+\nu)}{E} \tag{8.41}$$

In the implementation of the two-dimensional problem we use simple bilinear isoparametric quadrilateral elements, hence, for the approximation

$$\boldsymbol{\sigma} \approx \sum_{j=1}^{m} N_j \mathbf{a}_j \tag{8.42}$$

m is 4 and the shape functions are given by

$$N_j = \tfrac{1}{4}(1+\xi_j\xi)(1+\eta_j\eta), \qquad j=1,2,3,4 \tag{8.43}$$

where ξ_j, η_j are the natural co-ordinates of the nodes as shown in Figure 8.2. and are given by

$$\left.\begin{aligned}\xi_j &= -1, 1, 1, -1\\ \eta_j &= -1, -1, 1, 1\end{aligned}\right\} \qquad j=1,2,3,4$$

Using standard isoparameteric concepts (see, for example, Reference 13) we can compute the derivatives of the shape functions with respect to x_1, x_2 for all specified ξ, η.

The arrays defined by (8.18), (8.20), (8.23), and (8.24) are now given by

$$\mathbf{G}_i = \sum_e \int_{\Gamma_1^e} N_i \begin{Bmatrix} u_1 \cos\theta \\ u_2 \sin\theta \\ \hat{u}_1 \sin\theta + \hat{u}_2 \cos\theta \end{Bmatrix} \mathrm{d}\Gamma \tag{8.44}$$

$$\mathbf{K}_{sij} = \sum_e \int_{\Omega^e} \begin{bmatrix} N_{i,1}N_{j,1} & 0 & N_{i,1}N_{j,2} \\ 0 & N_{i,2}N_{j,2} & N_{i,2}N_{j,1} \\ N_{i,2}N_{j,1} & N_{i,1}N_{j,2} & (N_{i,1}N_{j,1}+N_{i,2}N_{j,2}) \end{bmatrix} \mathrm{d}\Omega \tag{8.45}$$

$$\mathbf{F}_{si} = \sum_e \int_{\Omega_e} \begin{Bmatrix} N_{i,1}b_1 \\ N_{i,2}b_2 \\ N_{i,1}b_2 + N_{i,2}b_1 \end{Bmatrix} \mathrm{d}\Omega \tag{8.46}$$

$$\mathbf{Q}_{0i} = \sum_e \int_{\Omega^e} N_i \begin{Bmatrix} C_{11}\sigma_{11}^0 + C_{12}\sigma_{22}^0 \\ C_{12}\sigma_{11}^0 + C_{11}\sigma_{22}^0 \\ 2(C_{11}-C_{12})\sigma_{12}^0 \end{Bmatrix} \mathrm{d}\Omega \tag{8.47}$$

and

$$\mathbf{K}_{cij} = \sum_e \int_{\Omega^e} N_i \begin{bmatrix} C_{11} & C_{12} & 0 \\ C_{12} & C_{11} & 0 \\ 0 & 0 & 2(C_{11}-C_{12}) \end{bmatrix} N_j \, \mathrm{d}\Omega \tag{8.48}$$

respectively.

For each element we will evaluate the integrals on Ω^e and Γ^e numerically. It is at this stage that considerable care is needed since we must generate matrices which satisfy the condition that $\mathbf{K}_c + \alpha\mathbf{K}_s$ is non-singular (i.e. possesses no zero energy modes). These conditions are usually not difficult to satisfy. Another practical requirement is that we be able to compute viable solutions over a range of large α is often much more difficult to obtain. Selection of numerical quadrature formulae to satisfy all the above requirements is extremely delicate. Often certain classes of elements must

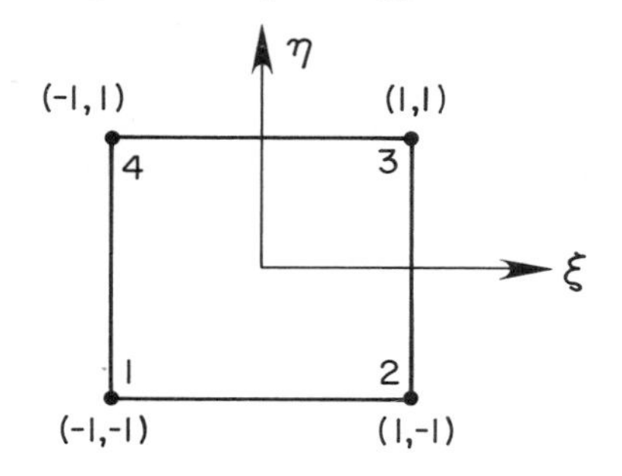

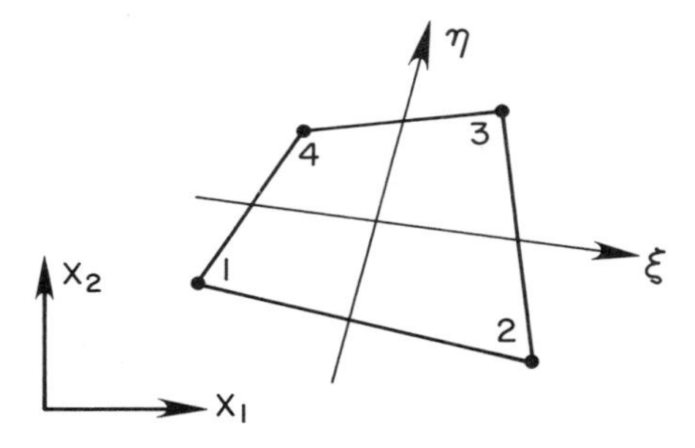

Figure 8.2 Typical four-node isoparametric quadrilateral element

not be used since no viable rule is available (e.g. see Reference 14 in which serendipity elements are shown to perform poorly for thin plate analyses where conditions are imposed by penalty functions). As a general rule the penalty terms are evaluated using a reduced order of numerical integration to produce the singular $\mathbf{K}_s$ matrix. The remaining terms (complementary energy terms here) can be integrated with any order formula which satisfies the other conditions—this may also be by reduced integration but usually not. For the present study iterated one-dimensional Gauss quadrature formulae are used. For the four-node isoparametric element we use a 2×2 formula for the complementary term and a one-point formula for the equilibrium penalty function. In a large mesh we will introduce three new unknowns with each element and two penalty function equilibrium constraints, hence, we have available about one degree of freedom per element to satisfy the constitutive equations. From numerical experiments (see next section) we have been able to obtain quite good results using very few elements. We did explore briefly the use of quadratic elements (nine-node Lagrange elements) using 2×2 reduced integration for the penalty terms and 3×3 integration for the remaining terms. This implies twelve new variables per element with, at best, eight penalty function constraints and thus there are four (or more) degrees of freedom per element to satisfy the constitutive equation. We found, however, that zero energy modes were still present and that further constraints are necessary. While we have not pursued this element further at this time we consider the use of the five-point formula of Irons[12] as a good choice. This will mean that two degrees of freedom remain to satisfy the constitutive equation which is much better than what the four-node element has.

8.6 RECOVERING A CONTINUOUS DISPLACEMENT FIELD

In Section 8.4 we have shown how to recover the displacement field from the penalty function analysis. In practice equilibrium will only be approximately satisfied at the quadrature points where $\mathbf{K}_s$ is evaluated, hence when we multiply by the large value of α we will get viable displacements only at these quadrature locations. In the case of the plane four-node isoparametric element this implies that displacements are obtained only at one point in each element. Often we may wish to obtain displacements at other points and in particular at boundaries. In the present study we have projected the element displacements using the notions of conjugate approximation introduced in Reference 15. As such we must compute the solution to the problem

$$\int_{\Omega} \delta\mathbf{u}^{\mathrm{T}}(\mathbf{u}-\mathbf{u}^{h})\,\mathrm{d}\Omega=0 \tag{8.49}$$

where $\mathbf{u}$ is the projected displacement solution and $\mathbf{u}^h$ is the displacement field computed from the penalty function equilibrium method. The projected displacement field is approximated here by

$$\mathbf{u}\approx\sum_{j=1}^{m} N_j\mathbf{c}_j \tag{8.50}$$

where $\mathbf{c}_j$ are nodal displacements. Thus, (8.49) becomes

$$\sum_e \int_{\Omega^e} N_i\begin{bmatrix}1 & 0\\ 0 & 1\end{bmatrix}N_j\,\mathrm{d}\Omega\begin{bmatrix}C_{1j}\\ C_{2j}\end{bmatrix}=\sum_e\int_{\Omega^e} N_i\begin{bmatrix}u_1^h\\ u_2^h\end{bmatrix}\mathrm{d}\Omega \tag{8.51}$$

The computation of each component of displacement uncouples and we are required to solve a problem of size equal to the number of nodal points with two right-hand sides (one for u_1, the second for u_2). This is a problem (in two dimensions) about one-third the size with one-third the bandwidth of that solved to obtain the stresses and hence costs about one-twenty-seventh as much—a small added cost indeed.

In (8.49) we assumed that $\mathbf{u}^h$ is constant over each element with a value of that computed at the constraint point. We then evaluated the integrals using a 2×2 Gauss formula and solved the resulting equation. We did not impose boundary conditions on $\mathbf{u}$ as these are weakly contained in $\mathbf{u}^h$. Some improvement would undoubtedly result but would require two different solutions as the boundary conditions for u_1 and u_2 are imposed at different locations. Even so the results we obtained are surprisingly good—often comparable to those obtained from a displacement solution. We also tried a 2×2 quadrature formula which samples only at nodes. This produces a diagonal (lumped) coefficient matrix which is then trivial to solve.† This procedure produced answers comparable to those of a consistent implementation except that the boundary conditions near boundaries were usually slightly more in error (see Table 8.1). For a large size problem this degradation is not serious and the relative cost difference between the lumped and consistent solution may be the deciding factor.

8.7 NUMERICAL EXAMPLES

In order to demonstrate the type of results which can be obtained using the method described above we consider two sample problems. The first problem is a cantilever beam subjected to either pure bending or uniform normal load on the top surface. The second problem is a cylindrical region with non-zero initial stress and a stress-free circular opening. Details of the two problems are given below.

8.7.1 Cantilever beam

We consider first a rather short cantilever beam subjected to two different loading states—pure bending and uniform normal loading on the top surface. The material properties are

$$E = 100$$
$$\nu = 0.0$$

A coarse mesh of six four-node rectangular elements is considered and shown in Figure 8.3(a). The complementary energy terms are evaluated using 2×2 Gauss quadrature (which gives exact results for the rectangular elements) and the penalty function terms are evaluated using one-point quadrature at the element centre. The penalty constant α was assigned a value of 10^4. The penalty functions terms are about $\alpha E/A^e$ (A^e is the element area) larger than those of the complementary energy. An α value of 10^6 gave stresses identical to the five significant digits printed and produced displacements with at least three digits the same. All calculations were carried out on a CDC 6400, which has about 15 significant digits.

† This was originally proposed for smoothing stresses by Lee, Gresho, and Sani.[16]

The beam was loaded in pure bending by a linearly varying σ_{11} stress as shown in Figure 8.3(a). The stresses computed at midspan are shown in Figure 8.4. The vertical displacement along the x_1 axis computed using the smoothing techniques described above are shown in Table 8.1.

A more refined mesh consisting of 60 four-node rectangular elements was constructed and is shown in Figure 8.3(b). The results for the σ_{11} stresses at midspan are also plotted in Figure 8.4, and the vertical displacement along the x_1 axis computed using the smoothing techniques are given in Table 8.1. The consistent method of projecting the displacements produced results which more nearly satisfy the zero displacement condition at the fixed end for both the coarse and fine mesh solutions. The results of the fine mesh analysis indicate, however, that there is not much difference between the consistent and lumped results. In the consistent method there is more tendency for the displacements to oscillate as they approach boundaries. This

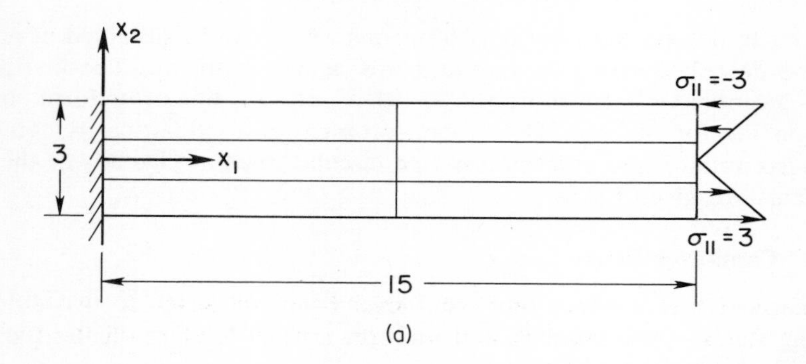

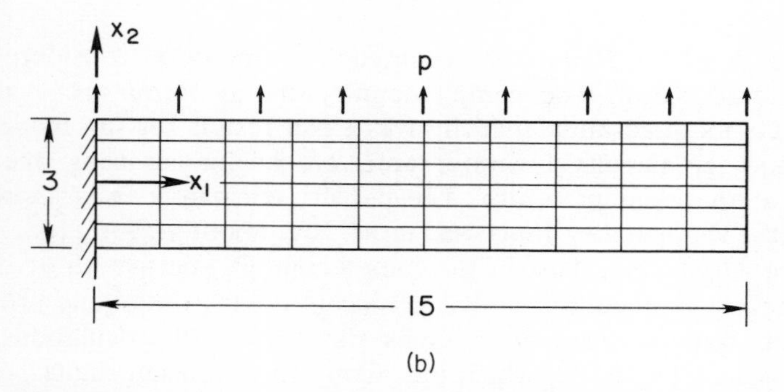

Figure 8.3 Mesh and loadings for beam analysis problem. (a) Coarse mesh. (b) Fine mesh

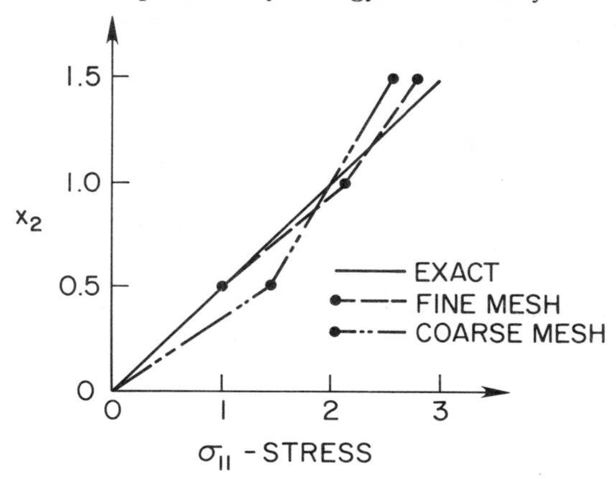

Figure 8.4 Stress σ_{11} at midspan of cantilever beam—pure bending load

Table 8.1 Vertical displacement of cantilever beam—pure bending $E = 100$, $\nu = 0$, $\alpha = 10^4$

		Coarse mesh, projected solution		Fine mesh, projected solution	
x_1	Exact	Consistent	Lumped	Consistent	Lumped
0	0	−0.004	0.314	0.0003	0.001
1.5	0.0225			0.033	0.034
3.0	0.09			0.098	0.103
4.5	0.2025			0.218	0.218
6.0	0.36			0.382	0.378
7.5	0.5625	0.937	0.938	0.596	0.585
9.0	0.81			0.852	0.837
10.5	1.1025			1.169	1.136
12.0	1.44			1.497	1.481
13.5	1.8225			1.982	1.872
15.0	2.25	1.862	1.563	2.114	2.079

is somewhat surprising since the consistent method uses all the data to project the displacements while the lumped method uses results only in the elements attached to each node. This tendency was also observed by Lee, Gresho, and Sani in Reference 16.

It was expected initially that the stresses in this problem would be exactly recovered since this is a possible solution for the four-node element used. In retrospect, however, if this had happened it would not have been possible to recover the displacement field as shown in Section 8.4. For coarse meshing the stress errors are quite pronounced but as finer meshing is employed convergence is rapidly occurring to the correct linear result.

A second loading of a uniform normal stress σ_{32} applied to the top surface of the fine mesh as shown in Figure 8.3(b) was also considered. The computed stress σ_{11} along the top boundary as shown in Figure 8.5. For purposes of comparison the solution obtained allowing warpage due to shear stresses at the support is shown (same as beam solution). The comparison is quite good. In Figure 8.6 we show the vertical displacement along the x_1-axis for the projection methods and compare with the solution permitting warpage but with a zero mean of the u_1 displacement along the x_2 axis. Again we notice that the consistent projected stresses have a greater tendency to oscillate near the free end of the beam.

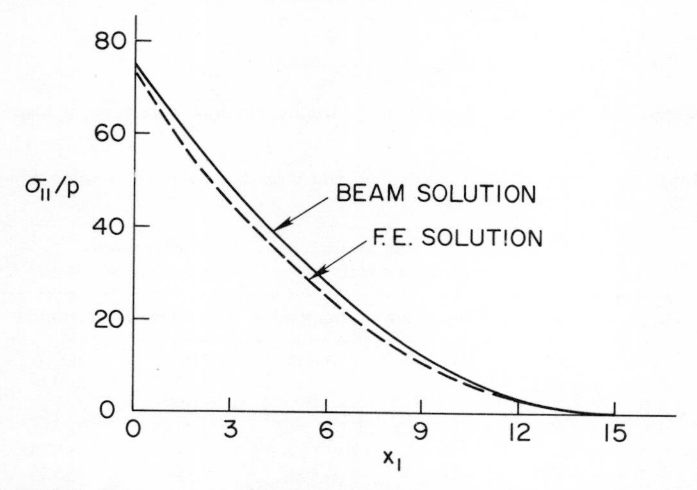

Figure 8.5 Stress σ_{11} along top of uniformly loaded cantilever beam

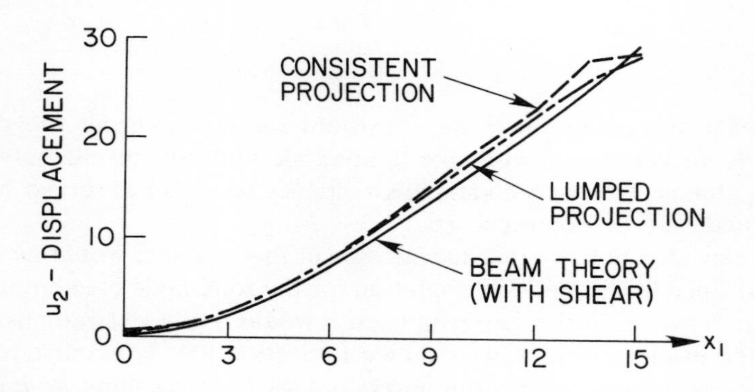

Figure 8.6 Centreline vertical displacement u_2 for uniformly loaded cantilever beam

8.7.2 Initially stressed cylindrical region with a stress free circular opening

As a second example we consider an initially stressed plane stress region in which a circular opening is made. The circular boundary of the opening is stress-free and it is assumed that a circular boundary with ten times the radius of the hole undergoes no displacement. For homogeneous initial stresses the exact displacement and stress field can be computed.

The initial stresses are given by

$$\sigma_{11}^0 = \sigma_{22}^0 = \sigma^0 \qquad \text{(constant)}$$
$$\sigma_{12}^0 = 0$$

and the material has a Poisson ratio of 0.3.

The solution for a stress-free opening of radius R with the displacements at radius $10R$ set zero is given in polar components as

$$\frac{Eu_r}{R} = -\frac{0.91}{71.3}\left(\rho - \frac{100}{\rho}\right)\sigma^0$$
$$\sigma_{rr} = \frac{70}{71.3}\left(1 - \frac{1}{\rho^2}\right)\sigma^0$$
$$\sigma_{\theta\theta} = \frac{70}{71.3}\left(1 + \frac{1}{\rho^2}\right)\sigma^0$$

where $\rho = r/R$. While the stress differ by less than 2 per cent from those of the infinite sheet, the displacements of the infinite sheet at ten times the opening are about 10 per cent of those at the opening, which cannot be ignored in assessing the overall accuracy. A finite element analysis using the stress method has been performed using a mesh of 35 quadrilateral four-node elements as shown in Figure 8.7. For purpose of comparison a conventional displacement solution has also been performed using the same mesh. The results are plotted using polar components in Figures 8.8 and 8.9. The displacements shown in Figure 8.9 for the stress method were computed using the consistent projection method. Considering the coarseness of the mesh the results are quite good.

A second loading state for this problem which gives stresses with steeper gradients near the hole is

$$\sigma_{11}^0 = \sigma^0 \qquad \text{constant}$$
$$\sigma_{12}^0 = \sigma_{22}^0 = 0$$

The exact solution for an infinite region is used for the comparison and a condition of plane stress is assumed.[11]

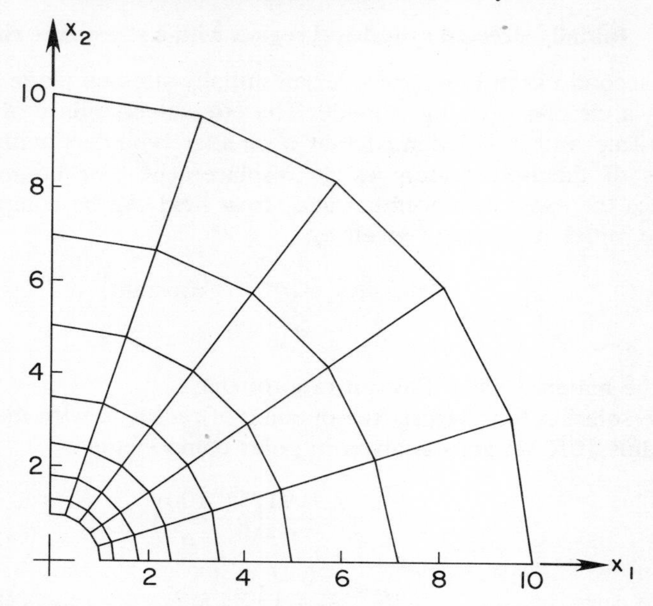

Figure 8.7 Mesh for stress analysis of initially stressed sheet with a circular hole

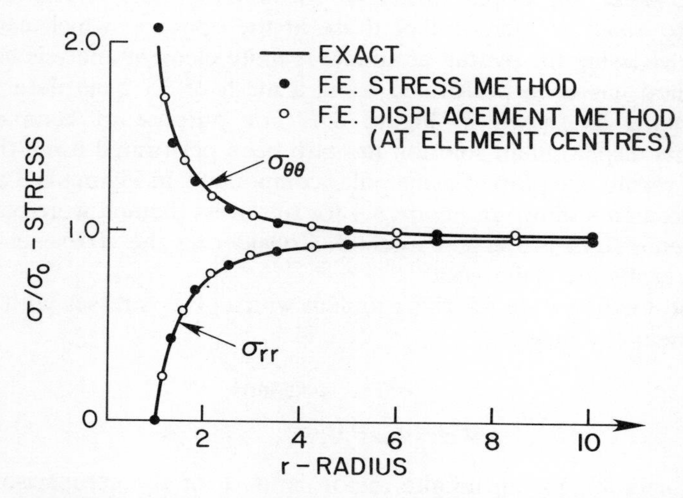

Figure 8.8 Stresses in sheet with circular hole—initial stress $\sigma_{11}^0 = \sigma_{22}^0 = \sigma_0$

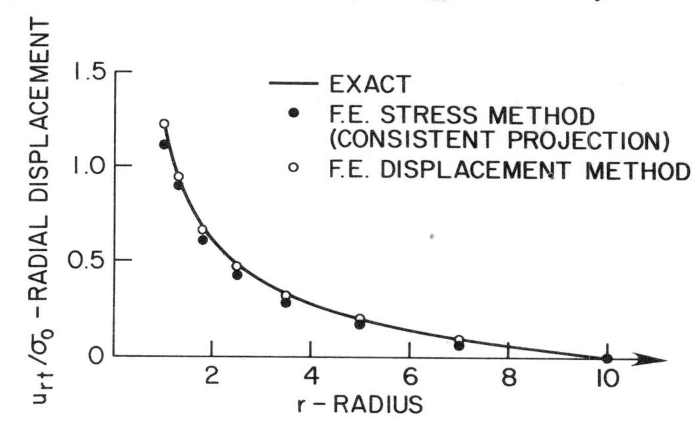

Figure 8.9 Radial displacement in sheet with circular hole—initial stress $\sigma_{11}^0 = \sigma_{22}^0 = \sigma_0$

Both finite element stress and displacement method solutions have been computed. The displacement distribution around the hole for both solutions is shown in Figure 8.10 and compared with the infinite solution. In Reference 17 the amount of error due to replacing the infinite region by a finite region are shown to be about half the discrepancy shown in Figure 8.10. In Figure 8.11 we show the tangential stress computed around the opening while in Figure 8.12 the normal stresses to the axes are shown. The stress method in all these cases has faithfully captured the character of the stress distribution using very few elements.

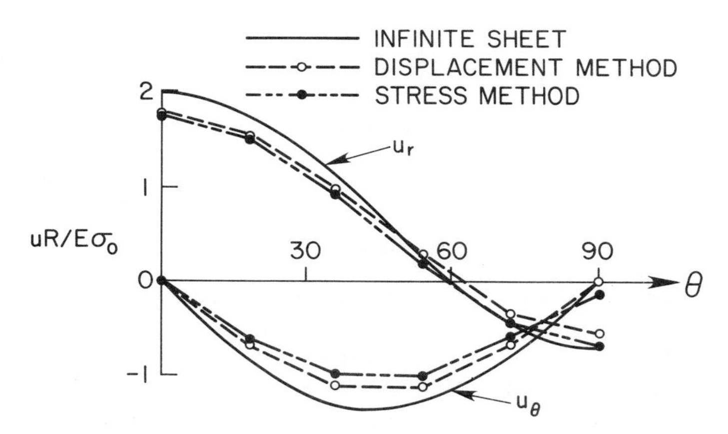

Figure 8.10 Displacements around circular hole in a sheet with initial stress $\sigma_{11}^0 = \sigma_0$

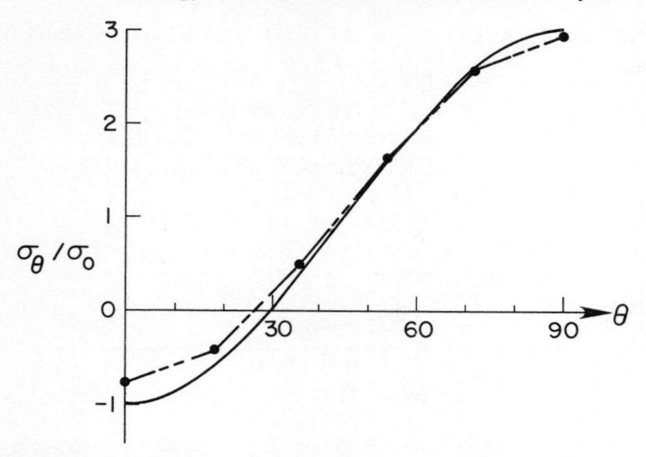

Figure 8.11 Tangential stress around circular hole in a sheet with initial stress $\sigma_{11}^0 = \sigma_0$

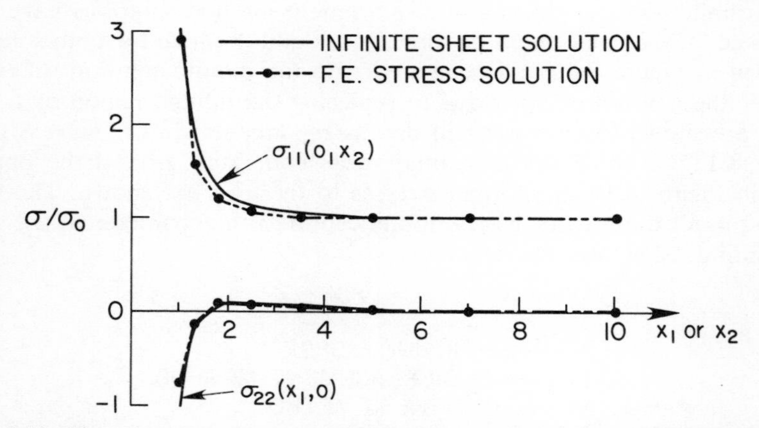

Figure 8.12 Normal stresses along co-ordinate axes for a sheet with circular hole and initial stress $\sigma_{11}^0 = \sigma_0$

8.8 CLOSURE

In this paper we have presented a stress method with equilibrium imposed by a penalty function. Numerical results for two-dimensional linear elastic problems have also been included to demonstrate the types of results which can be obtained using the method. The extension to non-linear problems which are formulated in terms of stresses appears to be straightforward; analyses which impose constraints on stresses such as the constraining of stresses on the yield surface for plasticity problems should also be easily adapted to the method and thus overcome difficulties which are encountered in keeping stresses on the yield surface in displacement analyses. Finally, for problems which require an accurate evaluation of stresses on boundaries the method is extremely attractive.

ACKNOWLEDGEMENT

The numerical results were performed using the CDC 6400 computer at the Computer Center of the University of California, Berkeley. We gratefully acknowledge the support and computer time furnished by the Computer Center.

REFERENCES

1. B. Fraeijs de Veubeke, 'Upper and lower bounds in matrix structural analysis', *AGARDograph*, Vol. 72, Pergamon Press, 1964, p. 165.
2. B. Fraeijs de Veubeke, Displacement and equilibrium models in the finite element method', *Stress Analysis* (ed. O. C. Zienkiewicz and G. S. Holister), Wiley, London, 1965, p. 145.
3. G. Sander, 'Application of the dual analysis principle', *Proc. IUTAM Conf. on High Speed Computing of Elastic Structs.*, Liège, 1970, p. 16.
4. B. Fraeijs de Veubeke, 'Stress function approach', *World Congress on Finite Element Method in Structural Mechanics*, Bournemouth, 1975.
5. J. Robinson, *Integrated Theory of Finite Element Methods*, Wiley, London, 1973.
6. T. H. H. Pian and P. Tong, 'Basis of finite element methods for solid continua', *Int. J. Num. Meth. Engr.*, **1,** 3 (1969).
7. L. R. Herrmann, 'Finite element bending analysis of plates', *Proc. Am. Soc. Civ. Eng.*, **94,** EM5, 13 (1968).
8. O. C. Zienkiewicz, D. R. J. Owen, and K. N. Lee, 'Least square finite element for elasto-static problems. Use of reduced integration', *Int. J. Num. Meth. Engr.*, **8,** 341 (1974).
9. O. C. Zienkiewicz, 'Constrained variational principles and penalty function methods in finite element analysis', *Lecture Notes in Mathematics*, No. 363, Springer-Verlag, 1974, p. 207.
10. D. S. Malkus and T. J. R. Hughes, 'Mixed finite element methods—Reduced and selective integration techniques: A unification of concepts', *Comp. Meth. Appl. Mech. Eng.*, **15,** 1–12 (1978).
11. I. S. Sokolnikoff, *Mathematical Theory of Elasticity*, McGraw-Hill, New York, 1956.
12. B. Irons, 'The SemiLoof shell element', *Finite Elements for Thin Shells and Curved Members* (D. G. Ashwell and R. H. Gallagher, Eds.), Wiley, London, 1976, p. 197.
13. O. C. Zienkiewicz, *The Finite Element Method*, 3rd edn, McGraw-Hill, London, 1977.
14. E. D. L. Pugh, E. Hinton, and O. C. Zienkiewicz, 'A study of quadrilateral plate bending elements with reduced integration', *Int. J. Num. Meth. Eng.*, **12,** 1059 (1978).
15. J. T. Oden and J. N. Reddy, *An Introduction to the Mathematical Theory of Finite Elements*, Wiley, New York, 1976.

16. R. L. Lee, P. M. Gresho, and R. L. Sani, 'Numerical smoothing techniques applied to some finite element solutions of Navier–Stokes equations', *Report UCRL 80127, Lawrence Livermore Laboratories*, Livermore, California, 1977.
17. O. C. Zienkiewicz, R. L. Taylor, and G. N. Pande, Quasi-plane strain in the analysis of geological problems, *Proc. Conf. on Computer Methods in Tunnel Design*, Institution of Civil Engineers, London (to appear).

Chapter 9

Dual Finite Element Models for Second Order Elliptic Problems

P. A. Raviart and J. M. Thomas

9.1 INTRODUCTION

Finite element methods based on the standard variational principle of minimum potential energy have been extensively studied from both practical and theoretical points of view. The mathematical theory of such methods is now firmly established and may be considered as a classical part of the numerical analysis of partial differential equations (see, for instance, the books of Strang and Fix[1] and Ciarlet[2]). On the other hand, considerable efforts have been made in the last few years in order to give a complete mathematical analysis of other classes of finite element methods (equilibrium methods, mixed methods, hybrid methods, etc.) based on different variational principles (complementary energy principle, Hellinger–Reissner principle, etc.). The mathematical theory of such finite element methods is now almost complete and it is the purpose of this paper of expository nature to discuss some aspects of this theory.

For the sake of simplicity, we shall restrict ourselves to the following simple second order elliptic model problem

$$\begin{cases} -\Delta u = f \text{ in } \Omega \\ u = 0 \text{ on } \Gamma \end{cases} \tag{9.1}$$

where Ω is a bounded domain in the Euclidean plane with a sufficiently smooth boundary Γ. Moreover we shall only discuss finite element methods based on dual models which generalize the equilibrium model considered by Fraeijs de Veubeke.[3,4]

An outline of the paper is as follows. We give in Section 9.2 an abstract variational principle and we derive as in[5] the dual mixed-hybrid formulation of problem (9.1). Section 9.3 is devoted to the construction of general methods of approximation based on the abstract principle of Section 9.2 and to the derivation of error estimates. In Section 9.4, we study the mixed-hybrid generalization of the equilibrium finite element method of Fraeijs de Veubeke. Section 9.5 contains an analysis of a mixed-hybrid generalization of a stress-assumed hybrid model of Pian type.

For the analysis of primal hybrid methods for second order elliptic problems, we refer to Babuška, Oden, and Lee[6] and to Raviart and Thomas[7] where non-conforming methods are also studied.

Similar but technically different analyses have been also developed for fourth order elliptic problems. We refer to Quarteroni[8] for a discussion of primal hybrid methods, to Brezzi[9] and Brezzi and Marini[10] for dual hybrid methods; a survey of such methods may also be found in Brezzi.[23]

Mixed models have been mathematically studied by Brezzi and Raviart,[11] Ciarlet and Raviart,[12] Johnson,[13] Miyoshi.[14] For a discussion of equilibrium methods, see Brezzi and Raviart[11] and Raviart.[15]

For a related work, we refer to the book of Oden and Reddy.[16]

9.2 AN ABSTRACT VARIATIONAL PRINCIPLE

We begin by introducing a general abstract variational principle. Let X and M be two (real) Hilbert spaces with norms $\|\cdot\|_X$ and $\|\cdot\|_M$. Let X' and M' be the dual spaces of X and M respectively. We denote by $\langle\cdot,\cdot\rangle$ the pairing between the spaces X' and X or between the spaces M' and M.

We introduce two continuous bilinear forms

$$a(\cdot,\cdot): X\times X\to\mathbb{R},$$
$$b(\cdot,\cdot): X\times M\to\mathbb{R}.$$

We consider the following problem: *Given* $l\in X'$ *and* $\chi\in M'$, *find a pair* $(p,\lambda)\in X\times M$ *such that*

$$a(p,q)+b(q;\lambda)=\langle l,q\rangle \qquad \forall q\in X \tag{9.2}$$

$$b(p;\mu)=\langle\chi,\mu\rangle \qquad \forall\mu\in M \tag{9.3}$$

We set

$$\left.\begin{aligned} &V(\chi)=\{q\in X;\, b(q;\mu)=\langle\chi,\mu\rangle\forall\mu\in M\}\\ &V=V(0)\end{aligned}\right\} \tag{9.4}$$

Then we have the fundamental result (cf. Babuška,[17] Brezzi[9]).

Theorem 9.1 *Assume that there exist two constants* α *and* $\beta>0$ *such that*

$$a(q,q)\geq\alpha\|q\|_X^2 \qquad \forall q\in V \tag{9.5}$$

$$\sup_{q\in X}\frac{b(q;\mu)}{\|q\|_X}\geq\beta\|\mu\|_M \qquad \forall\mu\in M \tag{9.6}$$

Then problem (9.2), (9.3) *has a unique solution* $(u,\lambda)\in X\times M$.

Clearly, the first argument p of the solution of problem (9.2), (9.3) may be characterized as the unique element of $V(\chi)$ which satisfies the equation

$$a(p, q) = \langle l, q \rangle \qquad \forall q \in V. \tag{9.7}$$

Remark 9.1 When the bilinear form $a(\cdot, \cdot)$ is symmetric (i.e., $a(p, q) = a(q, p)$ for all $p, q \in X$), we introduce the quadratic functionals:

$$I : q \in X \to I(q) = \tfrac{1}{2} a(q, q) - \langle l, q \rangle$$
$$\mathcal{L} : (q; \mu) \in X \times M \to \mathcal{L}(q; \mu) = I(q) + b(q; \mu) - \langle \chi, \mu \rangle$$

If in addition the bilinear form $a(\cdot, \cdot)$ is semi-positive-definite on X (i.e. $a(q, q) \geq 0$ for all $q \in X$), the solution (p, λ) of problem (9.2), (9.3) is the unique saddle-point of the functional $\mathcal{L}$ over $X \times M$, i.e.

$$\mathcal{L}(p, \lambda) = \operatorname*{Min}_{q \in X} \operatorname*{Max}_{\mu \in M} \mathcal{L}(q; \mu) = \operatorname*{Max}_{\mu \in M} \operatorname*{Min}_{q \in X} \mathcal{L}(q; \mu) \tag{9.8}$$

On the other hand, p is the unique element of $V(\chi)$ which minimizes the functional I:

$$I(p) = \operatorname*{Min}_{q \in V(\chi)} I(q) \tag{9.9}$$

Hence, in that case, λ appears to be the Lagrange multiplier associated with the constraint $p \in V(\chi)$. ■

Let us illustrate the previous result by introducing a mixed-hybrid formulation of the model problem (9.1). We first need some notation. For any integer $m \geq 0$, we define $H^m(\Omega)$ to be the usual Sobolev space of functions which are square integrable on Ω together with all the derivatives $\partial^\alpha v = \partial^{|\alpha|} v / \partial x_1^{\alpha_1} \partial x_2^{\alpha_2}$ of order $|\alpha| \leq m$. We provide $H^m(\Omega)$ with the norm

$$\|v\|_{m,\Omega} = \left(\sum_{|\alpha| \leq m} \int_\Omega |\partial^\alpha v|^2 \, dx \right)^{1/2}$$

and the semi-norm

$$|v|_{m,\Omega} = \left(\sum_{|\alpha| = m} \int_\Omega |\partial^\alpha v|^2 \, dx \right)^{1/2}.$$

We denote by $H^{1/2}(\Gamma)$ the space of the traces on Γ of all functions of $H^1(\Omega)$ equipped with the norm

$$\|\psi\|_{1/2,\Gamma} = \inf \{ \|v\|_{1,\Omega};\ v \in H^1(\Omega),\ v|_\Gamma = \psi \}$$

and we set:

$$H_0^1(\Omega) = \{ v \in H^1(\Omega);\ v|_\Gamma = 0 \}$$

Next, we introduce the space

$$H(\text{div}; \Omega) = \{\mathbf{q} \in (L^2(\Omega))^2; \text{div}\, \mathbf{q} \in L^2(\Omega)\}$$

normed by

$$\|\mathbf{q}\|_{H(\text{div};\Omega)} = \{\|\mathbf{q}\|^2_{0,\Omega} + \|\text{div}\, \mathbf{q}\|^2_{0,\Omega}\}^{1/2}$$

Note that, for $\mathbf{q} \in H$ (div; Ω), we may define $\mathbf{q} \cdot \mathbf{n}$ as an element of the dual space $H^{-1/2}(\Gamma)$, of $H^{1/2}(\Gamma)$, $\mathbf{n}$ being the unit outward normal to Γ. Moreover, we have the Green's formula.

$$\int_\Omega \{\mathbf{q} \cdot \mathbf{grad}\, v + v \,\text{div}\, \mathbf{q}\}\, \text{d}x = \int_\Gamma \mathbf{q} \cdot \mathbf{n} v \,\text{d}s \qquad \forall v \in H^1(\Omega) \tag{9.10}$$

where the integral $\int_\Gamma$ represents the duality between the spaces $H^{-1/2}(\Gamma)$ and $H^{1/2}(\Gamma)$.

Consider a general 'triangulation' $\mathcal{T}_h$ of $\bar{\Omega}$ with elements K of diameters $\leq h$. If v is any function defined on Ω, we denote by v_K its restriction to the element K and by $v_{\partial K}$ its restriction to the boundary ∂K of K.

With this partition of $\bar{\Omega}$, we associate the following spaces:

$$Q = \{\mathbf{q} \in (L^2(\Omega))^2, \mathbf{q}_K \in H(\text{div}; K), K \in \mathcal{T}_h\}, \tag{9.11}$$

normed by

$$\|\mathbf{q}\|_Q = \left(\sum_{K \in \mathcal{T}_h} \|\mathbf{q}_K\|^2_{H(\text{div};K)} \right)^{1/2} \tag{9.12}$$

and

$$\Psi = \left\{ \psi \in \prod_{K \in \mathcal{T}_h} H^{1/2}(\partial K); \right. \\ \textit{there exists a function} \\ \left. v \in H^1_0(\Omega) \textit{ such that } v = \psi \textit{ on } \bigcup_{K \in \mathcal{T}_h} \partial K \right\} \tag{9.13}$$

with the norm

$$\|\psi\|_\Psi = \inf \left\{ |v|_{1,\Omega};\, v \in H^1_0(\Omega),\, v = \psi \text{ on } \bigcup_{K \in \mathcal{T}_h} \partial K \right\} \tag{9.14}$$

We introduce the continuous bilinear forms $a(\cdot, \cdot): Q \times Q \to \mathbb{R}$ and $b(\cdot; \cdot, \cdot): Q \times (L^2(\Omega) \times \Psi) \to \mathbb{R}$ given by

$$a(\mathbf{p}, \mathbf{q}) = \int_\Omega \mathbf{p} \cdot \mathbf{q}\, \text{d}x, \qquad \mathbf{p}, \mathbf{q} \in Q \tag{9.15}$$

and

$$b(\mathbf{q}; v, \psi) = \sum_{K \in \mathcal{T}_h} \left(\int_K \text{div}\, \mathbf{q}_K v_K \,\text{d}x - \int_{\partial K} \mathbf{q}_K \cdot \mathbf{n}_K \psi_{\partial K} \,\text{d}s \right) \tag{9.16}$$

$\mathbf{q} \in Q$, $v \in L^2(\Omega)$, $\psi \in \Psi$, where $\mathbf{n}_K$ is the unit outward normal to ∂K. Then we can state

Theorem 9.2 *Assume that the function f belongs to the space* $L^2(\Omega)$. *There exists a unique triple* $(\mathbf{p}, u, \phi) \in Q \times L^2(\Omega) \times \Psi$ *solution of the equations*

$$a(\mathbf{p}, \mathbf{q}) + b(\mathbf{q}; u, \phi) = 0 \qquad \forall \mathbf{q} \in Q \tag{9.17}$$

$$b(\mathbf{p}; v, \psi) = -\int_\Omega fv \, dx \qquad \forall (v, \psi) \in L^2(\Omega) \times \Psi \tag{9.18}$$

In addition, u is the solution of problem (9.1) *and*

$$\mathbf{p} = \mathbf{grad}\, u, \qquad \phi = u \text{ on } \bigcup_{K \in \mathcal{T}_h} \partial K \tag{9.19}$$

Proof: Let us briefly check that Theorem 9.1 applies with $X = Q$, $M = L^2(\Omega) \times \Psi$, $\mu = (v, \psi)$, $l = 0$ and $\langle \chi, \mu \rangle = -\int_\Omega fv \, dx$. In fact, we have

$$V = \{\mathbf{q} \in (L^2(\Omega))^2; \operatorname{div} \mathbf{q} = 0 \text{ in } \Omega\}$$

so that the condition (9.5) holds with $\alpha = 1$. On the other hand, given $\mu = (v, \psi) \in L^2(\Omega) \times \Psi$, we define w to be the function of $H_0^1(\Omega)$ which satisfies for all $K \in \mathcal{T}_h$

$$\Delta w_K = \varepsilon v_K \text{ in } K$$

$$w_K = -\psi_{\partial L} \text{ on } \partial K$$

where $\varepsilon > 0$ is a parameter. By setting $\mathbf{q} = \mathbf{grad}\, w$ and choosing ε small enough, we obtain

$$b(\mathbf{q}; v, \psi) \geq c(\|v\|_{0,\Omega}^2 + |w|_{1,\Omega}^2) \geq \beta \|\mathbf{q}\|_Q (\|v\|_{0,\Omega}^2 + |w|_{1,\Omega}^2)^{1/2}$$

Hence we get by (9.13)

$$\frac{b(\mathbf{q}; v, \psi)}{\|\mathbf{q}\|_Q} \geq \beta(\|v\|_{0,\Omega}^2 + \|\psi\|_\Psi^2)^{1/2}$$

so that (9.6) holds.

Therefore we may apply Theorem 9.1 and problem (9.17), (9.18) has a unique solution $(\mathbf{p}, u, \phi)$. It remains only to check that the triple $(\mathbf{p}, u, \phi)$, where u is the solution of (9.1) and $\mathbf{p}$ and ϕ are given by (9.19), is indeed the solution of problem (9.17), (9.18). This is an easy consequence of the Green's formula (9.10) used in each $K \in \mathcal{T}_h$. ■

Remark 9.2 Define the quadratic functional

$$\mathcal{L}(\mathbf{q}; v, \psi) = \tfrac{1}{2} a(\mathbf{q}, \mathbf{q}) + b(\mathbf{q}; v, \psi) + \int_\Omega fv \, dx \tag{9.20}$$

Then, as a consequence of Remark 9.1, the triple $(\mathbf{p}, u, \phi)$ may be characterized by

$$\mathscr{L}(\mathbf{p}; u, \phi) = \underset{\mathbf{q}\in Q}{\text{Min}} \underset{(v,\psi)\in L^2(\Omega)\times\Psi}{\text{Max}} \mathscr{L}(\mathbf{q}; v, \psi) = \underset{(v,\psi)\in L^2(\Omega)\times\Psi}{\text{Max}} \underset{\mathbf{q}\in Q}{\text{Min}} \mathscr{L}(\mathbf{q}; v, \psi) \tag{9.21}$$

On the other hand, we have

$$V(\chi) = V^f = \{\mathbf{q} \in (L^2(\Omega))^2; \operatorname{div} \mathbf{q} + f = 0 \text{ in } \Omega\} \tag{9.22}$$

Hence $\mathbf{p}$ is the unique function of V^f such that

$$a(\mathbf{p}, \mathbf{q}) = 0 \qquad \forall \mathbf{q} \in V \tag{9.23}$$

or equivalently such that

$$I(\mathbf{p}) = \underset{\mathbf{q}\in V^f}{\text{Min}}\, I(\mathbf{q}), \qquad I(\mathbf{q}) = \tfrac{1}{2}\|\mathbf{q}\|^2_{0,\Omega} \tag{9.24}$$

we obtain here the equilibrium formulation of problem (1.1) based on the complementary energy principle (9.25). ■

Remark 9.3 By eliminating ϕ in (9.17), (9.18), we get the dual mixed formulation of (9.1): the pair $(\mathbf{p}, u) \in H(\text{div}; \Omega) \times L^2(\Omega)$ is the unique solution of the equations

$$\left.\begin{aligned} &a(\mathbf{p}, \mathbf{q}) + \int_\Omega u \operatorname{div} \mathbf{q}\, dx = 0 && \forall \mathbf{q} \in H(\text{div}; \Omega) \\ &\int_\Omega v(\operatorname{div} \mathbf{p} + f)\, dx = 0 && \forall v \in L^2(\Omega) \end{aligned}\right\} \tag{9.25}$$

Similarly, eliminating u in (9.17), (9.18) gives the dual hybrid formulation of problem (9.1): setting

$$Q^f = \{\mathbf{q} \in Q; \operatorname{div} \mathbf{q} + f = 0 \text{ in each } K \in \mathscr{T}_h\} \tag{9.26}$$

the pair $(\mathbf{p}, \phi) \in Q^f \times \Psi$ is the unique solution of the equations

$$\left.\begin{aligned} &a(\mathbf{p}, \mathbf{q}) - \sum_{K\in\mathscr{T}_h} \int_{\partial K} \mathbf{q}_K \cdot \mathbf{n}_K \phi_{\partial K}\, ds = 0 && \forall \mathbf{q} \in Q^0 \\ &\sum_{K\in\mathscr{T}_h} \int_{\partial K} \mathbf{p}_K \cdot \mathbf{n}_K \psi_{\partial K}\, ds = 0 && \forall \psi \in \Psi \end{aligned}\right\} \tag{9.27}$$ ■

9.3 A GENERAL APPROXIMATION SCHEME

In order to approximate the solution of the abstract problem (9.2), (9.3), it may be useful to give another characterization of this solution. Thus we are given two other (real) Hilbert spaces $\tilde{X}$ and $\tilde{M}$ with norms $\|\cdot\|_{\tilde{X}}$ and $\|\cdot\|_{\tilde{M}}$ such that

$$\tilde{X} \subset X, M \subset \tilde{M} \text{ with continuous imbeddings} \tag{9.28}$$

$$M \text{ is dense in } \tilde{M} \tag{9.29}$$

Again we denote by $\langle \cdot, \cdot \rangle$ the pairing between $\tilde{M}$ and its dual space $\tilde{M}'$.

We assume that the bilinear form $b(\cdot; \cdot)$ can be extended by continuity as a continuous bilinear form on $\tilde{X} \times \tilde{M}$ and that χ' belongs to $\tilde{M}'$. Then we consider the following problem: *Find a pair* $(u, \lambda) \in \tilde{X} \times \tilde{M}$ *such that*

$$a(p, q) + b(q; \lambda) = \langle l, q \rangle \qquad \forall q \in \tilde{X} \tag{9.30}$$

$$b(p; \mu) = \langle \chi, \mu \rangle \qquad \forall \mu \in \tilde{M} \tag{9.31}$$

This problem is not well posed at least in general. However we have

Theorem 9.3 *Assume that there exists a constant* $\tilde{\beta} > 0$ *such that*

$$\sup_{q \in \tilde{X}} \frac{b(q; \mu)}{\|q\|_{\tilde{X}}} \geqslant \tilde{\beta} \|\mu\|_{\tilde{M}} \qquad \forall \mu \in \tilde{M} \tag{9.32}$$

and that the first argument p of the solution (p, λ) *of problem* (9.2), (9.3) *belongs to the space* $\tilde{X}$. *Then* (p, λ) *is the unique solution of problem* (9.30), (9.31).

Proof: Let (p, λ) be the solution of (9.2), (9.3). If p belongs to the space $\tilde{X}$, it follows from the property (9.29) that (p, λ) is also a solution of (9.30), (9.31). The uniqueness of this solution is a consequence of the inequalities (9.5) and (9.32). ■

In order to illustrate Theorem 9.3, we go back to the mixed-hybrid formulation (9.17), (9.18) of the model problem (1.1). We introduce the spaces

$$\tilde{Q} = \{\mathbf{q} \in (L^2(\Omega))^2; \mathbf{q}_K \in H(\text{div}; K), \mathbf{q}_K \cdot \mathbf{n}_K \in L^2(\partial K), K \in \mathcal{T}_h\} \tag{9.33}$$

normed by

$$\|\mathbf{q}\|_{\tilde{Q}} = \left\{ \sum_{K \in \mathcal{T}_h} (\|\mathbf{q}_K\|^2_{H(\text{div}; K)} + h_K \|\mathbf{q}_K \cdot \mathbf{n}_K\|^2_{0, \partial K}) \right\}^{1/2} \tag{9.34}$$

and

$$\tilde{\Psi} = \left\{ \psi \in \prod_{K \in \mathcal{T}_h} L^2(\partial K); \psi_{\partial K_1} = \psi_{\partial K_2} \text{ on } \partial K_1 \cap \partial K_2 \text{ and } \psi_{\partial K} = 0 \text{ on } \partial K \cap \Gamma \right\} \tag{9.35}$$

with the (strange but natural) norm

$$\|\psi\|_{\tilde{\Psi}} = \left\{ \sum_{K \in \mathcal{T}_h} h_K^{-1} \left(\inf_{c \in \mathbb{R}} \|\psi_{\partial K} + c\|_{0,\partial K}^2 \right) \right\}^{1/2} \tag{9.36}$$

where h_K is the diameter of K.

Clearly, the bilinear form $b(\cdot, \cdot)$ given in (9.16) is continuous on $\tilde{Q} \times (L^2(\Omega) \times \tilde{\Psi})$ and Theorem 9.3 applies with $\tilde{X} = \tilde{Q}$ and $\tilde{M} = L^2(\Omega) \times \tilde{\Psi}$. Therefore we get

Theorem 9.4 *Assume that the function f belongs to* $L^2(\Omega)$ *such that the solution u of* (9.1) *satisfies* $\mathbf{grad}\, u \in \tilde{Q}$. *Then the triple* $(\mathbf{p}, u, \phi)$ *given by* (9.19) *is the unique solution of the equations*

$$a(\mathbf{p}, \mathbf{q}) + b(\mathbf{q}; u, \phi) = 0 \qquad \forall \mathbf{q} \in \tilde{Q} \tag{9.37}$$

$$b(\mathbf{p}; v, \psi) = -\int_\Omega fv \,\mathrm{d}x \qquad \forall (v, \psi) \in L^2(\Omega) \times \tilde{\Psi} \tag{9.38}$$

In many cases, problem (9.30), (9.31) is easier to approximate numerically than problem (9.2), (9.3) itself. Thus, let us give a general method of approximation of problem (9.30), (9.31). We introduce two finite-dimensional spaces X_h and M_h such that

$$X_h \subset \tilde{X}, \qquad M_h \subset \tilde{M} \tag{9.39}$$

We consider the approximate problem: *Find a pair* $(p_h, \lambda_h) \in X_h \times M_h$ *such that*

$$a(p_h, q_h) + b(q_h; \lambda_h) = \langle l, q_h \rangle \qquad \forall q_h \in X_h \tag{9.40}$$

$$b(p_h; \mu_h) = \langle \chi, \mu_h \rangle \qquad \forall \mu_h \in M_h \tag{9.41}$$

We set:

$$\left. \begin{aligned} &V_h(\chi) = \{q_h \in X_h;\, b(q_h; \mu_h) = \langle \chi, \mu_h \rangle \,\forall \mu_h \in M_h\} \\ &V_h = V_h(0) \end{aligned} \right\} \tag{9.42}$$

Note that in general V_h is not contained in V.

Concerning the existence and uniqueness of the solution (9.40), (9.41), one can trivially prove the following result.

Theorem 9.5 *The properties*

$$\{q_h \in V_h;\, a(q_h, q_h) = 0\} = \{0\} \tag{9.43}$$

$$\{\mu_h \in M_h;\, b(q_h; \mu_h) = 0 \,\forall q_h \in X_h\} = \{0\} \tag{9.44}$$

are a necessary and sufficient condition for problem (9.40), (9.41) *to have a unique solution.*

In most applications, the hypothesis (9.43) is very easily checked. On the other hand, (9.44) expresses a compatibility condition between the spaces X_h and M_h: in fact, the practical construction of finite-dimensional spaces X_h and M_h such that (9.45) holds is never obvious.

In order to derive abstract bounds for the errors $p-p_h$ and $\lambda-\lambda_h$, we need a stronger form of the hypotheses (9.43) and (9.44). We have (cf. References 9 and 18)

Theorem 9.6 *Assume that there exists a constant* $\alpha_\star > 0$

$$a(q_h, q_h) \geqslant \alpha_\star \|q_h\|_X^2 \qquad \forall q_h \in V_h \tag{9.45}$$

we have for some constant $C_1 = C_1(\alpha_\star) > 0$

$$\|p-p_h\|_X \leqslant C_1 \left\{ \inf_{q_h \in V_h(\chi)} \|p-q_h\|_X + \inf_{\mu_h \in M_h} \sup_{q_h \in V_h} \frac{b(q_h; \lambda-\mu_h)}{\|q_h\|_X} \right\} \tag{9.46}$$

If in addition, there exists a constant $\beta_\star > 0$ *such that*

$$\sup_{q_h \in X_h} \frac{b(q_h; \mu_h)}{\|q_h\|_{\tilde{X}}} \geqslant \beta_\star \|\mu_h\|_{\tilde{M}} \qquad \forall \mu_h \in M_h \tag{9.47}$$

we get for some constant $C_2 = C_2(\beta_\star) > 0$

$$\|\lambda-\lambda_h\|_{\tilde{M}} \leqslant C_2 \left\{ \|p-p_h\|_X + \inf_{\mu_h \in M_h} \|\lambda-\mu_h\|_{\tilde{M}} \right\} \tag{9.48}$$

As a consequence of the inequality (9.46) we obtain

$$\|p-p_h\|_X \leqslant C_1 \inf_{q_h \in V_h(\chi)} \|p-q_h\|_X \quad \text{if} \quad V_h \subset V \tag{9.46a}$$

and

$$\|p-p_h\|_X \leqslant C_1 \left\{ \inf_{q_h \in V_h(\chi)} \|p-q_h\|_X + \inf_{\mu_h \in M_h} \|\lambda-\mu_h\|_M \right\} \tag{9.46b}$$

9.4 A MIXED-HYBRID GENERALIZATION OF FRAEIJS DE VEUBEKE'S EQUILIBRIUM METHOD

Let us apply the abstract results of Theorems 9.5 and 9.6 to the approximation of the solution u of (9.1) using the mixed-hybrid variational formulation (9.37), (9.38). We construct finite-dimensional subspaces Q_h, L_h and Ψ_h with

$$Q_h \subset \tilde{Q}, \qquad L_h \subset L^2(\Omega), \qquad \Psi_h \subset \tilde{\Psi} \tag{9.49}$$

and we consider the discretized problem: *Find a triple* $(p_h, u_h, \phi_h) \in$

$Q_h \times L_h \times \Psi_h$ *such that*

$$a(\mathbf{p}_h, \mathbf{q}_h) + b(\mathbf{q}_h; u_h, \phi_h) = 0 \qquad \forall \mathbf{q}_h \in Q_h \tag{9.50}$$

$$b(\mathbf{p}_h; v_h, \psi_h) = -\int_\Omega f v_h \,\mathrm{d}x \qquad \forall (v_h, \psi_h) \in L_h \times \Psi_h \tag{9.51}$$

Just to be specific, we restrict ourselves to the case of a polygonal plane domain $\bar{\Omega}$ and of a triangulation $\mathcal{T}_h$ of $\bar{\Omega}$ made up with triangular elements K whose diameters are $\leqslant h$. Given an integer $l \geqslant 0$ and a triangle $K \in \mathcal{T}_h$, we define:

- $P_l(K)$ = space of the restrictions to K of all polynomials of degree $\leqslant l$ in the the two variables x_1, x_2; (9.52)
- $Q_l(K)$ = subspace of $(P_{l+1}(K))^2$ of functions $\mathbf{q} = (q_1, q_2)$ of the form

$$\begin{aligned} q_1 &= \tilde{q}_1 + \sum_{i=0}^{l} \alpha_i x_1^{l+1-i} x_2^i \\ q_2 &= \tilde{q}_2 + \sum_{i=0}^{l} \alpha_i x_1^{l-i} x_2^{i+1} \end{aligned} \tag{9.53}$$

where $\tilde{q}_i \in P_l(K), \qquad i = 1, 2;$

- $S_l(\partial K)$ = space of all functions defined on ∂K whose restrictions to any side of K are polynomials of degree $\leqslant l$; (9.54)
- $T_l(\partial K)$ = subspace of $S_l(\partial K)$ of functions which are continuous on ∂K. (9.55)

Clearly $T_l(\partial K)$ is the space of the restrictions to ∂K of all polynomials of $P_l(K)$. On the other hand, we have div $Q_l(K) = P_l(K)$ and $S_l(\partial K)$ is exactly the space of functions of the form $\mathbf{q} \cdot \mathbf{n}_K$, $q \in Q_l(K)$.

For any triangle $K \in \mathcal{T}_h$, we introduce a subspace Q_K of the space $\{\mathbf{q} \in H(\mathrm{div}; K); \mathbf{q} \cdot \mathbf{n}_K \in L^2(\partial K)\}$ and we define

$$Q_h = \{\mathbf{q} \in \tilde{Q}; \mathbf{q}_K \in Q_K, K \in \mathcal{T}_h\} \tag{9.56}$$

$$L_h = \{v \in L^2(\Omega); v_K \in P_l(K), K \in \mathcal{T}_h\} \tag{9.57}$$

$$\Psi_h = \{\psi \in \tilde{\Psi}; \psi_{\partial K} \in S_l(\partial K), K \in \mathcal{T}_h\} \tag{9.58}$$

Theorem 9.7 Let there be given spaces Q_h, L_h *and* Ψ_h *defined as in* (9.56)–(9.58). *Assume that the following inclusion*

$$Q_l(K) \subset Q_K, \; K \in \mathcal{T}_h \tag{9.59}$$

holds. Then problem (9.50), (9.51) *has a unique solution.*

Proof: We apply Theorem 9.5 with $X_h = Q_h$ and $N_h = L_h \times \Psi_h$. Since $a(\mathbf{q}, \mathbf{q}) = \|\mathbf{q}\|_{0,\Omega}^2$, condition (9.43) clearly holds. On the other hand, let $(v, \psi) \in L_h \times \Psi_h$

be such that $b(\mathbf{q}; v, \psi) = 0$ for all $\mathbf{q} \in X_h$. Using (9.59), we have for all $\mathbf{q} \in Q_l(K)$

$$\int_K v_K \operatorname{div} \mathbf{q} \, dx - \int_{\partial K} \mathbf{q} \cdot \mathbf{n}_K \psi_{\partial K} \, ds = 0$$

First, we may choose $\mathbf{q} \in Q_l(K)$ so that $\operatorname{div} \mathbf{q} = 0$ and $\mathbf{q} \cdot \mathbf{n}_K$ is an arbitrary function of $S_l(\partial K)$ with $\int_{\partial K} \mathbf{q} \cdot \mathbf{n}_K \, ds = 0$. Hence, there exists for all $K \in \mathcal{T}_h$ a constant c_K such that

$$\psi_{\partial K} = c_K \text{ on } \partial K$$

and therefore

$$\int_K v_K \operatorname{div} \mathbf{q} \, dx - \int_{\partial K} \mathbf{q} \cdot \mathbf{n}_K \Psi_{\partial K} \, ds = \int_K (v_K - c_K) \operatorname{div} \mathbf{q} \, dx = 0 \qquad \forall \mathbf{q} \in Q_l(K)$$

Next, we may choose $\mathbf{q} \in Q_l(K)$ such that $\operatorname{div} \mathbf{q} = v_K - c_K$ in K. We obtain $v_K = c_K$ in K. Now, since ψ belongs to the space $\tilde{\Psi}$, we have $c_K = 0$ for all $K \in \mathcal{T}_h$. Thus, we have $v = 0$, $\Psi = 0$ so that (9.45) holds and the desired result follows from Theorem 9.5. ■

Consider the very important case where $Q_K = Q_l(K)$. Then the first argument $\mathbf{p}_h$ of the solution of problem (9.50), (9.51) may be characterized as the solution of an equilibrium method. In fact, defining

$$V_h^f = \left\{\mathbf{q}_h \in Q_h ; b(\mathbf{q}_h ; v_h, \psi_h) = -\int_\Omega f v_h \, dx \; \forall (v_h, \psi_h) \in L_h \times \Psi_h\right\} \tag{9.60}$$

we have

$$V_h^f = \{\mathbf{q}_h \in Q_h \cap H(\operatorname{div}; \Omega); \operatorname{div} \mathbf{q}_h + f_h = 0 \text{ in } \Omega\} \tag{9.61}$$

where f_h is the $L^2(\Omega)$-projection of f upon L_h. Hence $V_h \subset V$ and $\mathbf{p}_h$ is the unique function of V_h^f such that

$$a(\mathbf{p}_h, \mathbf{q}_h) = 0 \qquad \forall \mathbf{q}_h \in V_h \tag{9.62}$$

or equivalently which minimizes the quadratic functional $I(\mathbf{q}) = \frac{1}{2}\|\mathbf{q}\|_{0,\Omega}^2$ over V_h^f. This equilibrium finite method has been introduced by Fraeijs de Veubeke (cf. References 3, 4, and 19) and mathematically analysed in References 18 and 20.

Similarly, if we eliminate φ_h in (9.50), (9.51) we find that the pair $(\mathbf{p}_h, u_h) \in (Q_h \cap H(\operatorname{div}; \Omega)) \times L_h$ is the unique solution of the equations

$$\left.\begin{aligned} & a(\mathbf{p}_h, \mathbf{q}_h) + \int_\Omega u_h \operatorname{div} \mathbf{q}_h \, dx = 0 && \forall \mathbf{q}_h \in Q_h \cap H(\operatorname{div}; \Omega) \\ & \int_\Omega v_h (\operatorname{div} \mathbf{p}_h + f) \, dx = 0 && \forall v_h \in L_h \end{aligned}\right\} \tag{9.63}$$

We obtain a mixed finite element method introduced and analysed in References 18 and 21.

We now turn to error estimates. We consider a *regular family* $(\mathcal{T}_h)$ of triangulations of $\bar{\Omega}$ in that there exists a constant $\sigma > 0$ independent of h such that

$$h_K \leqslant \sigma \rho_K, \; K \in \mathcal{T}_h \tag{9.64}$$

where h_K is the diameter of K and ρ_K is the diameter of the inscribed circle in K.

Concerning the approximation properties of the space Q_h, we have (cf. References 18 and 21).

Lemma *Assume that the inclusion* (9.59) *holds and that* $(\mathcal{T}_h)$ *is a regular family of triangulations of* $\bar{\Omega}$. *Then there exists a linear operator* $\Pi_h \in \mathcal{L}((H^1(\Omega))^2; \, Q_h \cap H(\text{div}; \Omega))$ *such that*:

(i) *for all* $\mathbf{p} \in (H^1(\Omega))^2$, $\text{div}\, \Pi_h \mathbf{p}$ *is the* $L^2(\Omega)$-*projection of* $\text{div}\, \mathbf{p}$ *upon* L_h;

(ii) *for all* $\mathbf{p} \in (H^{l+1}(\Omega))^2$ *with* $\text{div}\, \mathbf{p} \in H^{l+1}(\Omega)$, *we have for some constant* $C > 0$ *independent of* h

$$\|\mathbf{p} - \Pi_h \mathbf{p}\|_{H(\text{div};\Omega)} \leqslant C h^{l+1} (|\mathbf{p}|_{l+1,\Omega} + |\text{div}\, \mathbf{p}|_{l+1,\Omega}) \tag{9.65}$$

We are now able to state

Theorem 9.8 *Let there be given spaces* Q_h, I_h *and* Ψ_h *defined as in* (9.56)–(9.58) *which are associated with a regular family* $(\mathcal{T}_h)$ *of triangulations of* $\bar{\Omega}$. *We assume that*

$$Q_K = Q_l(K), \qquad K \in \mathcal{T}_h \tag{9.66}$$

and that the solution u *of* (9.1) *belongs to* $H^{l+2}(\Omega)$ *with* $\Delta u \in H^{l+1}(\Omega)$. *Then we have for some constant* $C > 0$ *independent of* h *the following error bounds:*

$$\|\mathbf{p} - \mathbf{p}_h\|_{H(\text{div};\Omega)} + \|u - u_h\|_{0,\Omega} \leqslant C h^{l+1} (|u|_{l+1,\Omega} + |u|_{l+2,\Omega} + |\Delta u|_{l+1,\Omega}) \tag{9.67}$$

$$\|\phi - \phi_h\|_{\Psi} \leqslant C h^l (|u|_{l+1,\Omega} + |\Delta u|_{l,\Omega}) \tag{9.68}$$

Proof: We begin by estimating $\|\mathbf{p} - \mathbf{p}_h\|_{H(\text{div};\Omega)}$. We apply the first part of Theorem 9.6 with $X_h = Q_h$, $M_h = L_h \times \Psi_h$ and $V_h(\chi) = V_h^f$. Since in this case $V_h \subset V$, we may use the inequality (9.46a) and we get ($\alpha_\star = 1$)

$$\|\mathbf{p} - \mathbf{p}_h\|_{H(\text{div};\Omega)} \leqslant C_1 \inf_{\mathbf{q}_h \in V_h^f} \|\mathbf{p} - \mathbf{q}_h\|_{H(\text{div};\Omega)}$$

Now, if $\mathbf{p} = \mathbf{grad}\, u \in (H^{l+1}(\Omega))^2$, we have by the property (i) of the previous lemma

$$\text{div}\, \Pi_h \mathbf{p} + f_h = 0$$

so that $\Pi_h \mathbf{p} \in V_h^f$. Therefore it follows from (9.65) that

$$\|\mathbf{p}-\mathbf{p}_h\|_{H(\text{div};\Omega)} \leqslant C_1\|\mathbf{p}-\Pi_h\mathbf{p}\|_{H(\text{div};\Omega)} \leqslant Ch^{l+1}(|\mathbf{p}|_{l+1,\Omega}+|\text{div}\,\mathbf{p}|_{l+1,\Omega}) \tag{9.69}$$

In order to derive a bound for $\|u-u_h\|_{0,\Omega}$, we observe that the pair $(\mathbf{p}_h, u_h)$ is the unique solution of the equations (9.63). Thus we apply the second part of Theorem 9.6 with

$$X = H(\text{div};\, \Omega),\, M = \tilde{M} = L^2(\Omega)$$

$$a(\mathbf{p}, \mathbf{q}) = \int_\Omega \mathbf{p}\cdot\mathbf{q}\, \mathrm{d}x,\, b(\mathbf{q};\, v) = \int_\Omega v\, \text{div}\,\mathbf{q}\, \mathrm{d}x$$

$$l = 0,\, \langle \chi, v\rangle = -\int_\Omega fv\, \mathrm{d}x$$

and

$$X_h = Q_h \cap H(\text{div};\, \Omega),\, M_h = L_h$$

Since, by Reference 18 and 21, there exists a constant $\beta_\star > 0$ independent of h such that

$$\sup_{\mathbf{q}_h\in Q_h\cap H(\text{div};\Omega)} \frac{\int_\Omega v_h\, \text{div}\,\mathbf{q}_h\, \mathrm{d}x}{\|\mathbf{q}_h\|_{H(\text{div};\Omega)}} \geqslant \beta_\star \|v_h\|_{0,\Omega}$$

it follows from the inequality (9.48) that

$$\|u-u_h\|_{0,\Omega} \leqslant C_2\left\{\|\mathbf{p}-\mathbf{p}_h\|_{H(\text{div};\Omega)} + \inf_{v_h\in L_h} \|u-v_h\|_{0,\Omega}\right\} \tag{9.70}$$

Using a standard result in finite element approximation theory (see, for example, References 1 and 2) we have if $u \in H^{l+1}(\Omega)$

$$\inf_{v_h\in L_h} \|u-v_h\|_{0,\Omega} \leqslant Ch^{l+1}|u|_{l+1,\Omega} \tag{9.71}$$

The inequality (9.67) follows from (9.69), (9.70) and (9.71).

Using a similar method of proof, one can derive the error estimate (9.69). For a detailed analysis, see Reference 18. ■

9.5 A MIXED-HYBRID FINITE ELEMENT METHOD OF PIAN TYPE

In this section, we use directly the mixed-hybrid variational formulation (9.17), (9.18) in order to construct a method of approximation of the model problem (9.1). We consider again the finite-dimensional space $Q_h \subset Q$ and

$L_h \subset L^2(\Omega)$ introduced in Section 9.4 and we define the following subspace Ψ_h of Ψ by

$$\Psi_h = \{\psi \in \Psi;\ \psi_{\partial K} \in T_{l+1}(\partial K),\ K \in \mathcal{T}_h\} \tag{9.72}$$

Theorem 9.9 Let there be given spaces Q_h, L_h, *and* Ψ_h *defined as in* (9.56), (9.57) *and* (9.72). *Assume that the inclusion* (9.59) *holds. Then the corresponding problem* (9.50), (9.51) *has a unique solution.*

Proof: We again apply Theorem 9.5 with $X_h = Q_h$ and $M_h = L_h \times \Psi_h$. Let $(v, \psi) \in L_h \times \Psi_h$ be such that $b(\mathbf{q}; v, \psi) = 0$ for all $\mathbf{q} \in X_h$. Given $K \in \mathcal{T}_h$, we denote by $\tilde{\psi}_{\partial K}$ the $L^2(\partial K)$-projection of $\psi_{\partial K}$ upon $S_l(\partial K)$. Exactly as in the proof of Theorem 9.7, we get for some constant c_K

$$\tilde{\psi}_{\partial K} = c_K \text{ on } \partial K$$

Next we notice that $\Psi_{\partial K}$ and $\tilde{\Psi}_{\partial K}$ coincide at the $(l+1)$ Gauss–Legendre points of each side of the triangle K. Therefore, when l is an even integer, we find

$$\psi_{\partial K} = c_K \text{ on } \partial K$$

and the end of the proof parallels that of Theorem 9.7.

On the other hand, when the integer l is added, we get

$$\psi_{\partial K} = \begin{cases} c_K & \text{at the } (l+1) \text{ Gauss–Legendre points of each side of } K \\ d_K & \text{at the vertices of } K \end{cases}$$

where d_K is another constant. Now, since $\psi_{\partial K}$ belongs to the space Ψ, we necessarily have $c_K = d_K = 0$ for all $K \in \mathcal{T}_h$. Therefore we obtain $\psi = 0$ and consequently $v = 0$. Again condition (9.44) holds and the result follows from Theorem 9.5. ■

By choosing the space Ψ_h as in (9.72), we cannot have $V_h \subset V$ and we never get an equilibrium method. On the other hand, if we eliminate u_h in (9.50), (9.51), we obtain a stress-assumed hybrid method of Pian type which has been analysed in References 18 and 22.

Concerning error estimates, we first state

Theorem 9.10 Let there be given spaces Q_h, L_h, *and* Ψ_h *defined as in* (9.56), (9.57), (9.72) *which are associated with a regular family* $(\mathcal{T}_h)$ *of triangulations of* $\bar{\Omega}$. *We assume that the inclusion* (9.59) *holds and that the solution* u *of* (9.1) *belongs to* $H^{l+2}(\Omega)$ *with* $\Delta u \in H^{l+1}(\Omega)$. *Then we have*

$$\|\mathbf{p} - \mathbf{p}_h\|_Q + \|u - u_h\|_{0,\Omega} \leqslant Ch^{l+1}(|u|_{l+1,\Omega} + |u|_{l+2,\Omega} + |\Delta u|_{l+1,\Omega}) \tag{9.73}$$

Proof: Let us estimate $\|\mathbf{p}-\mathbf{p}_h\|_Q$. Applying the inequality (9.46b) with $X_h = Q_h$ and $M_h = L_h \times \Psi_h$ gives

$$\|\mathbf{p}-\mathbf{p}_h\|_Q \leqslant c_1\left(\inf_{\mathbf{q}_h \in V_h^f} \|\mathbf{p}-\mathbf{q}_h\|_Q + \inf_{v_h \in L_h} \|u - v_h\|_{0,\Omega} + \inf_{\psi_h \in \Psi_h} \|\phi - \psi_h\|_\Psi \right)$$

where V_h^f is defined as in (9.60). Clearly $\Pi_h \mathbf{p} \in V_h^f$ so that

$$\inf_{\mathbf{q}_h \in V_h^f} \|\mathbf{p}-\mathbf{q}_h\|_Q \leqslant \|\mathbf{p}-\Pi_h \mathbf{p}\|_{H(\text{div};\Omega)}$$

Hence, using (9.65), (9.71) and the inequality

$$\inf_{\psi_h \in \Psi_h} \|\phi - \psi_h\|_\Psi \leqslant Ch^{l+1}|u|_{l+2,\Omega}$$

which follows from a classical result in finite element approximation theory, we get

$$\|\mathbf{p}-\mathbf{p}_h\|_Q \leqslant Ch^{l+1}(|u|_{l+1,\Omega} + |u|_{l+2,\Omega} + |\Delta u|_{l+1,\Omega})$$

The bound for $\|u-u_h\|_{0,\Omega}$ is obtained as in the proof of Theorem 9.8. ■

Finally, in order to obtain an optimal error estimate for $\|\phi - \phi_h\|_\Psi$, we need to choose the space Q_K in a precise way. Given a triangle $K \in \mathcal{T}_h$, we denote by $\chi_i = \chi_i(x)$, $1 \leqslant i \leqslant 3$, the barycentric coordinates of a point x with respect to the vertices of K and we introduce the function

$$\omega_K = (\chi_1 - \chi_2)(\chi_2 - \chi_3)(\chi_3 - \chi_1)((\chi_1\chi_2)^{(l-1)/2} + (\chi_2\chi_3)^{(l-1)/2} + (\chi_3\chi_1)^{(l-1)/2}) \tag{9.74}$$

Then we have the following result whose proof can be found in References 18 and 22.

Theorem 9.11 *Assume the hypotheses of Theorem* 9.10. *Assume in addition that*

$$Q_K \supset \begin{cases} Q_l(K) \text{ when } l \text{ is even} \\ Q_l(K) \oplus \textbf{curl}\ \omega_K \text{ when } l \text{ is odd} \end{cases} \tag{9.75}$$

Then we have

$$\|\phi - \phi_h\|_\Psi \leqslant Ch^{l+1}(|u|_{l+1,\Omega} + |u|_{l+2,\Omega} + |\Delta u|_{l+1,\Omega}) \tag{9.76}$$

REFERENCES

1. Strang, G. and Fix, G. J., *An Analysis of the Finite Element Method*, Prentice-Hall, Englewood Cliffs, 1973.

2. Ciarlet, P. G., *The Finite Element Method for Elliptic Problems*, North-Holland, Amsterdam, 1977.
3. Fraeijs de Veubeke, B., 'Displacement and equilibrium models in the finite element method', in *Stress Analysis*, (Zienkiewicz and Holister, Eds.), Wiley, London, 1965.
4. Fraeijs de Veubeke, B., 'Diffusive Equilibrium Models', University of Calgary, Lecture Notes, 1973.
5. Oden, J. T., and Lee, J. K., 'Dual-mixed hybrid finite element method for second-order elliptic problems', in *Mathematical Aspects of Finite Element Methods, Rome* 1975, (Galligani and Magenes, Eds.); *Lecture Notes in Mathematics*, **606,** 275–291 (1977).
6. Babuška, I., Oden, J. T., and Lee, J. K., 'Mixed-hybrid finite element approximations of second order elliptic boundary-value problems', *Comput. Methods Appl. Mech. Engrg.*, **11,** 175–206 (1977).
7. Raviart, P. A., and Thomas, J. M., 'Primal hybrid finite element methods for 2nd order elliptic equations', *Math. of Comput.*, **31,** 138, 391–413 (1977).
8. Quarteroni, A., 'Primal hybrid finite element methods for 4th order elliptic equations' (to appear).
9. Brezzi, F., 'On the existence, uniqueness and approximation of saddle-point problems arising from Lagrangian multipliers', *R.A.I.R.O., Analyse Numérique*, **R2,** 129–151 (1974).
10. Brezzi, F., and Marini, L. D., 'On the numerical solution of plate bending problems by hybrid methods', *R.A.I.R.O., Analyse Numérique*, **R3,** 5–50 (1975).
11. Brezzi, F., and Raviart, P. A., 'Mixed finite element methods for 4th order elliptic equations', in *Topics in Numerical Analysis*, Vol. III (J. J. Miller, Ed.), Academic Press, New York, 1977, pp. 315–338.
12. Ciarlet, P. G., and Raviart, P. A., 'A mixed finite element method for the biharmonic equation', in *Mathematical Aspects of Finite Elements in Partial Differential Equations* (C. de Boor, Ed.), Academic Press, New York, 1974.
13. Johnson, C., 'On the convergence of a mixed finite element method for plate bending problems', *Numer. Math.*, **21,** 43–62 (1973).
14. Miyoshi, T., 'A finite element method for the solution of fourth order partial differential equations', *Kuamoto J., Math.* **9,** 87–116 (1973).
15. Raviart, P. A., 'Mixed and equilibrium methods for 4th order problems', from Japan Seminars on Functional and Numerical Analysis, Tokyo and Kyoto, Sept. 20–28 1976.
16. Oden, J. T., and Reddy, C. T., *Variational Methods in Theoretical Mechanics*, Springer-Verlag, Heidelberg, 1976.
17. Babuška, I., 'Error bounds for finite element method', *Numer. Math.*, **16,** 322–333 (1971).
18. Thomas, J. M., Thesis, Université P. and M. Curie, Paris, 1977.
19. Fraeijs de Veubeke, B., and Hogge, M. A., 'Dual analysis for heat conduction problems by finite elements', *Int. J. Num. Meth. Engrg.*, **5,** 65–82 (1972).
20. Thomas, J. M., 'Méthodes des éléments finis équilibre', in Journées des Eléments Finis, Mai 1975, Université de Rennes et INSA de Rennes.
21. Raviart, P. A., and Thomas, J. M., 'A mixed finite element method for 2nd order elliptic problems', in *Mathematic Aspects of Finite Element Methods, Rome* 1975, (Galligani and Magenes, Eds.); *Lecture Notes in Mathematics*, **606,** 292–315 (1977).

22. Thomas, J. M., 'Méthodes des éléments finis hybrides duaux pour les problèmes elliptiques du second ordre', *R.A.I.R.O., Analyse Numérique,* **10,** 12, 51–79 (1976).
23. Brezzi F., Chapter 10 of this book.

Chapter 10

Non-standard Finite Elements for Fourth Order Elliptic Problems

F. Brezzi

10.1 INTRODUCTION

The aim of this chapter is to present a brief survey of the different finite element methods for solving boundary value problems for fourth order linear elliptic operators, with particular emphasis on the so-called 'non-standard' methods. We shall take as a model problem the homogeneous Dirichlet problem for the biharmonic operator. However, extension of the theory to more general problems will be immediate. The terminology to be used is taken (mainly for historical reasons) from the 'model problem' in relation with the linear bending of plates; nevertheless, on the one hand the theory itself can be regarded as purely mathematical and, on the other hand, applications to completely different physical problems are also of great interest, as we shall briefly see later on.

From an historical point of view, non-standard finite element methods were introduced first by engineers for solving linear and nonlinear elasticity problems.[1–20] A few years later appeared the first attempts at synthesis of the different methods in a single framework and, besides, the mathematical analysis of some of them was carried out.[21–57] At the present moment this kind of work cannot yet be considered completed. However, in the author's opinion, the essential tools to be used in the analysis have been pointed out, at least for the linear cases; the nonlinear cases, as usual, are much more complicated, since each problem has almost to be treated individually and the lack of a general theory is, in some sense, 'physiological'. However, many interesting applications to specific problems have already been studied[58–72] and the field seems to be very promising.

The content of the chapter is as follows: in Section 10.2 the different methods will be presented, in a very schematic way, in their essential aspects. Little mathematics and no functional analysis will be used at this level. In Section 10.3 we shall give the main abstract theorems to be used, in order to carry out a correct analysis of the methods from the mathematical point of view. In Section 10.4 the abstract theorems will be applied to the different methods in order to derive sufficient conditions of convergence and proper error bounds.

We shall always suppose for simplicity that all finite elements are triangular and that the family of triangulations to be used is *quasi-uniform* in the sense that there exist two positive constants k_1, k_2 such that for each triangulation and for each triangle T in the triangulation we have

$$k_1 r(T) \leq d(T) \leq k_2 r(T)$$

with $d(T) =$ *diameter* of T and $r(T) =$ *radius* of the largest circle inscribed in T.

10.2 PRESENTATION OF THE VARIOUS MODELS

10.2.1 The model problem

Let Ω be a convex polygon in the x_1, x_2 plane and let p be a smooth given function in Ω. The model problem to be solved is:

$$\left.\begin{array}{l} \text{find } \psi \text{ such that} \\ \Delta^2 \psi = p \quad \text{in} \quad \Omega \\ \psi = \dfrac{\partial \psi}{\partial n} = 0 \quad \text{on} \quad \partial\Omega \end{array}\right\} \tag{10.1}$$

We shall assume that the solution ψ is smooth, although this is not always the case for a polygonal domain.

10.2.2 Notation and conventions

—The convention of summation of repeated indices (from 1 to 2) is assumed.
—The letters ψ, ϕ, χ will always denote functions: the letters u, v will always denote 2×2 symmetric tensors; the letters n and t will denote unit vectors, in general the (outward) normal and tangential directions to a given (boundary) line.

$$\phi_{/i} = \frac{\partial \phi}{\partial x_i} \quad (i = 1, 2); \qquad \phi_{/n} = \frac{\partial \phi}{\partial n}; \qquad \phi_{/t} = \frac{\partial \phi}{\partial t}$$

$$M_n(v) = v_{ij} n_i n_j; \qquad M_n(\phi) = \phi_{/ij} n_i n_j$$

$$M_{nt}(v) = v_{ij} n_i t_j; \qquad M_{nt}(\phi) = \phi_{/ij} n_i t_j$$

$$Q_n(v) = v_{ij/i} n_j; \qquad Q_n(\phi) = \frac{\partial \Delta \phi}{\partial n}$$

$$K_n(v) = \frac{\partial M_{nt}(v)}{\partial t} + Q_n(v); \qquad K_n(\phi) = \frac{\partial M_{nt}(\phi)}{\partial t} + Q_n(\phi)\dagger$$

† Even for smooth v or ϕ, K_n can include Dirac measures, on polygonal boundaries, owing to the jumps of the normal and tangential directions.

10.2.3 *A local stationary principle*

Let T be a triangle, let the boundary ∂T be composed of three connected parts:

$$\partial T = \partial T_c \cup \partial T_s \cup \partial T_f$$

and consider the problem:

$$\left.\begin{array}{l} \text{find } \phi \text{ such that:} \\ \Delta^2\phi = p \quad \text{in} \quad T \\ \phi = \bar{\phi}, \qquad \phi_{/n} = \bar{\phi}_n \quad \text{on} \quad \partial T_c \\ \phi = \bar{\phi}, \qquad M_n(\phi) = \bar{M}_n \quad \text{on} \quad \partial T_s \\ M_n(\phi) = \bar{M}_n, \quad K_n(\phi) = \bar{K}_n \quad \text{on} \quad \partial T_f \end{array}\right\} \tag{10.2}$$

where p, $\bar{\phi}$, $\bar{\phi}_n$, $\bar{M}_n$, $\bar{K}_n$ are prescribed quantities.

It is easy to see that the following functional

$$\begin{aligned} \Pi_T = & -\frac{1}{2}\int_T v_{ij}v_{ij}\,\mathrm{d}x + \int_T v_{ij}\chi_{/ij}\,\mathrm{d}x - \int_T p\chi\,\mathrm{d}x \\ & + \int_{\partial T_c} [(\chi - \bar{\phi})K_n(v) - (\chi_{/n} - \bar{\phi}_n)M_n(v)]\,\mathrm{d}l \\ & + \int_{\partial T_s} [(\chi - \bar{\phi})K_n(v) - \chi_{/n}\bar{M}_n]\,\mathrm{d}l \\ & + \int_{\partial T_f} [\chi\bar{K}_n - \chi_{/n}\bar{M}_n]\,\mathrm{d}l \; (= \Pi_T(v, \chi, \bar{\phi}, \bar{\phi}_n, \bar{M}_n, \bar{K}_n)) \end{aligned} \tag{10.3}$$

is *stationary* at the point $v_{ij} = \phi_{/ij}$, $\chi = \phi$ (ϕ being the unique solution of (10.2)) for all variations of v and χ. In the applications to plate bending problems this turns out to be the *Hellinger–Reissner principle.*

For a given decomposition of Ω into triangles T, problem (10.1) can be split into problems of type (10.2), provided that similar boundary conditions with compatible data are prescribed along the common boundary of adjacent elements. Of course the values of $\bar{\phi}$, $\bar{\phi}_n$, $\bar{M}_n$, $\bar{K}_n$ at the interelement boundaries should now be regarded as unknowns.

10.2.4 *A composite stationary principle*

For a given decomposition of Ω and a given choice between ∂T_c, ∂T_s, and ∂T_f for each interelement boundary, the previous remark leads us to consider a more general principle.

The functional

$$\Pi(v, \chi, \bar{\phi}, \bar{\phi}_n, \bar{M}_n, \bar{K}_n) = \sum_T \Pi_T(v, \chi, \bar{\phi}, \bar{\phi}_n, \bar{M}_n, \bar{K}_n) \tag{10.4}$$

is stationary at the point:

$$\begin{aligned} v_{ij} &= \psi_{/ij}; \qquad \chi = \psi; \qquad \bar{\phi} = \psi; \\ \bar{\phi}_n &= \psi_{/n}; \qquad \bar{M}_n = M_n(\psi); \qquad \bar{K}_n = K_n(\psi) \end{aligned} \tag{10.5}$$

(ψ being the solution of (10.1)) with respect to all the variations of v, χ, $\bar{\phi}$, $\bar{\phi}_n$, $\bar{M}_n$, $\bar{K}_n$ such that

—v is smooth inside each element T
—χ is smooth inside each element T
—$\bar{\phi}$ is an 'interelement' smooth function on $\bigcup_T (\partial T_c \cup \partial T_s)$
—$\bar{\phi}_n$ is an 'interelement' smooth function on $\bigcup_T (\partial T_c)$
—$\bar{M}_n$ is an 'interelement' smooth function on $\bigcup_T (\partial T_s \cup \partial T_f)$
—$\bar{K}_n$ is an 'interelement' smooth function on $\bigcup_T (\partial T_f)$ plus possible Dirac measures at the corners of each T

Moreover we shall require that the variations of $\bar{\phi}$ and $\bar{\phi}_n$ verify C^1-compatibility conditions at the vertices, and the prescribed boundary conditions ($\bar{\phi} = \bar{\phi}_n = 0$) on $\partial\Omega$. Moreover, the sum over T of the Dirac measures of $\bar{K}_n$ at a given internal vertex must be equal to zero.

Remark It is known that in problem (10.2) one cannot assume $\partial T = \partial T_f$ since, in that case, the problem itself will not be well posed. As a conséquence, in the choice of the type of the interelement boundaries, the following rule must be observed: the choice $\partial T \equiv \partial T_f$ for an interior element T will not be allowed unless the continuity of χ at the three corners is *a priori* required.

10.2.5 *Derivation of the different models from the composite stationary principle*

10.2.5.1 *Pure displacement model*

We assume that each interelement boundary is of type ∂T_f and we restrict the variations of χ to $C^1(\bar{\Omega})$ with $\chi = \chi_{/n} = 0$ on $\partial\Omega$. It is easy to see that Π

reduces to

$$\Pi(v, \chi) = \int_\Omega \left(-\tfrac{1}{2}v_{ij}v_{ij} + v_{ij}\chi_{/ij} - p\chi\right) dx = \text{stationary}$$

and variations over v give $v_{ij} = \chi_{/ij}$ at the stationary point. Therefore we obtain

$$\left.\begin{array}{c}\Pi(\chi) = \dfrac{1}{2}\displaystyle\int_\Omega \chi_{/ij}\chi_{/ij}\, dx - \int_\Omega p\chi\, dx = \text{minimum} \\ \text{for } \chi \text{ smooth in } \Omega \text{ and } \chi = \chi_{/n} = 0 \text{ on } \partial\Omega\end{array}\right. \qquad (10.6)$$

that is the *minimum potential energy principle.*

10.2.5.2 *Assumed displacements (primal) hybrid model*

We assume that each interelement boundary is of type ∂T_f and we restrict the variations of χ to be continuous at the vertices and zero on the vertices of $\partial\Omega$ (therefore we can forget the possible Dirac measures at the corners, for they will not appear in the final expression of the functional). It is easily seen that Π reduces to

$$\Pi(v, \chi, \bar{M}_n, \bar{K}_n) = \sum_T \left\{ \int_T \left(-\tfrac{1}{2}v_{ij}v_{ij} + v_{ij}\chi_{/ij} - p\chi\right) dx + \int_{\partial T} (\chi\bar{K}_n - \chi_{/n}\bar{M}_n)\, dl \right\}$$
$$= \text{stationary}$$

Variations over v give again $v_{ij} = \chi_{/ij}$ at the stationary point so that the final principle can be written:

$$\left.\begin{array}{l}\Pi(\chi, \bar{M}_n, \bar{K}_n) = \displaystyle\sum_T \left\{ \frac{1}{2}\int_T \chi_{/ij}\chi_{/ij}\, dx - \int_T \chi p\, dx + \int_{\partial T} (\chi\bar{K}_n - \chi_{/n}\bar{M}_n)\, dl \right\} = \\ \quad\text{stationary for } \chi \text{ smooth in each } T\text{, continuous at the vertices} \\ \quad\text{and zero at the vertices on } \partial\Omega\text{, and } \bar{M}_n,\ \bar{K}_n \text{ smooth at the} \\ \quad\text{interelement boundaries}\end{array}\right\} \qquad (10.7)$$

that is some *modified minimum potential energy principle.*[14,48]

10.2.5.3 *Pure equilibrium model*

We assume that each interelement boundary is of type ∂T_c and we restrict the variations of v to be C^1-continuous in Ω and satisfy the equilibrium equation

$$v_{ij/ij} - p = 0 \qquad (10.8)$$

We have then

$$\Pi(v,\chi)=\sum_T\left\{\int_T(-\tfrac{1}{2}v_{ij}v_{ij}+v_{ij}\chi_{/ij}-p\chi)\,\mathrm{d}x+\int_{\partial T}(\chi K_n(v)-\chi_{/n}M_n(v))\,\mathrm{d}l\right\}$$
$$=\text{stationary}$$

Integrating by parts on each T we obtain

$$\left.\begin{array}{l}\Pi(v)=-\dfrac{1}{2}\displaystyle\int_\Omega v_{ij}v_{ij}\,\mathrm{d}x=\text{maximum}\\ \text{for } v \text{ smooth and satisfying } v_{ij/ij}=p \text{ in } \Omega\end{array}\right\}\qquad(10.9)$$

that is the minimum complementary energy principle.

10.2.5.4 *Assumed stresses (dual) hybrid model*

We assume that each interelement boundary is of type ∂T_c and we restrict the variations of v to satisfy the equilibrium equation (10.8) *in each T*. Integrating again by parts on each T we obtain

$$\left.\begin{array}{l}\Pi(v,\bar{\phi},\bar{\phi}_n)=\displaystyle\sum_T\left\{-\frac{1}{2}\int_T v_{ij}v_{ij}\,\mathrm{d}x+\int_{\partial T}(\bar{\phi}_nM_n(v)-\bar{\phi}K_n(v))\,\mathrm{d}l\right\}=\\ \text{stationary for } v \text{ satisfying } v_{ij/ij}=p \text{ in each } T \text{ and } \bar{\phi},\bar{\phi}_n \text{ smooth}\\ \text{interelement functions satisfying the } C^1\text{-compatibility condi-}\\ \text{tions at the vertices and } \bar{\phi}=\bar{\phi}_n=0 \text{ on } \partial\Omega\end{array}\right\}\qquad(10.10)$$

that is the *modified minimum complementary energy principle.*[30,12]

10.2.5.5 *Mixed models*

We assume that each interelement boundary is of type ∂T_s and we restrict the variations of χ and v to satisfy the conditions $\chi=\bar{\phi}$ and $M_n(v)=\bar{M}_n$ at the interelement boundaries (this will imply that $\chi\in C^0(\Omega)$ and that $M_n(v)$ is continuous across the interelement boundaries. We obtain

$$\left.\begin{array}{l}\Pi(v,\chi)=\displaystyle\sum_T\left\{\int_T(-\tfrac{1}{2}v_{ij}v_{ij}+v_{ij}\chi_{/ij}-p\chi)\,\mathrm{d}x-\int_{\partial T}M_n(v)\chi_{/n}\,\mathrm{d}l\right\}=\\ \text{stationary for } \chi\in C^0(\Omega)\text{, smooth in each } T\text{, and zero on } \partial\Omega,\\ \text{and for } v \text{ smooth in each } T \text{ and with } M_n(v) \text{ continuous}\end{array}\right\}\qquad(10.11)$$

that is, some *modified Hellinger–Reissner principle.*

The principle (10.11) is the foundation of the so called *Hellan–Herrmann Johnson scheme*;[7,8,41] if one assumes more continuity on v, the principle is of course still valid, but the corresponding numerical scheme will change. For example, if in (10.11) we assume v to be *continuous* we obtain the *Herrmann–Miyoshi scheme.*[9,44] If, instead, the continuity requirement for $K_n(v)$ is added to (10.11) we find the *mixed-equilibrium* (or simply '*equilibrium*') *schemes of Fraeijs de Veubeke–Sanders.*[4,5]

10.3 ABSTRACT THEOREMS

10.3.1 *The general form*

As seen in Section 10.2, the previous non-standard models are mostly based on stationary principles. The abstract situation, in all those cases, is

$$\mathscr{L}(y, z) = \tfrac{1}{2}a(y, y) - \langle g, y\rangle + b(y, z) - \langle f, z\rangle = \text{stationary} \qquad (10.12)$$

for all variations of y and z in suitable spaces Y and Z respectively; however, the meaning of the forms $a(y, y)$ and $b(y, z)$ will change from one case to another together with the definitions of Y, Z, g, and f. In any case the following general frame can be assumed for the abstract functional (10.12):

$$\left.\begin{array}{l}
\text{—}Y, Z \text{ are reflexive (real) Banach spaces}\\
\text{—}a(\bar{y}, y) \text{ is a bilinear continuous symmetric form on } Y\times Y\\
\quad\text{with norm } \|a\|\\
\text{—}b(y, z) \text{ is a bilinear continuous form on } Y\times Z, \text{ with norm}\\
\quad \|b\|\\
\text{—}g \in Y' \text{ (dual space of } Y)\\
\text{—}f \in Z' \text{ (dual space of } Z)
\end{array}\right\} \qquad (10.13)$$

Under assumptions (10.13) the functional (10.12) is well defined. Moreover it is easy to see that the pair $\{\bar{y}, \bar{z}\} \in Y\times Z$ is a *stationary point* for $\mathscr{L}$ *if and only if* the following conditions hold:

$$\left.\begin{array}{ll}
a(\bar{y}, y) + b(y, \bar{z}) = \langle g, y\rangle & \forall y \in Y\\
b(\bar{y}, z) = \langle f, z\rangle \quad \forall z \in Z &
\end{array}\right\} \qquad (10.14)$$

In all the applications it is easy to see the existence of at least one stationary point, just starting from the existence of the solution ψ of (10.1) and making use of (10.5). The following conditions will guarantee the uniqueness of the solution of (10.14):

For any $y \in Y$, condition '$b(y, z) = 0\ \forall z \in Z$ and $a(y, y) = 0$' implies $y = 0$ (10.15)

For any $z \in Z$, condition '$b(y, z) = 0\ \forall y \in Y$' implies $z = 0$ (10.16)

The proof is trivial. In order to have abstract existence theorems more hypotheses (not always satisfied in the applications) should be added.[29]

Approximation of (10.14) are usually performed by means of two *finite dimensional* subspaces $Y_h \subset Y$ and $Z_h \subset Z$. The approximate solution (if any) $\{y_h, z_h\} \in Y_h \times Z_h$ will satisfy the following *linear* system of equations:

$$\left.\begin{aligned} &a(\bar{y}_h, y_h) + b(y_h, \bar{z}_h) = \langle g, y_h \rangle \qquad \forall y_h \in Y_h \\ &b(\bar{y}_h, z_h) = \langle f, z_h \rangle \qquad \forall z_h \in Z_h \end{aligned}\right\} \tag{10.17}$$

10.3.2 *Three cases of special interest*

In each particular case, some assumptions will be necessary in order to have existence and uniqueness of the solution of (10.16) and an abstract error bound. We shall consider, here, essentially three cases.

Case 1: We assume that

—Y, Z are Hilbert spaces

—$\exists \alpha > 0$ such that $a(y_h, y_h) \geq \alpha \, \|y_h\|_Y^2 \qquad \forall y_h \in K_h(0)$

where, in general, for $f \in Z'$ we define

$$K(f) = \{y \mid y \in Y;\ b(y, z) = \langle f, z \rangle \, \forall z \in Z\}$$
$$K_h(f) = \{y_h \mid y_h \in Y_h;\ b(y_h, z_h) = \langle f, z_h \rangle \, \forall z_h \in Z_h\}$$

—$\exists \beta > 0$ such that:

$$\sup_{y_h \in Y_h} \|y_h\|_Y^{-1} \, b(y_h, z_h) \geqslant \beta \, \|z_h\|_Z \qquad \forall z_h \in Z_h$$

Then problem (10.17) has a *unique solution* and

$$\|\bar{y} - \bar{y}_h\|_Y + \|\bar{z} - \bar{z}_h\|_Z \leqslant C(\|a\|, \|b\|, \alpha, \beta)(d_y(\bar{y}, Y_h) + d_Z(\bar{z}, Z_h)) \tag{10.18}$$

where, in general, if E is a Banach space, $F \subseteq E$, and $\eta \in E$, we define

$$d_E(\eta, F) = \operatorname*{Inf}_{\xi \in F} \|\eta - \xi\|_E$$

For the proof of this result, see Reference 29.

Case 2: We assume that

—Y, Z are Hilbert spaces

—L is a Hilbert space such that:

—$Y \subsetneq L$

—$a(\bar{y}, y)$ can be extended as a bilinear continuous form on $L \times L$ with norm $\|\tilde{a}\|$

—$\exists \alpha > 0$ such that $a(y_h, y_h) \geqslant \alpha \|y_h\|_L^2 \qquad \forall y_h \in Y_h$

—$\exists \beta > 0$ such that $\sup_{y_h \in Y_h} \|y_h\|_Y^{-1} b(y_h, z_h) \geqslant \beta \|z_h\|_Z \qquad \forall z_h \in Z_h$

Then problem (10.17) has a unique solution and

$$\|\bar{y} - \bar{y}_h\|_L + \|\bar{z} - \bar{z}_h\|_Z \leqslant C(\alpha, \beta, \|\tilde{a}\|, \|b\|)(d_y(\bar{y}, Y_h) + (1 + S(h))d_Z(z, Z_h)) \tag{10.19}$$

where

$$S(h) = \sup_{y_h \in Y_h} \|y_h\|_Y \|y_h\|_L^{-1}$$

For the proof see Reference 31.

Case 3: We assume that

—Y is a Hilbert space, Z a reflexive Banach space

—M is a Hilbert space such that $Z \subsetneq M$

—L is a Hilbert space such that

—$Y \subsetneq L$

—$a(\bar{y}, y)$ can be extended as a bilinear continuous form on $L \times L$ with norm $\|\tilde{a}\|$

—$\exists \alpha > 0$ such that $a(y_h, y_h) \geqslant \alpha \|y_h\|_L^2 \qquad \forall y_h \in Y_h$

—$\exists \beta > 0$ such that $\sup_{y_h \in Y_h} \|y_h\|_Y^{-1} b(y_h, z_h) \geqslant \beta \|z_h\|_M \qquad \forall z_h \in Z_h$

—$K_h(0) \subseteq K(0)$

Then problem (10.17) has a unique solution and

$$\|\bar{y} - \bar{y}_h\|_L \leqslant C(\|\tilde{a}\|, \alpha) d_L(\bar{y}, K_h(f)) \tag{10.20}$$

$$\|\bar{z} - \bar{z}_h\|_M \leqslant C(\|\tilde{a}\|, \beta)[\|\bar{y} - \bar{y}_h\|_L + \sup_{y_h \in Y_h} \inf_{z_h \in Z_h} b(y_h, \bar{z} - z_h) \|y_h\|_Y^{-1}] \tag{10.21}$$

For the proof see Reference 31. We observe that in (10.21) the Sup Inf part is taken in order to avoid the use of $\|b\|$ which, in some of the applications of this abstract case, depends on the decomposition.

10.4 MATHEMATICAL ANALYSIS OF THE VARIOUS MODELS

10.4.1 *Orientation*

We shall discuss briefly here the applications of the abstract theorems of Section 10.3 to the models described in Section 10.2. We shall restrict our analysis to the stationary principles, since the others can be considered as classical. For each model we shall present the proper functional frame, the various kinds of approximation, the error bounds and some general ideas of the computational approach.

10.4.2 *The primal hybrid model*

We choose

$Y=\{\phi \mid \phi \in H^2(T)\forall T,\ \phi$ continuous at the vertices and zero at the boundary vertices$\}$

$$\|\phi\|_Y^2=\sum_T \|\phi\|^2_{H^2(T)}$$

$Z=\{\{\lambda_1,\lambda_2\} \mid \lambda_1 \in (H^{1/2}_{00}(T')),\ \lambda_2 \in H^{-3/2}(T')$† for each edge T' of the decomposition, with λ_1 single valued at T' and λ_2 taking opposite values on the opposite sides of each $T'\}$

$$\|\{\lambda_1,\lambda_2\}\|_Z=\operatorname*{Inf}_{\phi\in R(\lambda_1,\lambda_2)} \|\phi\|_Y$$

$$R(\lambda_1,\lambda_2)=\{\phi \mid \phi\in Y,\ \Delta^2\phi\in L^2(T)\forall T,\ M_n(\phi)=\lambda_1 \quad K_n(\phi)=\lambda_2 \text{ on each } \partial T\}$$

$$a(\chi,\phi)=\sum_T \int_T \chi_{/ij}\phi_{/ij}\,\mathrm{d}x$$

$$b(\phi,(\lambda_1,\lambda_2))=\sum_T \int_{\partial T} (\phi\lambda_2-\phi_{/n}\lambda_1)\,\mathrm{d}l$$

$$g:\phi\to\int_\Omega p\phi\,\mathrm{d}x$$

$$f\equiv 0$$

Let now (r, m, n) be three integers such that:

$$r\geqslant 2,\qquad m\geqslant 0,\qquad n\geqslant -1,\qquad m\leqslant n+1$$
$$3(m+n+3)\geqslant (r+1)(r+2)/2$$

and define

$$Y_h=\{\phi_h \mid \phi_h\in Y,\ \phi_h|_T\in P_r\}‡$$

$$Z_h=\{\boldsymbol{\lambda}^h \mid \boldsymbol{\lambda}^h\in Z,\ \lambda^h_{1|T'}\in P_m,\ \lambda^h_{2|T'}\in P_n \text{ for each edge } T'\}$$

† For the Sobolev spaces H^s and their properties see, for example, Reference 73, Chapter I.
‡ P_k = polynomials of degree $\leqslant k$; $P_{-1}=\{0\}$.

It can be proved that, with the above choice, hypotheses of Case 1 are satisfied. Moreover if $\{\chi_h, \boldsymbol{\lambda}^h\}$ is the (unique) stationary point of (10.7) over $Y_h \times Z_h$ we have:

$$\|\psi - \chi_h\|_Y \leq C\,|h|^\nu$$

$$\|\{M_n(\psi), K_n(\psi)\} - \{\boldsymbol{\lambda}_1^h, \boldsymbol{\lambda}_2^h\}\|_Z \leq C\,|h|^\nu \quad (\text{for } n \geq 0)$$

with $|h|$ = mesh size and $\nu = \min(m+1, r-1)$.

For the proof see Reference 48. Once the stationary principle (10.7) is written on $Y_h \times Z_h$ as a linear system of equation, the resulting matrix takes the form

$$\begin{pmatrix} A & B^{\mathrm{T}} \\ -B & 0 \end{pmatrix} \tag{10.22}$$

this is not well featured for the application of most numerical methods for solving linear systems. Of course the Gaussian elimination with total pivoting will still work very well; on the other hand, different approaches can be used. For instance, one could define

$$K_h(0) = \{\phi_h \mid \phi_h \in Y_h, b(\phi_h, \boldsymbol{\lambda}^h) = 0\ \forall \boldsymbol{\lambda}^h \in Z_h\} \tag{10.23}$$

and solve

$$a(\chi_h, \phi_h) = \langle p, \phi_h \rangle \qquad \forall \phi_h \in K_h(0) \tag{10.24}$$

It can be proved (see Reference 48) that $K_h(0)$ comes out to be a space of *non-conforming finite elements* which pass different levels (depending on m, n) of patch test.

Therefore this approach could be viewed as a *non-conforming approximation* of the *pure displacement* model (10.6) (see References 74 and 54; see also References 6, 18, and 57 for similar ideas).

Another possibility could be to relax the continuity requirements on Y_h by means of *Lagrangian multipliers* at the vertices. The new field $\tilde{Y}_h$ obtained in that way, would be completely discontinuous and then could be eliminated *a priori* (the corresponding matrix A being block diagonal). In that case the method could be finally considered as a *force method* in a generalized sense.

10.4.3 *Dual hybrid model*

We choose first a tensor $\bar{v}$ such that

$$\bar{v}_{ij} \in L^2(\Omega); \qquad \bar{v}_{ij/ij} = p \text{ in each } T$$

and we choose

$$Y = \{v \mid v_{ij} \in L^2(\Omega),\ v_{ij/ij} = 0 \text{ in each } T\}$$

$$\|v\|_Y^2 = \int_\Omega v_{ij} v_{ij}\, dx$$

$$Z = \{\phi \mid \phi \in H_0^2(\Omega),\ \Delta^2\phi = 0 \text{ in each } T\}$$

$$\|\phi\|_Z = \|\phi\|_{H^2(\Omega)}$$

$$a(u, v) = \int_\Omega u_{ij} v_{ij}\, dx$$

$$b(v, \phi) = \sum_T \int_{\partial T} (K_n(v)\phi - M_n(v)\phi_{/n})\, dl$$

$$g : v \to a(\bar{v}, v)$$

$$f : \phi \to b(\bar{v}, \phi)$$

Let now (r, s, m) be three integers such that

$$r \geq 3, \qquad s \geq 1, \qquad m \geq k(r, s)$$

(where $k(r, s)$ is equal to $r-1$ if $r = s+1$ and s is *even* while $k(r, s) = \max(r-2, s-1)$ in all the other cases) and define

$$Y_h = \{v_h \mid v_h \in Y,\ v_{h|T} \in (P_m)^4\ \forall T\}$$

$$Z_h = \{\phi_h \mid \phi_h \in Z,\ \phi_{h|T'} \in P_r,\ \phi_{h/n|T'} \in P_s \text{ for each edge } T'\}$$

where $\phi_{h/n|T'}$ denotes the *normal derivative.*

It can be proved that, with the above choice, hypotheses of Case 1 are satisfied. Moreover let $\bar{Y}_h$ be defined as $Y_h + \{\bar{v}\}$ and let $(\bar{v}_h, \psi_h)$ be the (unique) stationary point of (10.10) over $\bar{Y}_h \times Z_h$; then we have

$$\|\{\psi_{/ij}\} - \bar{v}_h\|_Y + \|\bar{\psi} - \psi_h\|_Z \leq C\,|h|^\nu$$

where

$$\nu = \min(m+1, s, r-1)$$

and $\bar{\psi}$ is the unique element of Z such that $\psi - \bar{\psi} \in H_0^2(T)\forall T$. For the proof see Reference 30.

We observe that in fact Z and Z_h can be considered as spaces of functions $\{\bar{\phi}, \bar{\phi}_n\}$ defined only at the interelement boundaries. In fact for a given $\phi \in Z$, only the values of ϕ and $\phi_{/n}$ at the interelement boundaries are taken into account in the form $b(v, \phi)$ (and therefore in all computations).

The resulting linear system has the form:

$$\begin{cases} AV + B^T\Psi = G \\ \qquad\ \ BV = F \end{cases}$$

Since the elements of Y_h behave independently in each element, the matrix A is block diagonal. An *a priori* inversion of A gives

$$BA^{-1}B^{\mathrm{T}}\Psi = BA^{-1}G - F \qquad (10.25)$$

In this form the method can be considered as a *displacement method.* See Reference 26 for further developments of this idea. More details on the numerical approach can be found in References 75 and 76. A trick similar to (10.23), (10.24) could also be used, leading to a non-conforming approximation of the pure equilibrium model.

10.4.4 *Mixed models*

Let us consider first the simplest case, in which the C^0 continuity of the moment field v is assumed. In this case, integrating by parts in each T, the functional $\Pi(v, \chi)$ of (10.11) becomes

$$\Pi(v, \chi) = \sum_T \left\{ \int_T (-\tfrac{1}{2} v_{ij} v_{ij} - v_{ij/i} \chi_{/j} - p\chi)\, \mathrm{d}x \right\} \qquad (10.26)$$

We choose then:

$$Y = \{v \mid v_{ij} \in H^1(\Omega)\}, \qquad Z = H_0^1(\Omega), \qquad \text{(natural norms)}$$

$$a(u, v) = \int_\Omega u_{ij} v_{ij}\, \mathrm{d}x, \qquad b(v, \phi) = \int_\Omega v_{ij/i} \chi_{/j}\, \mathrm{d}x$$

$$g \equiv 0, \qquad f : \phi \to \int_\Omega p\phi\, \mathrm{d}x$$

Let k be an integer $\geqslant 2$, we take

$$L = \{v \mid v_{ij} \in L^2(\Omega)\}$$
$$Y_h = \{v_h \mid v_h \in Y, v_{h|T} \in (P_h)^4\ \forall T\}$$
$$Z_h = \{\phi_h \mid \phi_h \in Z, \phi_{h|T} \in P_k\ \forall T\}$$

It is easy to prove that, with the given choices, hypotheses of Case 2 are satisfied. Moreover if $\{u_h, \psi_h\}$ is the (unique) stationary point of (10.26) over $Y_h \times Z_h$ we obtain

$$\|\{\psi_{/ij}\} - u_h\|_L + \|\psi - \psi_h\|_Z \leqslant C\,|h|^{k-1}; \qquad \|\psi - \psi_h\|_{L^2(\Omega)} \leqslant C\,|h|^k$$

(for the proof see, for example, Reference 31). In the case $k = 1$ an error bound of the type $C\,|h|^{1/2}$ can also be proved, but only for special types of decomposition.[44]

The matrix of the corresponding linear system of equations has always the form (10.22) and the prescribed continuity of the moments makes difficult any transform to the type (10.25) (which is however obtainable relaxing the moment continuity by means of some additional Lagrangian multiplier). A very interesting variant of this method is the following one:[34,38,43]

$$\left.\begin{array}{l}\Pi(w,\chi)=\displaystyle\int_\Omega(-\tfrac{1}{2}w^2-w_{/i}\chi_{/i}-p\chi)\,\mathrm{d}x=\text{stationary}\\ \text{for all variations of } w \text{ in } Y=H^1(\Omega) \text{ and } \chi \text{ in } Z=H_0^1(\Omega)\end{array}\right\}\quad(10.27)$$

Of course the unique stationary point is $\{\Delta\psi, \psi\}$. The numerical analysis of (10.27) is almost identical to the one of (10.26).[31,38] However, from the computational point of view, the model can be regarded as solving $\Delta\psi = w$ and $\Delta w = p$ successively by means of classical displacement methods (but the problem is on the boundary conditions). This idea has been analysed extensively in the recent years (see References 33 and 39 and the references therein contained) with very good results.

We consider now the case in which in (10.11) the only continuity of $M_n(v)$ is required to the moment field v. Integrating by parts in each T, (10.11) can be equivalently written

$$\Pi(v,\chi)=\sum_T\left\{\int_T(-\tfrac{1}{2}v_{ij}v_{ij}-v_{ij/i}\chi_{/j}-p\chi)\,\mathrm{d}x+\int_{\partial T}M_{nt}(v)\chi_{/t}\,\mathrm{d}l\right\}\quad(10.28)$$

We choose then:

$$Y=\{v\mid v_{|T}\in(H^1(T))^4 T,\ M_n(v)\ \text{`continuous'}\ \text{at the interelement boundaries}\}$$

$$\|v\|_Y^2=\sum_T\|v\|^2_{(H^1(T))}4$$

$$Z=W_0^{1,q}(\Omega)\quad(q\ \text{to be chosen}>2)$$

$$a(u,v)=\int_\Omega u_{ij}v_{ij}\,\mathrm{d}x$$

$$b(v,\phi)=\sum_T\left\{\int_T v_{ij/i}\phi_{/j}\,\mathrm{d}x-\int_{\partial T}M_{nt}(v)\phi_{/t}\,\mathrm{d}l\right\}$$

$$g\equiv 0,\qquad f:\phi\to\int_\Omega p\phi\,\mathrm{d}x$$

Let now k be an integer $\geqslant 0$ and define

$$\left.\begin{array}{l}L=\{v\mid v_{ij}\in L^2(\Omega)\}\\ M=H_0^1(\Omega)\\ Y_h=\{v_h\mid v_h\in Y,\ v_{h|T}\in(P_k)^4\ \forall T\}\\ Z_h=\{\phi_h\mid \phi_h\in Z,\ \phi_{h|T}\in P_{k+1}\ \forall T\}\end{array}\right\}\quad(10.29)$$

It is easy to prove that, with this choice, hypotheses of Case 3 are satisfied. Moreover if $\{u_h, \psi_h\}$ is the (unique) stationary point of (10.28) on $Y_h \times Z_h$ we have:

$$\{\|\{\psi_{/ij}\} - u_h\|_L + \|\psi - \psi_h\|_M \leqslant C\,|h|^{k+1};\ \|\psi - \psi_h\|_{L^2(\Omega)} \leqslant C\,|h|^{k+2}\}$$

For the proof see, for example, Reference 31.

If the continuity of $K_n(v)$ is also required in (10.11) to the moment field, a similar analysis can be carried out. The only changes to be made to the previous frame are

$$Y = \{v \mid v_{|T} \in (H^2(T))^4\ \forall T,\ M_n(v) \text{ and } K_n(v) \text{ 'continuous' at each interelement edge}\}$$

$$Z_h = \{\phi_h \mid \phi_h \in Z,\ \phi_{h|T} \in P^*_{k+1}(T)\ \forall T\}$$

$$P^*_{k+1}(T) = P_1 \oplus \{\text{bubble functions of } P_{k+1} \text{ on } T\}$$

We remark that the change on Y implies a corresponding change in the elements of Y_h (which will still be formally defined by (10.29)). Again the abstract results of Case 3 are applicable; the corresponding error bound is

$$\|\{\psi_{/ij}\} - u_h\|_L \leqslant C\,|h|^{k+1}$$

$$\|\psi - \psi_h\|_M \leqslant C\,|h|; \qquad \|\psi - \psi_h\|_{L^2(\Omega)} \leqslant C\,|h|^2$$

For the proof see Reference 50.

In both the last models the resulting matrix is of the type (10.22). Lagrangian multipliers can be used in order to relax the continuity requirements on $M_n(v)$ and on $M_n(v)$ and $K_n(v)$ respectively. After that, a trick like (10.25) can be used. Both methods could be regarded, at this stage, as 'displacement methods' in a generalized sense.

It should also be noted that, while the previous methods can be extended to the case of rectangular elements without any difficulty, the extension of the two last mixed methods is less immediate; for some results in this direction see References 40 and 56.

REFERENCES

1. Backlund, J., 'Mixed finite element analysis of elasto-plastic plates in bending', *Arch. Mech. Sto.*, **24,** 319–335 (1972).
2. Fraeijs de Veubeke, B., 'Displacements and equilibrium models in the finite element method' in *Stress Analysis* (Zienkiewicz and Holister, Eds.), Wiley, London, 1965.
3. Fraeijs de Veubeke, B., and Zienkiewicz, O. C., 'Strain energy bounds in finite element analysis by slab analogy', *J. of Strain Analysis*, **2,** 4 (1967).
4. Fraeijs de Veubeke, B., Sander, G., and Beckers, P., 'Dual analysis by finite elements: linear and nonlinear applications', *USAF Technical Report*, AFFDL-TR-72-93, 1972.

5. Fraeijs de Veubeke, B., Diffusive equilibrium models, *Univ. Calgary Lecture Notes,* 1973.
6. Fraeijs de Veubeke. B., 'Variational principles and the patch test', *Int. J. Num. Meth. Eng.,* **8,** 783–801 (1974).
7. Hellan, K., 'Analysis of elastic plates in flexure by a simplified finite element method', *Acta Polytechnica Scandinavia,* Ci. Ser., **46** (1967).
8. Herrmann, L. R., 'Finite element bending analysis for plates', *J. Eng. Mech. Div.* ASCE EM5, **93,** 49–83 (1967).
9. Herrmann, L. R., 'A bending analysis for plates', *Proc. Conf. On Matrix Methods in Structural Mechanics,* AFFDL-TR-66-88, 577–604.
10. Idelsohn, S., Analyse statique et dynamique des coques par la méthode des éléments finis, *Ph.D. Thesis,* University of Liège, Aerospace Lab., Report SF-25, 1974.
11. Pian, T. H. H., 'Formulations of finite element methods for solid continua', *Recent Advances in Matrix Methods of Structural Analysis and Design* (Gallagher, Yamada, Oden, Eds), The University of Alabama Press, 1971, pp. 49–83.
12. Pian, T. H. H., 'Finite element formulation by variational principles with relaxed continuity requirements', *The Mathematical Foundations of the F.E.M. with Applications to P.D.E.* (*Aziz, Ed.*), *Academic, New York,* 1973, pp. 671–687.
13. Pian, T. H. H., and Tong, P., 'Basis of finite element methods for solid continua', *Int. J. Numer. Meth. Eng.,* **1,** 3–68 (1969).
14. Pian, T. H. H., 'Hybrid models', *Int. Symp. on Numerical and Computer Methods in Structural Mechanics,* Urbana, Illinois, Sept. 1971.
15. Girault, V., 'A mixed finite element method for the stationary Stokes equations', *SIAM J. Num. Anal.,* **15,** 534–555 (1978).
16. Sander, G., 'Application de la méthode des éléments finis à la flexion des plaques', *Coll. Publ. Fac. Sc. Appl.,* vol. 15, 1969, University of Liège.
17. Sander, G., 'Application of the dual analysis principle', *Proc. IUTAM Symp., High Speed Computing of Elastic Structures,* University Liège, Belgium, 1970.
18. Sander, G., and Beckers, P., 'The influence of the choice of connectors in the finite element method'. *Mathematical aspects of finite element methods,* Springer, Lecture notes in mathematics, N-606, 1977, pp. 316–342.
19. Tong, P., 'An assumed stress hybrid finite element method of an incompressible and near-incompressible material', *Int. J. Solids and Structures,* **5,** 455–461 (1969).
20. Zienkiewicz, O. C., *The Finite Element Method in Engineering Science,* McGraw-Hill, London, 1978.
21. Babuška, I., The finite element method with Lagrangian multipliers, *Num. Math.,* **20,** 179–192 (1973).
22. Babuška, I., 'Error bounds for finite element method', *Num. Math.,* **16,** 322–333 (1971).
23. Babuška, I., Oden, J. T., and Lee, J. K., 'Mixed-hybrid finite element approximations of second order elliptic boundary-value problems', *TICOM Report* 75–7, University of Texas at Austin, 1975.
24. Bercovier, M., 'The link between first order mixed quadrilateral finite elements and finite difference schemes for solving Stokes stationary problem', *Computation Center,* T. note 75/MB2, Université de Jérusalem, 1975.
25. Bercovier, M., *Thèse de Doctorat d'Etat,* Université de Rouen, 1976.
26. Brezzi, F., 'Hybrid method for fourth order elliptic equations', *Mathematical*

aspects of finite element methods, Lecture notes in Mathematics, Springer, N. 606, 1977, pp. 35–46.

27. Brezzi, F., 'Sur une méthode hybride pour l'approximation du problème de la torsion d'une barre élastique', *Rend. Ist. Lombardo Sci. Lett. A*, **108,** 274–300 (1974).
28. Brezzi, F., 'Sur la méthode des éléments finis hybrides pour le problème biharmonique', *Num. Math.*, **24,** 103–131 (1975).
29. Brezzi, F., 'On the existence uniqueness and approximation of saddle point problems arising from Lagrangian multipliers', *R.A.I.R.O.*, **8–R2,** 129–151 (1974).
30. Brezzi, F., and Marini, L. D., 'On the numerical solution of plate bending problems by hybrid methods, *R.A.I.R.O.*, **R3,** 5–50 (1975).
31. Brezzi, F., and Raviart, P. A., 'Mixed finite element methods for 4th order elliptic equations', *Topics in Numerical Analysis*, Vol. III, (J. J. H. Miller, Ed.), Academic, London, 1976, pp. 33–36.
32. Ciarlet, P. G., *The Finite Element Method for Elliptic Problems*, North-Holland, Amsterdam, 1978.
33. Ciarlet, P. G., and Glowinski, R., 'Dual iterative techniques for solving a finite element approximation of the biharmonic equation', *Comp. Meth. Applied Mech. Eng.*, **5,** 277–295 (1975).
34. Ciarlet, P. G., and Raviart, P. A., 'A mixed finite element method for the biharmonic equation', *Mathematical Aspects of Finite Element in Partial Differential Equations* (C. de Boor, Ed.), Academic, New York, 1974, pp. 125–145.
35. Ciarlet, P. G., and Raviart, P. A., 'General Lagrange and Hermite interpolation in R^n with applications to finite element methods', *Arch. Rat. Mech. Anal.*, **46,** 177–199 (1972).
36. Crouzeix, M., and Raviart, P. A., 'Conforming and non-conforming finite element method for solving the stationary Stokes equations', *R.A.I.R.O.*, **R3,** 33–76 (1973).
37. Fix, G. J., 'Hybrid finite element methods', *SIAM Review*, **18,** 3, 460–484 (1976).
38. Glowinski, R., 'Approximations externes par éléments finis d'ordre un et deux du problème de Dirichlet pour Δ^2', *Topics in Numerical Analysis*, Vol. I (J. J. H. Miller, Ed.), Academic, London, 1973, pp. 123–171.
39. Glowinski, R., and Pironneau, O., 'Numerical methods for the first biharmonic equation and for the two-dimensional Stokes problem', *SIAM Review*, (to appear).
40. Johnson, C., 'Convergence of another mixed finite element method for plate bending problems', *Report No. 27, Department of Mathematics, Chalmers Institute of Technology and the University of Goteborg*, 1972.
41. Johnson, C., 'On the convergence of some mixed finite element methods for plate bending problems', *Num. Math.*, **21,** 43–62 (1973).
42. Kikuci, F., and Ando, Y., 'On the convergence of a mixed finite element scheme for plate bending, *Nucl. Eng. and Design*, **24,** 357–373 (1973).
43. Mercier, B., 'Numerical solution of the biharmonic problem by mixed finite elements of class C^0', *Boll. Unione Matematica Italiana*, (4), **10,** 133–149 (1974).
44. Miyoshi, T., 'A finite element method for the solution of fourth order partial differential equations', *Kunamoto J. Sci.* (*Math.*), **9,** 87–116 (1973).

45. Oden, J. T., *Finite Elements on Nonlinear Continua*, McGraw Hill, New York, 1972.
46. Oden, J. T., 'Some contributions to the mathematical theory of mixed finite element approximations', in *Theory and Practice in Finite Element Structural Analysis*, 3–23, University of Tokyo Press, 1973.
47. Oden, J. T., and Reddy, J. N., 'On dual-complementary variational principles in mathematical physics', *Int. J. Eng. Sci.*, **12,** 1–29 (1974).
48. Quarteroni, A., Primal hybrid methods for fourth order elliptic equations, to appear in *Calcolo.*
49. Raviart, P. A., Cours de troisème cycle de l'Université de Paris VI, 1971–1972.
50. Raviart, P. A., 'Mixed and equilibrium methods for 4th order problems', *Communication au Sèminaire Franco–Japonais sur l'Analyse Fonctionnelle et Numérique, Tokyo et Kyoto,* September 1976.
51. Raviart, P. A., and Thomas, J. M., 'A mixed finite element method for 2nd order elliptic problems', in *Mathematical Aspects of Finite Element Methods* Rome, 1975, Lecture Notes in Mathematics, Springer-Verlag (to appear).
52. Raviart, P. A., and Thomas, J. M., 'Primal hybrid finite element methods for 2nd order elliptic equations', *Math. of Comp.*, **31,** 138, 391–913 (1977).
53. Scholz, R., 'Approximation von Sattelpunkten mit finiten Elementen', *Bonner Mathematishe Schriften,* **89,** 53–66 (1976).
54. Strang, G., 'Variational crimes in the finite element methods', *The Mathematical Foundations of the F.E.M. with Applications to P.D.E.* (Aziz, Ed.), Academic, New York, 1973, pp. 689–710.
55. Strang, G., and Fix, G. J., *An Analysis of the Finite Element Method,* Prentice-Hall, Englewood Cliffs, 1973.
56. Scapolla, T., 'On a mixed finite element method for the biharmonic problem' (to appear).
57. Thomas, J. M., 'Sur l'analyse numérique des méthodes d'éléments finis hybrides et mixtes'. *Thèse de doctorat d'Etat,* Université de Paris VI, 1977.
58. Brezzi, F., Johnson, C., and Mercier, B., 'Analysis of a mixed finite element method for elasto-plastic plates', *Math. of Comp.* (to appear).
59. Falk, R. S., and Mercier, B., 'Estimation d'erreur en élasto-plasticité', *C.R. Acad. Sci.* Paris, T., **282,** série A, 645–648 (1976).
60. Fortin, M., 'Résolution numérique des équations de Navier–Stokes par des éléments finis de type mixte', *Rapport de Recherche N°. 184,* IRIA, August 1976.
61. Glowinski, R., 'Introduction to the approximation of elliptic variational inequalities', Université de Paris VI, Laboratoire d'Analyse Numérique, *Rapport N°. 76006* (see also, Presses de l'Université de Montréal (to appear)).
62. Johnson, C., 'A mixed finite element method for plasticity with hardening', *SIAM J. Numer. Analysis* (to appear).
63. Johnson, C., 'Existence theorems for plasticity problems', *J. de Math. Pures et Appl.*, **55,** 431–444 (1976).
64. Johnson, C., 'On finite element methods for plasticity problems', *Num. Math.*, **26,** 79–84 (1976).
65. Johnson, C., 'On plasticity with hardening' (to appear).
66. Johnson, C., 'A finite element method for consolidation of clay' (to appear).
67. Johnson, C., 'An elasto-plastic contact problem', (to appear).
68. Johnson, C., 'Some equilibrium finite element methods for two-dimensional elasticity problems' (to appear).

69. Mercier, B., 'Sur la théorie et l'analyse numérique de problèmes de plasticité', *Thèse de doctorat d'Etat,* Université de Paris VI, 1977.
70. Miyoshi, T., 'A mixed finite element method for the solution of the Von Kármán equations', *Num. Math.,* **26,** 255–269 (1976).
71. Strang, G., 'The finite element method—linear and nonlinear applications', *International Congress of Mathematicians,* Vancouver, Canada, 1974.
72. Strang, G., 'Some recent contributions to plasticity theory', *J. of the Franklin Institute* (to appear).
73. Lions, J.-L., and Magenes, E., *Problèmes aux limites non-homogènes,* vol. 1, Dunod, Paris, 1968.
74. Lascaux, P., and Lesaint, P., 'Some non-conforming finite elements for the plate bending problem', *Revue Française d'Automatique, Informatique, et Recherche Opérationnelle,* Avril 1975, Série Analyse Numérique, 9–54.
75. Marini, L. D., 'Implementation of hybrid finite element methods and associated numerical problems', Part I, *Publicazione N.136, LAN-CNR, Pavia,* 1976.
76. Marini, L. D., 'Implementation of hybrid finite element methods and associated numerical problems', part II (to appear).

Chapter 11

Some Equilibrium Finite Element Methods for Two-Dimensional Problems in Continuum Mechanics

C. Johnson and B. Mercier

11.1 INTRODUCTION

We shall describe in this chapter some results on mixed finite element methods of equilibrium type for problems in elasticity and plasticity, using piecewise linear approximation of the stresses and displacements. We also apply these methods after some modification to Navier–Stokes equations. Equilibrium methods for elasticity problems were first developed by Fraeijs de Veubeke;[1] in fact, one of the methods considered below is very similar to a method first introduced in Reference 1.

Our motivation to develop these methods came from plasticity where conventional displacements methods cannot be applied due to lack of smoothness of the displacements. As suggested by Fortin[2] methods of this type could also be useful for solving the Navier–Stokes equation in the case of high Reynold's numbers.

We shall first describe these methods in the context of an elasticity problem and then briefly discuss the application to problems in plasticity and fluid mechanics. The methods are mixed in the sense that independent finite element approximations are used for the stresses and the displacements (velocities). By an equilibrium method we shall mean a method where the continuity requirements will be put on the stresses and moreover the approximate stress field will satisfy the equilibrium equation exactly for loads f of particular form, e.g. $f=0$. Similar methods for plate bending problems have been introduced by Hellan and Herrman and analysed by Johnson[3] and Brezzi and Raviart.[4] Further, Raviart and Thomas[5] have developed equilibrium methods for Poisson's equation.

Below we shall use the summation convention: repeated indices indicate summation from 1 to 2.

11.2 A VARIATIONAL FORMULATION OF THE ELASTICITY PROBLEM

The elasticity problem in two dimensions (plane stress or plane strain) in the case of an isotropic body can be formulated as follows: Given the *load*

$f=(f_1, f_2)$ find a symmetric *stress tensor* $\sigma=\{\sigma_{ij}\}$, $i, j=1, 2$, and a *displacement* $u=(u_1, u_2)$, satisfying certain boundary conditions such that

$$\varepsilon(u)=-\lambda\bar{\sigma}\delta+\mu\sigma^d \quad \text{in } \Omega \tag{11.1}$$

$$\operatorname{div}\sigma+f=0 \quad \text{in } \Omega \tag{11.2}$$

where $\lambda \geqslant 0$, $\mu>0$, Ω is a bounded region in the plane,

$$\varepsilon(u)=\{\varepsilon_{ij}(u)\},\ \varepsilon_{ij}(u)=\frac{1}{2}\left\{\frac{\partial u_i}{\partial x_j}+\frac{\partial u_j}{\partial x_i}\right\}$$

is the *deformation tensor,*

$$\operatorname{div}\sigma=(\sigma_{1j,j}, \sigma_{2j,j})$$

and the pressure $\bar{\sigma}$ and *stress deviatoric* σ^d are given by σ through the decomposition

$$\sigma=\bar{\sigma}\delta+\sigma^d,$$

$$\bar{\sigma}=-(\sigma_{11}+\sigma_{22})/2$$

$$\delta=\{\delta_{ij}\},\ \delta_{ij}=\begin{cases}1 & \text{if} \quad i=j\\ 0 & \text{if} \quad i\neq j\end{cases} \qquad i, j=1, 2$$

Here (11.1), is the so-called *constitutive relation* relating stress and deformation and (11.2) is the *equilibrium equation.* Concerning the boundary conditions we have the following different possibilities on different parts of the boundary Γ of Ω (in the case of homogeneous conditions):

(i) $u\cdot n=0, \quad u\cdot t=0$
(ii) $\sigma_{nn}=0, \quad \sigma_{nt}=0,$
(iii) $u\cdot n=0, \quad \sigma_{nt}=0,$
(iv) $\sigma_{nn}=0, \quad u\cdot t=0$

where $n=(n_1, n_2)$ is an outward unit normal and $t=(t_1, t_2)=(n_2, -n_1)$ a tangent to Γ, $\sigma_{nn}=\sigma_{ij}n_in_j$ and $\sigma_{nt}=\sigma_{ij}n_it_j$ are the normal and tangential components of the stress σ, respectively. We shall here for simplicity consider the first condition, i.e.

$$u=0 \quad \text{on} \quad \Gamma \tag{11.3}$$

a choice which is however not essential in order to be able to apply our methods; the advantage of using (11.3) is only that the proof of the error estimate in that case will be somewhat simpler.

Remark 11.1 An incompressible (compressible) material corresponds to taking $\lambda=0$ ($\lambda>0$) in (11.1). Note also that with $\lambda=0$ in (11.1) we obtain the stationary *Stokes equations* with $1/\mu$ representing the viscosity. In this case (11.1) implies that $\operatorname{div} u=u_{i,i}=0$ in Ω.

In order to obtain a variational formulation of the elasticity problem (11.1)–(11.3), we introduce the function spaces

$$V = V(\Omega) = [L^2(\Omega)]^2$$
$$Y = Y(\Omega) = \{\tau = \{\tau_{ij}\} : \tau_{ij} \in L^2(\Omega),\ \tau_{ij} = \tau_{ij},\ i, j = 1, 2\}$$
$$H = H(\text{div}; \Omega) = \{\tau \in Y(\Omega) : \text{div}\ \tau \in V(\Omega)\}$$

and the scalar products

$$(\sigma, \tau) = \int_\Omega \sigma_{ij}\tau_{ij}\ \mathrm{d}x, \qquad \sigma, \tau \in Y$$

$$(u, v) = \int_\Omega u_i v_i\ \mathrm{d}x, \qquad u, v \in V$$

We also recall Green's formula:

$$(\tau, \varepsilon(v)) = \int_\Gamma (\tau \cdot n) v\ \mathrm{d}s - (\text{div}\ \tau, v) \tag{11.4}$$

where

$$\tau \cdot n = (\tau_{ij} n_j, \tau_{2j} n_j)$$

If (11.1) holds then for any $\tau \in H$, we have

$$(\varepsilon(u), \tau) = (\lambda\bar{\sigma}\delta + \mu\sigma^d, \bar{\tau}\delta + \tau^d) = a(\sigma, \tau)$$

where

$$a(\sigma, \tau) = \int_\Omega [2\lambda\bar{\sigma}\bar{\tau} + \mu\sigma_{ij}^d\tau_{ij}^d]\ \mathrm{d}x$$

so that using Green's formula (11.4) and the boundary condition $u = 0$ on Γ,

$$a(\sigma, \tau) + (u, \text{div}\ \tau) = 0, \qquad \tau \in H$$

Recalling (11.2), we are thus led to the following variational formulation of the elasticity problem: Find $(\sigma, u) \in H \times V$ such that

$$\begin{cases} a(\sigma, \tau) + (u, \text{div}\ \tau) = 0, & \tau \in H \qquad (11.5\text{a}) \\ (\text{div}\ \sigma, v) + (f, v) = 0, & v \in V \qquad (11.5\text{b}) \end{cases}$$

It is known (see, for example, Reference 6) that if $f \in [L^2(\Omega)]^2$, then this problem has a unique solution (σ, u), with $\bar{\sigma}$ unique only up to a constant in the case $\lambda = 0$. Further, if Γ is smooth one has for $\lambda > 0$ the regularity result

$$\|u\|_{m+2} + \|\sigma\|_{m+1} \leqslant C\|f\|_m$$

for $m = 0, 1, \ldots$, with a corresponding result for $\lambda = 0$. Here $\|\cdot\|_m$ denotes the norm in the space $[H^m(\Omega)]^n$, $n = 1, 2, \ldots$, where $H^m(\Omega)$ is the usual Sobolev space.

Remark 11.2 Note that the functions in V are not required to satisfy the boundary condition (11.3). This condition is in the above variational formulation implicitly contained in (11.5a); if we formally integrate by parts in (11.5a) and vary τ, we obtain (11.1) and (11.3).

Remark 11.3 If $(\sigma, u) \in H \times V$ satisfies (11.5), then σ is also the solution of the problem

$$\operatorname*{Inf}_{\tau \in E} [\tfrac{1}{2} a(\tau, \tau)]$$

where

$$E = \{\tau : (\operatorname{div} \tau, v) + (f, v) = 0, v \in V\}$$

i.e. σ is given by the *principle of minimum of the complementary energy.*

11.3 SOME FINITE ELEMENT METHODS FOR THE ELASTICITY PROBLEM

We shall consider finite element methods for the elasticity problem based on the variational formulation (11.5), i.e. methods of the form: Find $(\sigma_h, u_h) \in H_h \times V_h$ such that

$$a(\sigma_h, \tau) + (u_h, \operatorname{div} \tau) = 0, \qquad \tau \in H_h, \tag{11.6a}$$

$$(\operatorname{div} \sigma_h, v) + (f, v) = 0, \qquad v \in V_h, \tag{11.6b}$$

where H_h and V_h are finite-dimensional subspaces of H and V, respectively. Below these spaces will be constructed using piecewise linear functions defined on a finite element triangulation of Ω. In such a case the requirement $H_h \subset H$, i.e. $\operatorname{div} \tau \in V$ if $\tau \in H_h$, will be equivalent to requiring $\tau \cdot n$ to be continuous across interelement boundaries, i.e. for any finite elements K and K' with the common side S and $\tau \in H_h$, one has

$$\tau|_K \cdot n = \tau|_{K'} \cdot n \text{ on } S,$$

where n is a normal to S.

In order to be able to derive optimal error estimates the spaces H_h and V_h will be chosen to satisfy the following conditions:

Equilibrium condition: If $\tau \in H_h$ *and* $(\operatorname{div} \tau, v) = 0$ for all $v \in H_h$, then $\operatorname{div} \tau = 0$ in Ω. (11.7)

Interpolation condition: There exists an interpolation operator $\pi_h : H_h \to H$ such that for $\tau \in [H^1(\Omega)]^4$: (11.8)

(a) $(\operatorname{div} \pi_h \tau, v) = (\operatorname{div} \tau, v), \qquad v \in V_h,$
(b) $\|\tau - \pi_h \tau\| \leq C h^k \|\tau\|_k, \qquad k = 1, 2$
(c) $\|\operatorname{div} \pi_h \tau\| \leq C \|\operatorname{div} \tau\|$

where $\|\cdot\|$ denotes the L_2-norm. Moreover the space V_h will be chosen so that

$$\operatorname*{Inf}_{v \in V_h} \|w - v\| \leq Ch^2 \|w\|_2 \tag{11.9}$$

The basic result for finite element methods satisfying these conditions is the following in the case $\lambda > 0$ (for the simple proof see, for example, Reference 7).

Theorem 11.1 *If* (11.7)–(11.9) *hold then the discrete problem* (11.6) *has a unique solution* (σ_h, u_h) *and we have the following error estimates:*

$$\|\sigma - \sigma_h\| \leq \|\sigma - \pi_h \sigma\| \leq Ch^2 \|\sigma\|_2, \tag{11.10}$$

$$\|u - u_h\| \leq C\left\{ \operatorname*{Inf}_{v \in V_h} \|u - v\| + \|\sigma - \sigma_h\| \right\} \leq Ch^2(\|u\|_2 + \|\sigma\|_2) \tag{11.11}$$

Remark 11.4 In the case of Stoke's equation ($\lambda = 0$), the error estimate (11.10) will be changed into

$$\|\sigma^d - \sigma_h^d\| \leq Ch^2$$
$$\|\nabla\bar{\sigma} - \nabla\bar{\sigma}_h\| \leq Ch$$

We have only found two different choices of the spaces H_h and V_h using low degree polynomials which meet conditions (11.7) and (11.8). In both cases we use composite piecewise linear finite elements (macro elements) for the stresses, one triangular and one quadrilateral, together with piecewise linear discontinuous displacements. There seems to be a connection between these stress elements and the composite piecewise cubic plate bending elements of Hsieh–Clough–Tocher and Fraeijs de Veubeke (see Reference 8) via the Airy's stress function. The quadrilateral stress element has been first introduced by Fraeijs de Veubeke[1] for problems where $\operatorname{div} \sigma = 0$.

Let us now define the spaces H_h and V_h. For simplicity we shall then assume that the domain Ω is polygonal. Let $\{\mathscr{C}_h\}$ be a family of triangulations of Ω,

$$\Omega = \bigcup_{K \in \mathscr{C}_h} K$$

indexed by the parameter h representing the maximum diameter of the triangular or quadrilateral elements K. Each triangular (quadrilateral) element will itself be subdivided into 3 (4) subtriangles T_i. We shall assume that $\mathscr{C}_h$ is regular in the following sense: All angles of the elements $K \in \mathscr{C}_h$ are bounded away from zero and π uniformly in h and there is a positive constant α such that the length of any side of any $K \in \mathscr{C}_h$ is at least αh.

We can now define

$$V_h = \{v \in V : v \text{ is linear on each } K, K \in \mathscr{C}_h\},$$
$$H_h = \{\tau \in H : \tau|_K \in H_K, K \in \mathscr{C}_h\}$$

where H_k is a finite-dimensional subspace of $H(\text{div}; K)$ defined in the following way in the case of triangular and quadrilateral elements, respectively.

11.3.1 The triangular composite equilibrium element

Each triangle K is divided into three subtriangles T_i, $i = 1, 2, 3$, by connecting the centre of gravity of K with the three vertices of K (see Figure 11.1)

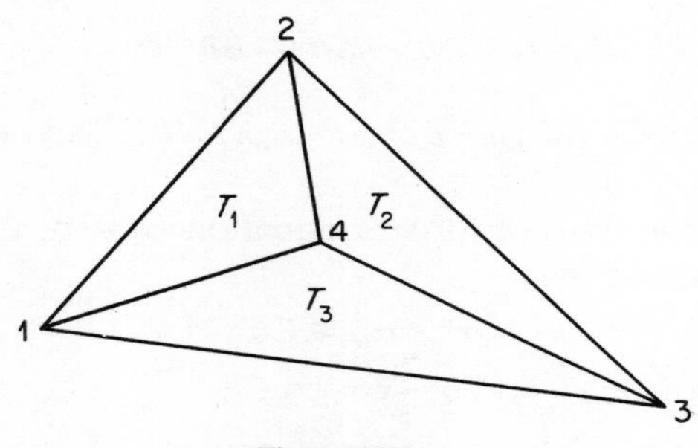

Figure 11.1

We then define

$$H_K = \{\tau \in H(\text{div}; K) : \tau|_{T_i} \text{ is linear}, \ i = 1, 2, 3\}$$

Note that the requirement $\tau \in H(\text{div}; K)$ means that $\tau \cdot n$ has to be continuous across the subtriangle boundaries 1–4, 2–4, and 3–4 (see Figure 11.1). We have (see Reference 7):

Proposition 11.1 An element $\tau \in H_K$ is uniquely determined by the following 15 degrees of freedom:

(i) the value of $\tau \cdot n$ at two points of each side of K, n being a normal to ∂K

(ii) $\int_K \tau_{ij} \text{d}x, \quad i, j = 1, 2$

Moreover, if $\tau \in H_K$ satisfies

$$\int_K v \cdot \text{div}\, \tau \, \text{d}x = 0 \text{ for } \quad v \text{ linear} \tag{11.12}$$

then $\text{div}\, \tau = 0$ in K.

11.3.2 The quadrilateral composite composite equilibrium element

In this case the four subtriangles are generated by the two diagonals of K (see Figure 11.2)

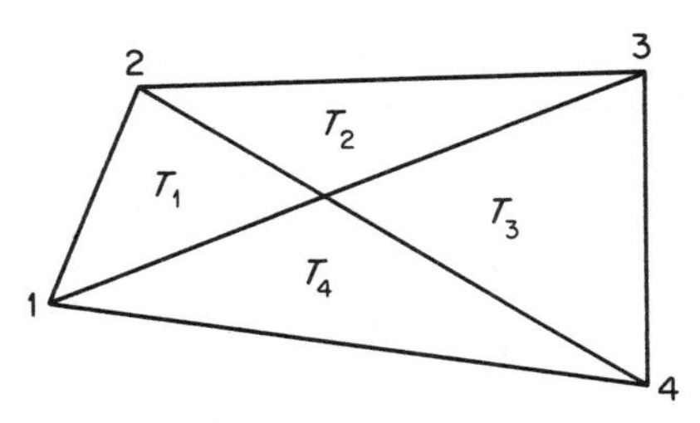

Figure 11.2

To define H_K we first introduce

$$R_K = \{\tau \in H(\text{div}; K) : \tau \text{ is linear on } T \text{ and } \tau = 0 \text{ on } K \backslash T\}$$
$$R'_K = \{\tau \in H(\text{div}; K) : \tau \text{ is linear on } T' \text{ and } \tau = 0 \text{ on } K \backslash T'\}$$
$$S_K = \{\tau \in H(\text{div}; K) : \tau \text{ is linear on } K\},$$

where $T = T_1 \cup T_2$, $T' = T_1 \cup T_4$ and then we define

$$H_K = R_K \oplus R'_K \oplus S_K$$

In analogy with Proposition 11.1, we have (see Reference 7)

Proposition 11.2 An element $\tau \in H_K$ is uniquely determined by the following 19 degrees of freedom:

(i) the value of $\tau \cdot n$ at two points at each side of K

(ii) $\int_K \tau_{ij} \, dx, \qquad i, j = 1, 2$

Moreover, if $\tau \in H_K$ satisfies

$$\int_K v \, \text{div} \, \tau \, dx = 0 \quad \text{for} \quad v \text{ linear} \tag{11.13}$$

then $\text{div} \, \tau = 0$ on K.

Let us now check that the spaces H_h and V_h will have the desired properties. From Proposition 11.1 and 11.2 it follows that the degrees of freedom for an element $\tau \in H_h$ can be chosen in the following way:

(i) the value of $\tau \cdot n$ at two points at each side of S of $\mathscr{C}_h$, n being a normal to S

(ii) the value of $\int_K \tau_{ij}\, dx$, $\quad i, j = 1, 2$, for each $K \in \mathscr{C}_h$

Since $\tau \cdot n$ is linear, condition (i) will guarantee that $\tau \cdot n$ is continuous across interelement boundaries, i.e. $\operatorname{div} \tau \in V$ if $\tau \in H_h$ so that $H_h \subset H$. Moreover, (11.12) and (11.13) imply that the equilibrium condition (11.7) will be satisfied. An interpolation operator π_h satisfying the interpolation condition (11.8) can be defined as follows: For $\tau \in [H^1(\Omega)]^4$ we define $\pi_h \tau$ to be the unique element of H_h satisfying

$$\int_S v \cdot (\tau - \pi_h \tau) \cdot n \, ds = 0 \quad \text{for} \quad v \text{ linear} \tag{11.14}$$

for any side S of $\mathscr{C}_h$, n being a normal to S, and

$$\int_K (\tau - \pi_h \tau)\, dx = 0, \qquad K \in \mathscr{C}_h \tag{11.15}$$

Let us now check that (11.8a) holds: Using Green's formula (11.4) on each K, (11.14), (11.15) and the fact that $\varepsilon(v)$ is constant on K if $v \in V_h$, we obtain

$$\begin{aligned}
(v, \operatorname{div} \tau) &= \sum_K \int_K v \cdot \operatorname{div} \tau \, dx = \sum_K \left\{ \int_{\partial K} v \cdot \tau \cdot n \, ds - \int_K \varepsilon(v) \cdot \tau \, dx \right\} \\
&= \sum_K \left\{ \int_{\partial K} v \cdot \pi_h \tau \cdot n \, ds - \int_K \varepsilon(v) \cdot \pi_h \tau \, dx \right\} \\
&= \sum_K \int_K v \cdot \operatorname{div} \pi_h \tau \, dx = (v, \operatorname{div} \pi_h \tau)
\end{aligned}$$

which proves (11.8a). It is easy to see that also (11.8b), (11.8c), and (11.9) hold. Thus, with our choice of H_h and V_h the conditions of Theorem 11.1 are satisfied.

11.4 APPLICATIONS TO PLASTICITY PROBLEMS

For simplicity we shall here consider a stationary plasticity problem corresponding to Henky's law. The quasi-static evolution problem can be handled in a analogous way (see Reference 9). The stationary plasticity problem can be formulated as follows (see Reference 6, 10, and 11): Find $(\sigma, u) \in P \times V$ such that

$$a(\sigma, \tau - \sigma) + (u, \operatorname{div} \tau - \operatorname{div} \sigma) \geq 0, \qquad \tau \in P \tag{11.16a}$$

$$(\operatorname{div} \sigma, v) + (f, v) = 0, \qquad v \in V \tag{11.16b}$$

where P denotes the set of plastically admissible stresses:

$$P=\{\tau \in H : \tau(x) \in B \text{ a.e. in } \Omega\}$$

and B is a closed convex set in $\mathbb{R}^4$. As an example we mention the analogue in two dimensions of the von Mises' yield criterion:

$$B=\{\tau \in \mathbb{R}^4 : |\tau^d| \leq g\}$$

with g a positive constant. Under a 'safe load hypothesis' one can prove existence of a solution (σ, u) of (11.16), where σ but not necessarily u is uniquely determined (see References 6, 10, and 11). In the obvious way we can now formulate a finite element method for the plasticity problem in the following way: Find $(\sigma_h, u_h) \in P_h \times V_h$, where $P_h = H_h \cap P$ such that

$$a(\sigma_h, \tau-\sigma_h)+(u_h, \operatorname{div} \tau-\operatorname{div} \sigma_h) \geq 0, \qquad \tau \in P_h \tag{11.17a}$$

$$(\operatorname{div} \sigma_h, v)+(f, v)=0, \qquad v \in V_h \tag{11.17b}$$

Under suitable assumptions this problem will have a solution and one can prove that σ_h will converge to σ in the L_2-norm. For details we refer to Reference 12 where an analogous situation in the case of elastoplastic plates is considered.

11.5 APPLICATIONS TO FLUID MECHANICS

We recall the stationary Navier–Stokes equations:

$$\varepsilon(u)=\mu\sigma^d \quad \text{in} \quad \Omega \tag{11.18a}$$

$$\operatorname{div} u=0 \quad \text{in} \quad \Omega \tag{11.18b}$$

$$-u \cdot \nabla u+\operatorname{div} \sigma+f=0 \quad \text{in} \quad \Omega \tag{11.18c}$$

$$u=0 \quad \text{on} \quad \Gamma \tag{11.18d}$$

where for simplicity we have considered a particular boundary condition. In general one can have (homogeneous) boundary conditions of the same form as those given for the elasticity problem in Section 11.1.

To formulate a finite element method for Navier–Stokes equations we shall need to modify the method for the Stokes equations introduced above. The reason is that since no continuity requirements are put on the functions in V_h, we would, using this space, have difficulties with the nonlinear term $u \cdot \nabla u$. A natural idea is to change the space V_h into

$$\tilde{V}_h=\{v \in V_h : \operatorname{div} v \in L^2(\Omega), v \cdot n=0 \text{ on } \Gamma\}$$

We note that for $v \in V_h$ the condition $\operatorname{div} v \in L^2(\Omega)$ is equivalent to requiring that $v \cdot n$ is continuous across interelement boundaries. As degrees of freedom for $v \in \tilde{V}_h$ we can choose the value of $v \cdot n$ at two points on each side of $\mathscr{C}_h$. Further, we shall seek independent approximations of the stress deviatoric and the pressure in the spaces

$$H_h^d = \left\{ \tau \in H_h : \int_K \bar{\tau}\, dx = 0, K \in \mathscr{C}_h \right\},$$

$$W_h = \left\{ q \in L^2(\Omega) : q|_K \text{ is constant, } K \in \mathscr{C}_h \right\}$$

In the case of Stokes equations, i.e. omitting the term $u \cdot \nabla u$ in (11.18c), the method is the following: Find $(\sigma_h^d, \bar{\sigma}_h, u_h) \in H_h^d \times W_h \times \tilde{V}_h$ such that

$$\mu(\sigma_h^d, \tau) + (u_h, \operatorname{div} \tau) = 0, \qquad \tau \in H_h^d, \tag{11.19a}$$

$$(\operatorname{div} u_h, q) = 0, \qquad q \in W_h, \tag{11.19b}$$

$$(\bar{\sigma}_h, \operatorname{div} v) + (\operatorname{div} \sigma_h^d, v) + (f, v) = 0, \qquad v \in \tilde{V}_h \tag{11.19c}$$

We note that (11.19b) will imply that

$$\operatorname{div} u_h = 0 \quad \text{in} \quad \Omega$$

and that the term $(\bar{\sigma}_h, \operatorname{div} v)$ in (11.19c) comes from integrating by parts in the term $(\operatorname{div} \bar{\sigma}, v)$. Further, we observe that the method is no longer an equilibrium method in the sense that (11.19c) will not imply that $\operatorname{div} \sigma_h \equiv 0$ if $f \equiv 0$. Because of this we will not get an optimal error estimate; what we can prove is the following:

$$\|\sigma^d - \sigma_h^d\| \leq Ch$$
$$\|u - u_h\| \leq Ch$$

Let us now consider the Navier–Stokes equations. To handle the nonlinear term we shall use a method proposed by Fortin producing an 'upwind' dissipative scheme (see Reference 2). In this method, given $u_h \in \tilde{V}_h$ for each $K \in \mathscr{C}_h$ we distinguish between the part ∂K_- of the boundary of K where the flow is entering,

$$\partial K_- = \{x \in \partial K : u_h \cdot n(x) \leq 0\}$$

and the part ∂K_+ where the flow is sorting,

$$\partial K_+ = \{x \in \partial K : u_h \cdot n(x) > 0\}$$

Here n is an outward normal to ∂K. We note that if $u_h \in \tilde{V}_h$ then $u_h \cdot n$ is continuous across interelement boundaries so that for two triangles K and K' with common side S one has $\partial K_- = \partial K'_+$ or $\partial K_+ = \partial K'_-$ on S.

We can now formulate the method for the stationary Navier–Stokes equations: Find $(\sigma_h, \bar{\sigma}_h, u_h) \in H_h^d \times W_h \times \tilde{V}_h$ such that

$$\mu(\sigma_h^d, \tau) + (u_h, \operatorname{div} \tau) = 0, \qquad \tau \in H_h^d \quad (11.20a)$$

$$(\operatorname{div} u_h, q) = 0, \qquad q \in W_h \quad (11.20b)$$

$$b^*(u_h, u_h, v) + (\bar{\sigma}_h, \operatorname{div} v) + (\operatorname{div} \sigma_h^d, v) + (f, v) = 0, \qquad v \in \tilde{V}_h \quad (11.20c)$$

where

$$b^*(u, u, v) = \sum_{K \in \mathscr{C}_h} \left\{ \int_K u_i u_j \frac{\partial v_i}{\partial x_j} \mathrm{d}x - \int_{\partial K} u \cdot n u_i v_i \, \mathrm{d}s \right\}$$

$$\tilde{u} = \begin{cases} \text{the trace of } u|_K \text{ on } \partial K_+ \\ \text{the trace of } u|_{K'} \text{ on } \partial K_- \end{cases}$$

and K' is a triangle $K' \neq K$, with $\partial K' \cap \partial K_- \neq \emptyset$. To motivate the expression for the nonlinear term $b(u, u, v)$ we note that multiplying (11.18) by v, we obtain the term

$$b(u, u, v) = -\sum_{K \in \mathscr{C}_h} \int_K u_i \frac{\partial u_j}{\partial x_i} v_j \, \mathrm{d}x$$

Using Green's formula on each $K \in \mathscr{C}_h$ we see that if $\operatorname{div} u = 0$, then

$$b(u, u, v) = \sum_K \left\{ \int_K u_i u_j \frac{\partial v_i}{\partial x_j} \mathrm{d}x - \int_{\partial K} u \cdot n u_j v_j \, \mathrm{d}x \right\}$$

Thus, the term b^* is obtained from b by replacing the 'interior trace' of u on ∂K_- by the 'exterior trace'.

One can prove (see Reference 13) that a subsequence of $\{(\sigma_h^d, u_h)\}$, $h > 0$, will converge to (σ^d, u), where (σ, u) is a solution of (11.18).

REFERENCES

1. Fraeijs de Veubeke, B., 'Displacement and equilibrium models in the finite element method', in *Stress Analysis* (ed. by O. C. Zienkiewicz and G. S. Holister,) Wiley, New York, 1965, pp. 145-197.
2. Fortin, M., 'Résolution numérique des équations de Navier–Stokes par des élément finis de type mixte', *2nd International Symposium on Finite Element Methods in Flow Problems*, S. Margherita Ligure, Italy, 1976.
3. Johnson, C., 'A mixed finite element method for plasticity with hardening', *Siam J. of Numer. Anal.*, **14** (4), 574–583 (1977).
4. Brezzi, F., and Raviart, P. A., 'Mixed finite element methods for 4th order elliptic equations', in *Topics in Numerical Analysis* III (J. J. H. Miller, Ed.) Academic Press, London, 1977, pp. 33–56.
5. Raviart, P. A., and Thomas, J. M., 'A mixed finite element method for 2nd order elliptic problems', in *Mathematical Aspects of Finite Element Methods* (I. Galligani, E. Magenes Eds.), Lecture Notes in Mathematics, Vol. 606, Springer-Verlag, 1977, pp. 292–315.
6. Duvaut, G., and Lions, J. L., *Les Inequations en mecanique et en physique*, Dunod, Paris, 1972.

7. Johnson, C., and Mercier, B., 'Some equilibrium finite element methods for two-dimensional elasticity problems', to appear.
8. Ciarlet, P. G., *The Finite Element Method for Elliptic Problems*, North-Holland, Amsterdam, 1978.
9. Johnson, C., 'On the convergence of a mixed finite element method for plate bending problems,' *Numer. Math.*, **21** 43–62 (1973).
10. Johnson C., 'Existence theorems for plasticity problems', *J. Math. pures et appl.*, **55** 431–444 (1976).
11. Mercier, B., 'Sur la theorie et l'analyse numérique des problems de plasticité', *These*, Paris, 1977.
12. Brezzi, F., Johnson, C., and Mercier, B., 'Analysis of a mixed finite element method for elasto-plastic plates', *Math. Comp.*, **31** (140), 809–817 (1977).
13. Johnson, C., 'A mixed finite element method for Navier–Stokes equations', *Research Report*, Chalmers Institute of Technology, 1977.

Chapter 12

Improved Displacement Finite Elements for Incompressible Materials

J. H. Argyris and P. C. Dunne

12.1 INTRODUCTION

In a recent paper[1] the authors introduced a method of improving the behaviour of the simplex constant strain finite elements TRIM3, TET4, and the axisymmetric element TRIAX3. The motivation for this work was the desire to retain the use of these simple elements in the incompressible regime where the unmodified elements are unsuitable, except for in the case of TRIM3 where a special crossed diagonal arrangement is suitable for small strain applications.[2–5]

The problem of the incompressible or nearly incompressible continuum discretized in finite elements may be regarded as a special case of those problems in which there are either a very large number of rigid constraints or in which the stiffnesses have a very wide range. Although some general[3] or special[5] methods have been developed to eliminate the constraints, which are often bedevilled by linear dependencies, these methods require, apart from the above mentioned exception, the use of the higher order finite elements and often the relaxation of point-wise incompressibility. The same applies to the nearly incompressible case where the counterpart to the relaxation of point-wise incompressibility is the use of reduced integration or of weighted residual methods with a reduced number of parameters representing the hydrostatic stress or strain.

The least number of incompressibility constraints that can be applied to a finite element is one. If, as is usually the case, the dilatational elastic modulus is considered constant within an element this implies a simple area constraint as the modulus becomes infinite. Linear strain elements with only one constraint sometimes give reasonable solutions. For example in Reference[6] the linear strain TRIM6 element has been applied to the incompressible fluid flow problem and in some but not all problems gave a better solution with simple area constraint than with full point-wise incompressibility. However, in the case of the constant strain simplex elements area constraint is synonomous with point-wise constraint and some other way of relaxing the constraints is necessary.

In Reference 1 this is done by using the orthodox simplex elements for the deviatoric part of the strain energy density and modified elements, in which the sides are allowed to bulge, for the dilatational energy. The total volume of each element remains invariant and convergence depends on the stiffness associated with the bulge coordinates which is easily determined for optimum average behaviour.

There are very good practical engineering reasons for working with the simplest elements in which the strains are constant within each element. Thus the main object of the method was to avoid the use of higher order elements. On the other hand it is possible that the principle of the method may prove useful in a more general context. Although there are some characteristics that the method shares in common with existing variational methods, in which stress or strain field are specified independently of the displacement field, the use of the stiffness associated with the bulge coordinates appears to be a new feature which does not fall out naturally from a variational statement expressed at continuum level. In the following sections we shall attempt to describe the method in a more general way in which the method of Reference 1 appears as a special case.

12.2 PARTITION OF STRAIN ENERGY DENSITY $\bar{\Phi}$

The strain energy density is supposed to be divided into two parts. Thus

$$\bar{\Phi} = \bar{\Phi}_d + \bar{\Phi}_v \tag{12.1}$$

where $\bar{\Phi}_d$ is the deviatoric or shear strain energy and $\bar{\Phi}_v$ the volumetric or dilational energy. Now $\bar{\Phi}_d$ is considered to be associated with an elastic continuum which is discretized as a pattern of finite elements with an $(N \times 1)$ nodal displacement vector $\mathbf{u}$. The global stiffness matrix due to $\bar{\Phi}_d$ is denoted by $\mathbf{K}_d$ and any external loading is reduced to forces in the directions of the displacement $\mathbf{u}$. The strain energy density $\bar{\Phi}_v$ is associated with another continuum connected to the above nodes. However, the deformation of the $\bar{\Phi}_v$ continuum is not entirely defined by $\mathbf{u}$. Thus we suppose that between nodes the $\bar{\Phi}_v$ continuum may bulge and that the amount of the bulging is represented by a vector $\boldsymbol{\varphi}$. Therefore there is a lack of conformity between the two continuums. When the material is actually incompressible the vector $\boldsymbol{\varphi}$ produces exchanges of volume between elements which, however, keep the volume of each element and the total volume constant. In the case of the constant strain simplex elements considered in Reference 1 the φ's where simple volume increments and it was unnecessary to associate with them a displacement mode within the element. For extension to higher order elements, in which more than a constant hydrostatic stress mode are to be considered, it is necessary to define a displacement mode corresponding to φ. Because of the stiffness assigned to the bulge coordinates φ the total average energy density will be increased by an amount $\bar{\Phi}_\varphi$. However, by proper choice of the stiffnesses this energy becomes negligible in relation to $\bar{\Phi}_d$ as the mesh is refined.

12.3 CALCULATION OF THE DILATATIONAL POTENTIAL ENERGY

The two weighted residual procedures used to relax the point-wise incompressibility in higher order finite elements are based either on the variational method of Herrman[7] or on that of Nagtegaal *et al.*[4] which was also previously advocated by Key.[8] The difference in the two methods is that the first uses an independent distribution of hydrostatic stress and the second uses one of dilational strain. In both cases the number of parameters representing the stress or strain determines the number of compressibility constraints when the bulk modulus becomes infinite. In spite of the claims made in Reference 4 that the second method has advantages in application there does not appear to be any essential difference between them and the methods become identical when the bulk modulus remains constant within each finite element. For our present purpose in which we shall suppose the elasticity to be constant within an element we choose the hydrostatic stress as our independent quantity.

The dilatational strain ε_h within an element may be expressed in terms of the nodal deflections u and the bulge coordinates φ. At element level we denote the u's by the vector $\boldsymbol{\rho}$.

Thus,

$$\varepsilon_h = [\boldsymbol{\omega}_\rho \boldsymbol{\omega}_\varphi]\{\boldsymbol{\rho}\ \boldsymbol{\varphi}\} \tag{12.2}$$

where the row matrix $[\boldsymbol{\omega}_\rho\ \boldsymbol{\omega}_\varphi]$ is composed of polynomials or other functions of position within the element. An alternative but equivalent procedure would use the natural deformation vector ρ_N instead of $\boldsymbol{\rho}$.

The hydrostatic stress in the gth element is represented by,

$$\sigma_h = \boldsymbol{\omega}_h \boldsymbol{\sigma}_{hg} \tag{12.3}$$

where $\boldsymbol{\omega}_h$ is a row matrix of polynomials and $\boldsymbol{\sigma}_{hg}$ a vector.

$$\boldsymbol{\sigma}_{hg} = \{\sigma_{h1}\ \sigma_{h2}\ \sigma_{h3}\}$$

Thus for elements with only one constraint $\omega_h = 1$ and $\sigma_{hg} = \sigma_{h1}$. For linear variation of σ_h in a membrane element we could take

$$\boldsymbol{\sigma}_{hg} = \{\sigma_{h0}\ \sigma_{h\xi}\ \sigma_{h\eta}\}$$

with

$$\boldsymbol{\omega}_h = [1\ \xi\ \eta] \tag{12.4}$$

where ξ, η are non-dimensional coordinates with respect to the centroid, or equivalently in a simplex TRIM element

$$\boldsymbol{\sigma}_{hg} = \{\sigma_{h1}\ \sigma_{h2}\ \sigma_{h3}\}$$

and

$$\boldsymbol{\omega}_h = [\zeta^1\ \zeta^2\ \zeta^3]$$

that is the nodal values and natural coordinates in the TRIM3 element containing the given TRIM element. In any case we would always include, explicitly or implicitly a constant term. Generally in simplex element the hydrostatic stress will be represented by the shape functions of the simplex element of one degree lower which means that the deviatoric and hydrostatic stresses are represented by the same degree polynomials as the deviatoric strains but the dilatational strains will contain higher degree terms.

If we regard the hydrostatic stresses as given quantities the potential energy of σ_h in an element of volume V is,

$$\int \sigma_h\, \varepsilon_h\, \mathrm{d}V$$

or with (12.2) and (12.3),

$$\boldsymbol{\sigma}_{hg}^t \left[\int \boldsymbol{\omega}_h^t [\boldsymbol{\omega}_\rho \boldsymbol{\omega}_\varphi]\, \mathrm{d}V \right] \{\boldsymbol{\rho}\ \boldsymbol{\varphi}\} \tag{12.5}$$

or

$$\boldsymbol{\sigma}_{hg}^t [\mathbf{h}_u^t \boldsymbol{\rho} + \mathbf{h}_\varphi^t \boldsymbol{\varphi}] \tag{12.6}$$

Thus, if the material is incompressible the point-wise incompressibility condition is not satisfied exactly but is replaced by the weighted mean conditions,

$$[\mathbf{h}_u^t \boldsymbol{\rho} + \mathbf{h}_\varphi^t \varphi] = \mathbf{0} \tag{12.7}$$

which, since $\boldsymbol{\omega}_h$ contains a constant term, always includes the condition of zero total volume change. The complete potential for the whole system including nodal forces may now be written

$$\bar{\Phi} = \tfrac{1}{2}\mathbf{u}^t \mathbf{K}_d \mathbf{u} + \tfrac{1}{2}\boldsymbol{\varphi}^t \mathbf{K}_\varphi \boldsymbol{\varphi} + \boldsymbol{\sigma}_h^t [\mathbf{H}_u^t \mathbf{u} + \mathbf{H}_\varphi^t \boldsymbol{\varphi}] - \mathbf{f}^t \mathbf{u} \tag{12.8}$$

where $\boldsymbol{\sigma}_h$ is the vector of all the $\boldsymbol{\sigma}_{hg}$. $\mathbf{H}_u$ and $\mathbf{H}_\varphi$ are the result of assembling all the terms of the form (12.6) and K_φ is a diagonal stiffness matrix of positive terms. When there are no bulge coordinates φ equation (12.8) reduces to the standard formulation.

On internal boundaries between elements quite small values of the elements of $\mathbf{K}_\varphi$ are sufficient to ensure rapid convergence. On unsupported external boundaries larger values of $\mathbf{K}_\varphi$ are necessary and in fact we usually assume $\varphi = 0$ on these boundaries whether supported or not. In Reference 1 the vector $\boldsymbol{\varphi}$ was taken as the actual bulge areas or volumes and the elements k_φ of $\mathbf{K}_\varphi$ were taken as

$$k_\varphi = \lambda G[V_\mathrm{I}^{-1} + V_\mathrm{II}^{-1}] \tag{12.9}$$

for solid elements of volumes V_I and V_II on each side of a boundary, and

$$k_\varphi = \lambda Gt[\Omega_\mathrm{I}^{-1} + \Omega_\mathrm{II}^{-1}] \tag{12.10}$$

for membrane elements of area Ω_I and Ω_II on each side of a boundary. G is the shear modulus, t the membrane thickness, and λ a numerical factor of the order unity for optimum convergence.

In the general case we may use

$$k_\varphi = \lambda G\left[\int \omega_{\varphi\mathrm{I}}^2 \,\mathrm{d}V_\mathrm{I} + \int \omega_{\varphi\mathrm{II}}^2 \,\mathrm{d}V_\mathrm{II}\right] \tag{12.11}$$

which reduces to (12.9) when φ are the bulge volumes or areas and $\omega_{\varphi\mathrm{I}} = V_\mathrm{I}^{-1}$ etc.

12.4 EQUILIBRIUM AND CONSTRAINT EQUATIONS FOR INCOMPRESSIBLE CASE

By variation of $\mathbf{u}$ and $\boldsymbol{\varphi}$ in (12.8) we obtain,

$$\begin{aligned} \mathbf{K}_d\mathbf{u} + \mathbf{H}_u\boldsymbol{\sigma}_h &= \mathbf{f} \\ \mathbf{K}_\varphi\boldsymbol{\varphi} + \mathbf{H}_\varphi\boldsymbol{\sigma}_h &= \mathbf{0} \end{aligned} \tag{12.12}$$

If the material is incompressible (12.7) holds and hence

$$\mathbf{H}_u^t\mathbf{u} + \mathbf{H}_\varphi^t\boldsymbol{\varphi} = \mathbf{0} \tag{12.13}$$

Equations (12.12), (12.13) may be reduced by the special procedure developed in Reference 3. However, the fact that $\mathbf{K}_\varphi$ is diagonal makes it simple to reduce the system to

$$\begin{bmatrix} \mathbf{K}_d & \mathbf{H}_u \\ \mathbf{H}_u^t & -\mathbf{F} \end{bmatrix} \begin{bmatrix} \mathbf{u} \\ \boldsymbol{\sigma}_h \end{bmatrix} = \begin{bmatrix} \mathbf{f} \\ \mathbf{0} \end{bmatrix} \tag{12.14}$$

with

$$\mathbf{F} = \mathbf{H}_\varphi^t\mathbf{K}_\varphi^{-1}\mathbf{H}_\varphi \tag{12.15}$$

Equation (12.14) is a sparse system and the most efficient method of solution depends on the structure of the matrix $\mathbf{F}$ which in turn depends on the choice of the bulge function $\boldsymbol{\varphi}$.

In the case of the constant strain simplex elements for which the method is especially suited or other elements with only a constant σ_h, the matrix $\mathbf{F}$ is singular only if no φ movements are allowed anywhere on the boundary. This will be so for a fixed boundary or if we assume all k_φ to be infinite on the boundary when the boundary is not completely fixed. However, we may always eliminate one of the $\boldsymbol{\sigma}_h$, say σ_{h1} and the corresponding constraint equation by writing,

$$\boldsymbol{\sigma}_h = \{\sigma_{h2}\sigma_{h3} \ldots \sigma_{hn}\} \tag{12.16}$$

Then

$$\mathbf{Ku} = \mathbf{f} \tag{12.17}$$

when the boundary is fixed, or

$$\begin{bmatrix} \mathbf{K} & \mathbf{L} \\ \mathbf{L}^t & 0 \end{bmatrix} \begin{bmatrix} \mathbf{u} \\ \sigma_{h1} \end{bmatrix} = \begin{bmatrix} \mathbf{f} \\ 0 \end{bmatrix} \tag{12.18}$$

when the boundary is free or partly free. In the above equations,

$$\mathbf{K} = [\mathbf{K}_d + \bar{\mathbf{H}}_u \bar{\mathbf{F}}^{-1} \bar{\mathbf{H}}_u^t] \tag{12.19}$$

where

$$\bar{\mathbf{F}} = \bar{\mathbf{H}}_\varphi^t \mathbf{K}_\varphi^{-1} \bar{\mathbf{H}}_\varphi \tag{12.20}$$

and $\bar{\mathbf{H}}_\varphi^t$ and $\bar{\mathbf{H}}_u^t$ are the result of omitting the rows corresponding to element 1 from H_φ^t and $\mathbf{H}_u^t$. The row matrix $\mathbf{L}^t$ is the result of adding all rows of $\mathbf{H}_u^t$ and

$$\mathbf{L}^t \mathbf{u} = 0 \tag{12.21}$$

may be interpreted as the constraint condition that the total area or volume contained by the boundary remains constant.

Equations (12.17), (12.18) although no longer sparse, are always well conditioned. Thus single precision calculations are generally adequate. In this respect the method compares very favourably with those using a high bulk modulus in higher degree elements.

The single constraint in equation (12.18) is easily dealt with. Thus we may eliminate one of the boundary displacements, say u_1, which enters strongly in equation (12.21). Then if

$$\bar{\mathbf{u}} = \{u_2 u_3 \ldots u_n\} \tag{12.22}$$

$$\mathbf{L} = \{l_1 l_2 \ldots l_n\} \tag{12.23}$$

$$\bar{\mathbf{L}} = -\left\{\frac{l_2}{l_1}\frac{l_3}{l_1} \ldots \frac{l_n}{l_1}\right\} \tag{12.24}$$

there follows,

$$\mathbf{u} = \mathbf{T}\bar{\mathbf{u}} \tag{12.25}$$

where,

$$\mathbf{T} = \begin{bmatrix} \bar{\mathbf{L}}^t \\ \mathbf{I}_{n-1} \end{bmatrix} \tag{12.26}$$

The equation of equilibrium in terms of $\bar{\mathbf{u}}$ is now

$$\bar{\mathbf{K}}\bar{\mathbf{u}} = \bar{\mathbf{f}} \tag{12.27}$$

where,

$$\bar{\mathbf{K}} = \mathbf{T}^t\mathbf{K}\mathbf{T} \tag{12.28}$$

and,

$$\bar{\mathbf{f}} = \mathbf{T}^t\mathbf{f} \tag{12.29}$$

If a lumped mass matrix is used with

$$\mathbf{M} = [m_1 m_2 \ldots m_n] \tag{12.30}$$

the equation of motion will be

$$\mathbf{M}\ddot{\mathbf{u}} + \mathbf{K}\mathbf{u} = \mathbf{f} \tag{12.31}$$

for fixed boundary, or,

$$\tilde{\mathbf{M}}\ddot{\bar{\mathbf{u}}} + \bar{\mathbf{K}}\bar{\mathbf{u}} = \bar{\mathbf{f}} \tag{12.32}$$

for free or partly free boundary, where,

$$\tilde{\mathbf{M}} = \mathbf{T}^t\mathbf{M}\mathbf{T} \tag{12.33}$$

In many problems u_1 may be chosen at a point having little movement and we may then take $m_1 = 0$. $\tilde{\mathbf{M}}$ then becomes the diagonal matrix,

$$\bar{\mathbf{M}} = [m_2 m_3 \ldots m_n] \tag{12.34}$$

The extension of this procedure to the higher degree elements with several hydrostatic stress modes requires a larger number of φ modes than is justified by the resulting increase in accuracy. For example, in the case of the linear strain element TRIM6 it is natural to choose only one φ mode per side which will turn out to be the anti-symmetrical cubic mode. This mode will not influence the total element area so that it couples only with the linear terms of equation (12.4). This means that the linear constraints corresponding to the constant term of $\boldsymbol{\sigma}_h$ will remain as in the standard TRIM6 element and the general method of Reference 3 will in any case have to be applied if we wish to arrive at an equation with all σ_h eliminated. To apply the parallel of equation (12.17) and (12.18) requires additional symmetrical φ modes on each side.

12.5 EQUILIBRIUM AND CONSTRAINT EQUATIONS IN COMPRESSIBLE CASE

When the material is compressible with bulk modulus E_h we require an expression for σ_h in terms of the strain. From the equivalence of the strain energy expressed in terms of σ_h and in terms of $\sigma_h \varepsilon_h$ we find, at element level,

$$\boldsymbol{\sigma}_{hg} = E_h \mathbf{h}_{\sigma g}^{-1}[\mathbf{h}_u^t \mathbf{h}_\varphi^t]\{\underline{\boldsymbol{\rho}}\boldsymbol{\varphi}\} \tag{12.35}$$

where

$$\mathbf{h}_{\sigma g} = \int \boldsymbol{\omega}_h^t \boldsymbol{\omega}_h \, \mathrm{d}V \tag{12.36}$$

The assembled vector $\boldsymbol{\sigma}_h$ is therefore,

$$\boldsymbol{\sigma}_h = E_h \mathbf{H}_\sigma^{-1}[\mathbf{H}_u^t \mathbf{H}_\varphi^t]\{\mathbf{u}\boldsymbol{\varphi}\} \tag{12.37}$$

where

$$\mathbf{H}_\sigma = [\boldsymbol{\eta}_{\sigma 1} \mathbf{h}_{\sigma 2} \cdots \mathbf{h}_{\sigma n}] \tag{12.38}$$

Substituting (12.37) in (12.12) we obtain,

$$\begin{aligned} &[\mathbf{K}_d + E_h \mathbf{H}_u \mathbf{H}_\sigma^{-1} \mathbf{H}_u^t]\mathbf{u} + E_h \mathbf{H}_u \mathbf{H}_\sigma^{-1} \mathbf{H}_\varphi^t \boldsymbol{\varphi} = \mathbf{f} \\ &E_h \mathbf{H}_\varphi \mathbf{H}_\sigma^{-1} \mathbf{H}_u^t \mathbf{u} + [\mathbf{K}_\varphi + E_h \mathbf{H}_\varphi \mathbf{H}_\sigma^{-1} \mathbf{H}_\varphi^t]\boldsymbol{\varphi} = \mathbf{0} \end{aligned} \tag{12.39}$$

In special cases the matrix $\mathbf{h}_\sigma$ may be diagonal. Thus if we have a linear distribution of σ_h expressed by,

$$\sigma_h = \sigma_{h0} + \sigma_{hx}(x/i_y) + \sigma_{hy}(y/i_x) \tag{12.40}$$

where i_x, i_y are the principal radii of gyration of a membrane element of area Ω and thickness t one obtains,

$$\mathbf{h}_\sigma = \Omega t \mathbf{I}_3 \tag{12.41}$$

For elements with only one hydrostatic stress, $h_\sigma = V$ for solid elements and $h_\sigma = \Omega t$ for membrane elements.

Equation (12.39) is sparse and may be used to give an improved performance in simple elements even when the material is not nearly incompressible. For the nearly incompressible case it may be better to use another form of (12.39) with $\boldsymbol{\varphi}$ eliminated. Thus,

$$[\mathbf{K}_d + \mathbf{H}_u[\mathbf{F} + E_h^{-1}\mathbf{H}_\sigma]^{-1}\mathbf{H}_u^t]\mathbf{u} = \mathbf{f} \tag{12.42}$$

where $\mathbf{F}$ is as in equation (12.15).

12.6 NUMERICAL EXAMPLES

All the numerical examples so far calculated have been for the simplex constant strain elements TRIM3 and TET4, and the axisymmetrical

TRIAX3 linear displacement ring element. Also, only small strain examples have been considered although it is hoped that the constant strain elements will prove especially advantageous in problems of incompressible large strain and plastic yield.

12.6.1 Symmetrically loaded equilateral triangle: TRIM3

This example has a central stress independent of Poisson's ratio and has been used previously as a test case.[2,3] Figure 12.1 shows the effect of the constant λ of equation (12.10) on the convergence for $N = 4$ and 8 elements

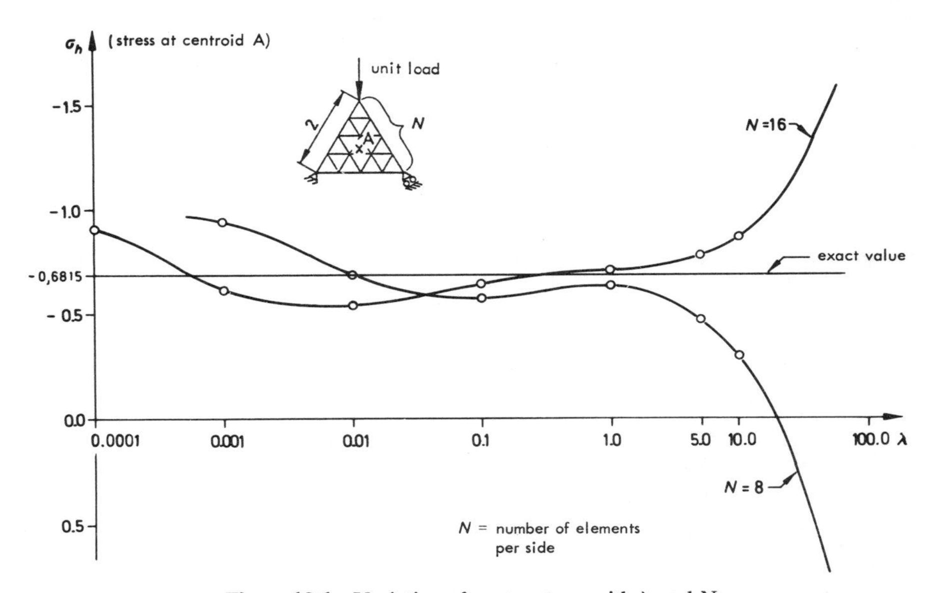

Figure 12.1 Variation of centre stress with λ and N

per side. Figure 12.2 compares the behaviour of the element with $\lambda = 1$ and increasing N with that of the standard compressible TRIM3 and of the nearly incompressible TRIM6.

12.6.2 Deep cantilever beam

Table 12.1 and Figure 12.3 show some results for a cantilever beam. We again see that the value $\lambda = 1$ gives engineering accuracy. The engineers' theory deflection is 41.6×10^{-3} but for the support conditions of the problem would be somewhat less. As λ becomes large we approach the ordinary TRIM3 result.

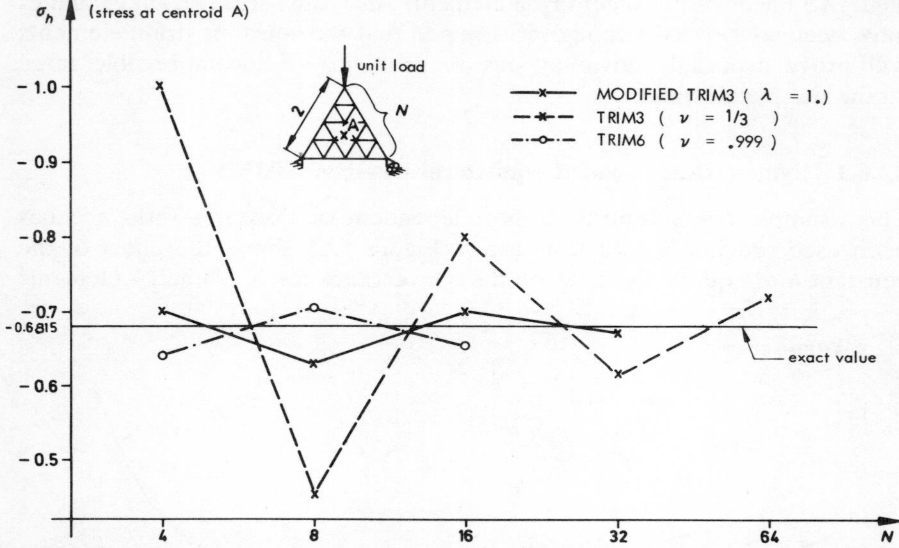

Figure 12.2 Variation of centre stress with n for TRIM3, TRIM6, and modified TRIM3 ($\lambda = 1.$)

Table 12.1 Variation of stresses and displacements with λ. (Modified trim 3)

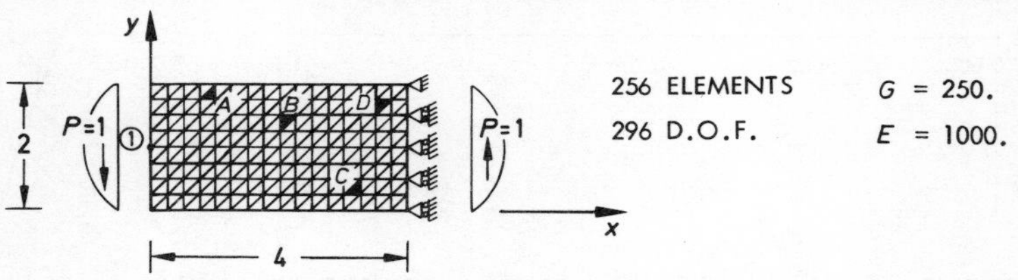

	ELEMENT A		ELEMENT B		ELEMENT C		ELEMENT D		Vertical Displacement at Node ①
λ	σ_{xx}	σ_{xy}	σ_{xx}	σ_{xy}	σ_{xx}	σ_{xy}	σ_{xx}	σ_{xy}	$(\times 10^3)$
.01	0.6158	0.2529	1.7812	0.3303	-3.6450	0.7820	4.3344	0.1117	-122.093
.1	0.9847	0.2411	1.5236	0.4998	-3.3815	0.5696	3.8855	0.3514	- 47.999
1.	1.0935	0.2534	1.3827	0.5228	-3.2537	0.5953	3.6874	0.3414	- 39.479
5.	0.9709	0.2655	1.4724	0.5219	-3.3066	0.5927	3.8325	0.3587	- 38.222
10.	0.7912	0.2750	1.6162	0.5187	-3.4158	0.5822	4.1197	0.3863	- 37.573
100.	-1.2533	0.3921	3.4656	0.4915	-4.7210	0.6097	7.6137	0.5980	- 30.880

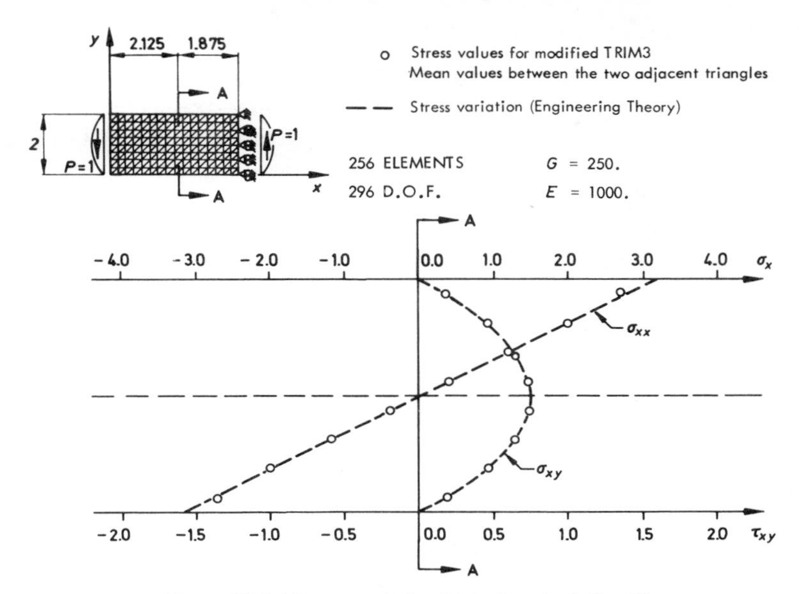

Figure 12.3 Stress variation in section A–A ($\lambda = 1$)

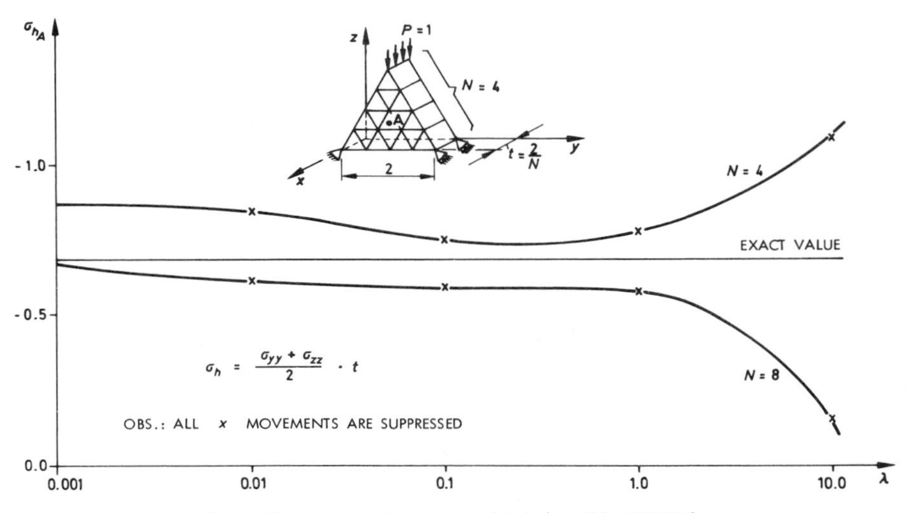

Figure 12.4 Variation of σ_h with λ (modified TET4)

12.6.3 Symmetrically loaded equilateral triangle: TET4

This is the same problem as Section 12.6.1 but using the TET4 element arranged in groups forming triangular prisms. Plane strain conditions are enforced with $\nu = 0,5$ so that the material is incompressible in the plane of the triangle. Figure 12.4 shows the variation with λ of the stress $(1/2)[\sigma_{yy} + \sigma_{zz}]$ at the centre. This stress is the mean of the values for the tetrahedrons composing the central prism and compares directly with the TRIM3 values in Figure 12.1.

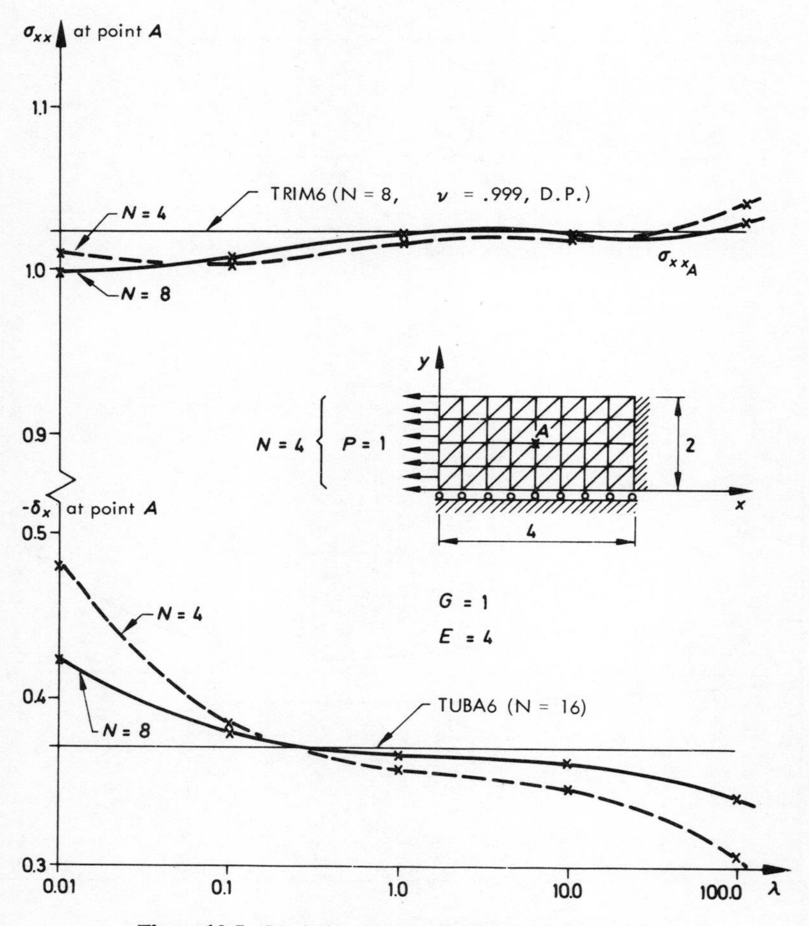

Figure 12.5 Variation of stress and displacement with λ

12.6.4 Built-in incompressible membrane with end loading

Only one half of the membrane is shown in Figures 12.5, 12.6, and 12.7, which are self-explanatory. Some comparative results using TRIM6 ($\nu =$ 0.999) and TUBA6 are also shown.

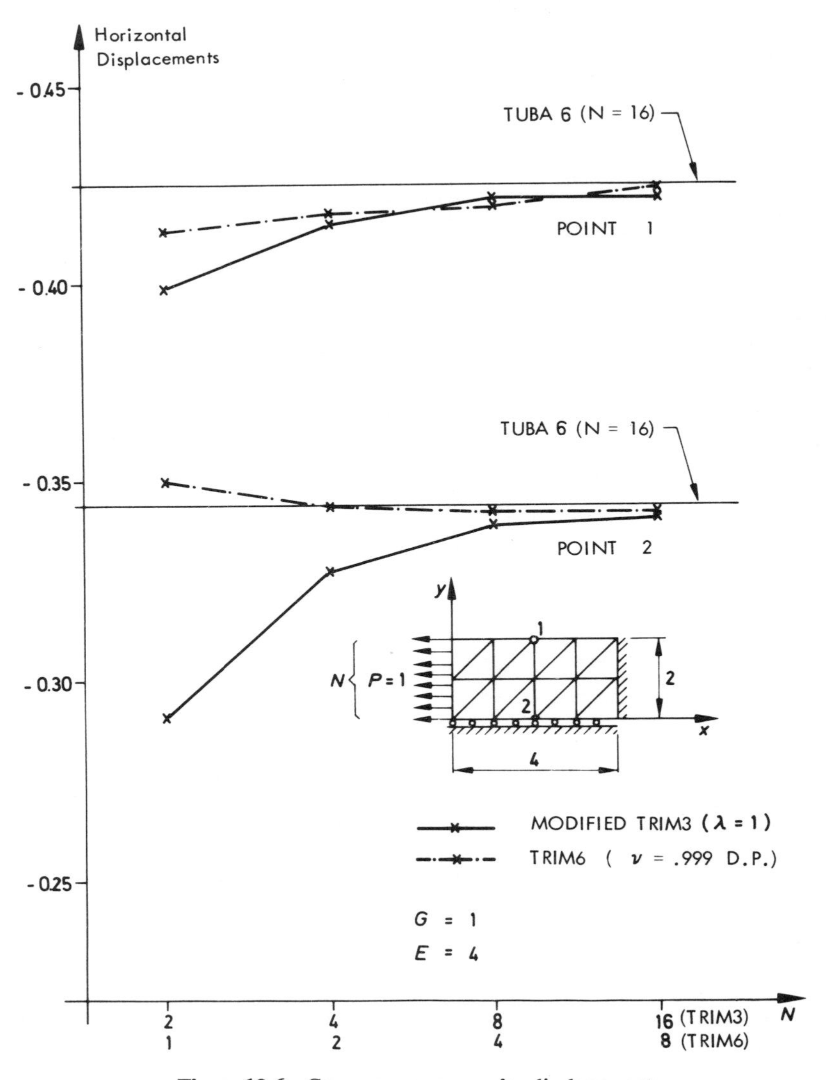

Figure 12.6 Convergence curves for displacements

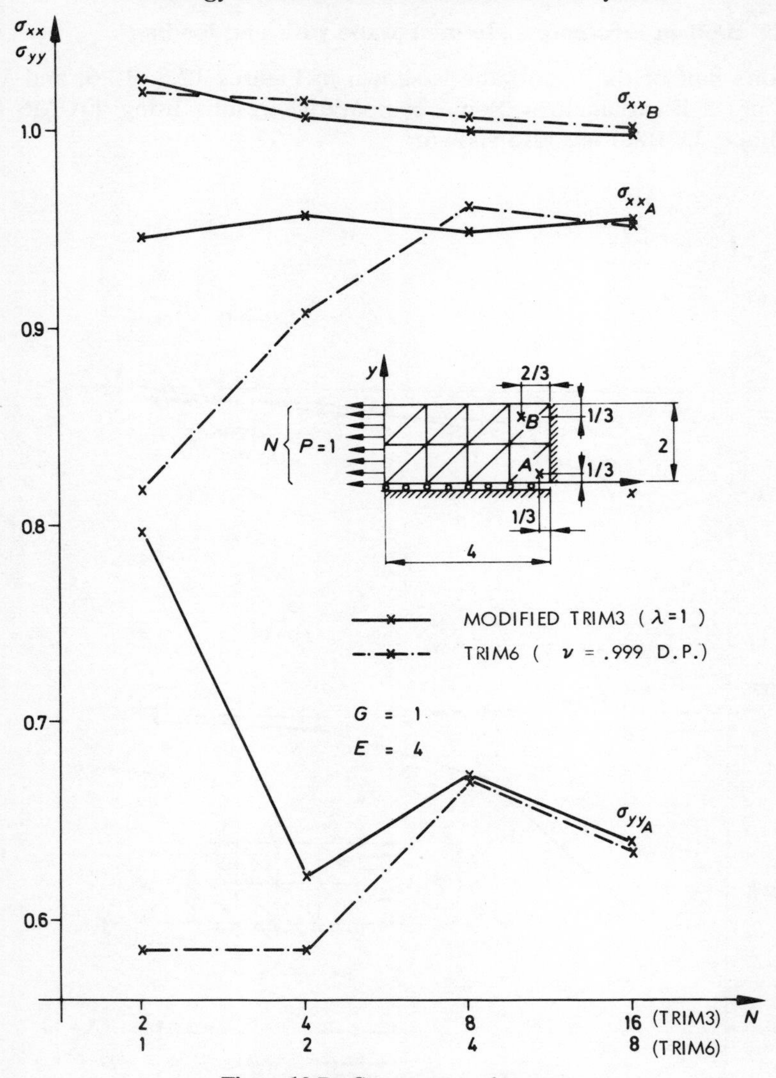

Figure 12.7 Convergence of stresses

The latter calculation uses a biharmonic equation of the displacement function ψ such that

$$\frac{\partial \psi}{\partial y} = u; \qquad \frac{\partial \psi}{\partial x} = -v \tag{12.43}$$

The TUBA6 deflections may be considered exact. Good agreement with

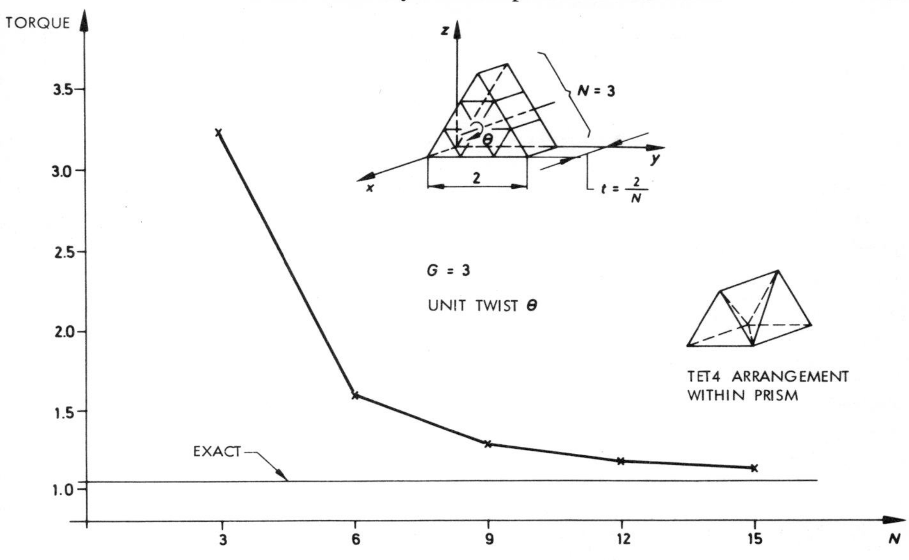

Figure 12.8 Convergence of torque with N (modified TET4, $\lambda = 1$)

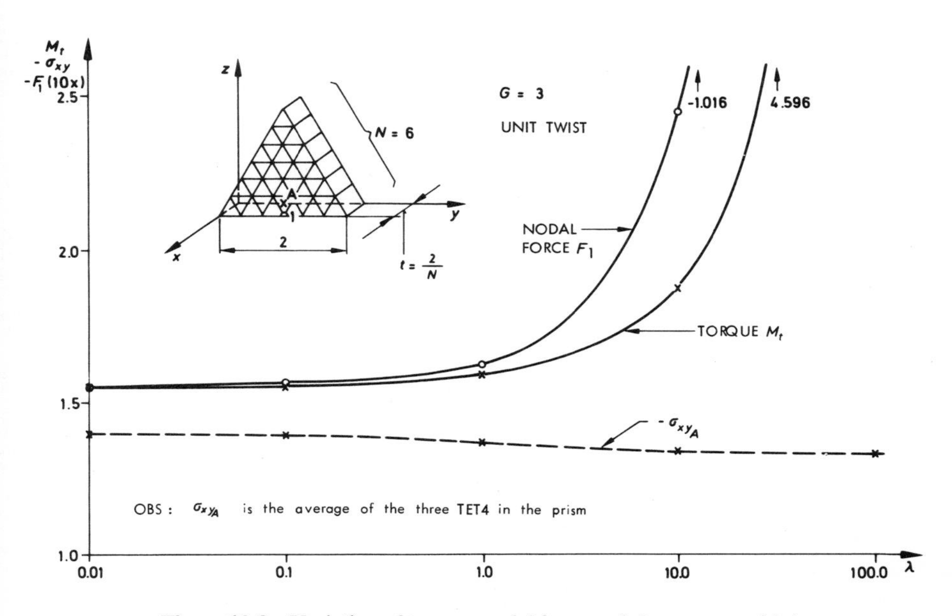

Figure 12.9 Variation of torque, nodal force and shear stress with λ

TUBA6 stresses was also obtained except near the top right-hand corner where a singularity exists.

12.6.5 Torsion of triangular bar

The problem was calculated with the same TET4 arrangement as in Section 12.6.3. To economize on the degrees of freedom the nodes on one face were fixed in the plane and on the other face received the movements corresponding to the rigid body rotation of the exact solution. The three corner points on one face were also constrained axially. All other points were free axially. The results in Figures 12.8, 12.9, and 12.10 show that the modified TET4 is capable of engineering accuracy whereas the ordinary TET4 is useless in incompressible conditions. Also, unlike in the case of TRIM3, no special arrangement of TET4's known to give a satisfactory constraint/freedom ratio.

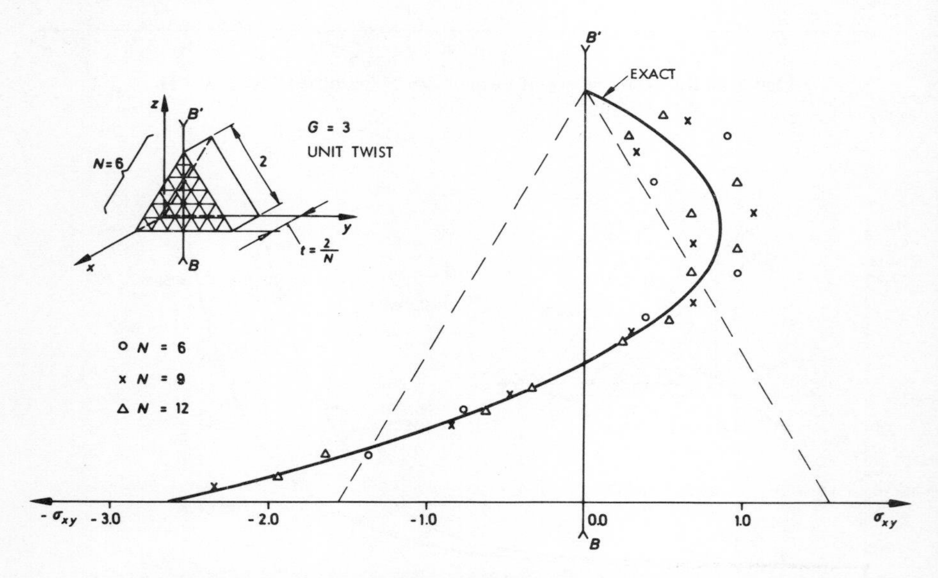

Figure 12.10 Stress variation along B–B′ (modified TET4, $\lambda = 1$)

REFERENCES

1. J. H. Argyris and P. C. Dunne, 'Constant strain finite elements for isochoric strain fields', Third International Symposium on Computing Methods in Applied Sciences and Engineering, Versailles, France, 1977.

2. J. H. Argyris, P. C. Dunne, T. Angelopoulos, and B. Bichat, 'Large natural strains and some special difficulties due to non-linearity and incompressibility in finite elements', *Comp. Meth. Appl. Eng.*, **4,** 220–278 (1974).
3. J. H. Argyris, P. C. Dunne, Th. L. Johnsen, and M. Müller, 'Linear systems with a large number of sparse constraints with applications to incompressible materials', *Comp. Meth. Appl. Mech. Eng.*, **10,** 105–132 (1977).
4. J. C. Nagtegaal, D. M. Parks, and J. R. Rice, 'On numerically accurate finite element solutions in fully plastic range', *Comp. Meth. Appl. Mech. Eng.*, **4,** 153–177 (1974).
5. A. Needleman and C. F. Shih, 'A finite element method for plane strain deformations of compressible solids', to be published in *Comp. Meth. Appl. Mech. Eng.*, **13,** (1978).
6. E. G. Thompson, 'Average and complete incompressibility in the finite element method', *Int. J. Num. Meth. Eng.*, **9,** 925–932 (1975).
7. L. R. Hermann, 'Elasticity equations for incompressible and nearly incompressible materials by a variational theorem,' *AIAAJ.*, **3,** 1896–1900 (1965).
8. S. W. Key, 'A variational principle for incompressible and nearly incompressible anisotropic elasticity,' *Int. J. Solids Structures*, **5,** 951–964 (1969).

Chapter 13

On Numerical Methods for the Stokes Problem

R. Glowinski and O. Pironneau

13.1 INTRODUCTION

A great deal of work has been done already for the numerical solution of the Stokes and Navier–Stokes equations. Since it is impossible to review all the papers on this subject, we shall mention only those which we feel are related to the methods developed in this chapter. For a more complete study we send the reader to Temam[1] and the bibliography therein.

The following study can be roughly divided into two parts. In the first part we shall review briefly the Stokes and Navier–Stokes equations and some classical methods for the solution of the stationary Stokes problem. *The cost of the numerical solution of the approximate problem* will be our point of view. In the second section we shall introduce a new method for the approximation of the Stokes problem; it is based upon a new variational formulation. This approach allows the use of *Lagrangian conforming elements of low order* (quadratic for the velocity and linear for the pressure). The errors of approximation are shown of optimal order. Then we shall describe several methods for the solution of the approximated problem which are based upon the very peculiar structure of the problem.

The main purpose of this study is to obtain an efficient 'Stokes solver' for an iterative solution of the Navier–Stokes equations.

This chapter follows the text of a lecture given at the VIIth Gatlinburg Meeting on Numerical Algebra and Optimization (Asilomar, California; 11 December 1977 – 17 December 1977).

13.2 THE STOKES AND THE NAVIER–STOKES EQUATIONS

Several *Sobolev spaces* will be used; for their definitions and properties see Adams,[2] Lions and Magenes,[3] Necas,[4] Oden–Reddy.[5]

Let Ω be an open set of $\mathbb{R}^N$ ($N=2$ or 3). Let $\Gamma=\partial\Omega$ be its boundary that we assume smooth. The non-stationary flows of incompressible viscous

Newtonian fluids are governed in Ω by the *Navier–Stokes equations:*

$$\left.\begin{aligned} &\frac{\partial \mathbf{u}}{\partial t} - \nu\Delta\mathbf{u} + (\mathbf{u}\cdot\nabla)\mathbf{u} + \nabla p = \mathbf{f} && \text{in} \quad \Omega \\ &\nabla\cdot\mathbf{u} = 0 && \text{in} \quad \Omega \\ &\mathbf{u}|_\Gamma = \mathbf{u}_\beta && \left(\text{with} \int_\Gamma \mathbf{u}_\beta\cdot\mathbf{n}\,d\Gamma = 0\right) \end{aligned}\right\} \tag{13.1}$$

In (13.1) and in a suitable system of units:

—$\mathbf{u}$ is the *velocity of the flow* and p is the *pressure* (which is defined up to a constant),
—ν (>0) is the (kinematic) *viscosity*,
—$\mathbf{n}$ is the unitary normal vector to Γ, exterior to Ω,
—$\mathbf{u}_\beta$ (given) is the velocity of the flow on Γ,
—$\mathbf{f}$ is the density of external forces,
—the condition $\nabla\cdot\mathbf{u}=0$ comes from the incompressibility of the fluid.

In this chapter we shall study the *homogeneous stationary Stokes problem:*

$$\left.\begin{aligned} -\nu\Delta\mathbf{u} + \nabla p &= \mathbf{f} && \text{in} \quad \Omega \\ \nabla\cdot\mathbf{u} &= 0 && \text{in} \quad \Omega \\ \mathbf{u}|_\Gamma &= \mathbf{0} \end{aligned}\right\} \tag{13.2}$$

The following results and methods are very easy to extend to the non-stationary and/or non-homogeneous flows (see Glowinski and Pironneau).[6]

Let us recall a theorem of existence whose proof and extention to the case Ω unbounded can be found in References 1 and 7:

Theorem 13.2.1 *If Ω is bounded (in one direction at least) and if* $\mathbf{f} \in (H^{-1}(\Omega))^N$ *then* (13.2) *has a unique solution in* $(H_0^1(\Omega))^N \times (L^2(\Omega)/\mathbb{R})$.

13.3 REVIEW OF SOME STANDARD NUMERICAL METHODS FOR THE STOKES PROBLEM

It follows from $\mathbf{v}|_\Gamma = 0$ that

$$\int_\Omega q\nabla\cdot\mathbf{v}\,dx = -\langle\nabla q, \mathbf{v}\rangle \qquad \forall q \in L^2(\Omega) \qquad \forall \mathbf{v} \in (H_0^1(\Omega))^N$$

where $\langle\cdot,\cdot\rangle$ stands for the duality pairing between $(H^{-1}(\Omega))^N$ and $(H_0^1(\Omega))^N$.

In other words,

$$-\nabla : L^2(\Omega) \to (H^{-1}(\Omega))^N \text{ is the } \textit{adjoint operator} \text{ to}$$
$$\nabla\cdot : (H_0^1(\Omega))^N \to L^2(\Omega)$$

This shows that (13.2) is of the form

$$\begin{pmatrix} A & B^t \\ B & 0 \end{pmatrix}\begin{pmatrix} \mathbf{u} \\ p \end{pmatrix} = \begin{pmatrix} \mathbf{f} \\ 0 \end{pmatrix} \tag{13.3}$$

In (13.3), $A \in \mathscr{L}(V, V')$, $B \in \mathscr{L}(H, H)$ where V (resp. H) is a Hilbert space whose dual is V' (resp. H' that we identify with H). Moreover A is *self-adjoint* and V-elliptic, i.e.

$$\langle Av, v \rangle \geqslant \alpha \|v\|_V^2 \qquad \forall v \in V$$

where $\langle \cdot, \cdot \rangle$ is the duality pairing between V' and V.

For the Stokes problem (13.2) we have:

$$A = -\nu\Delta, \qquad B = -\nabla \cdot, \qquad B^t = \nabla$$

$$H = L^2(\Omega), \qquad V = (H_0^1(\Omega))^N, \qquad V' = (H^{-1}(\Omega))^N$$

It is desirable that this structure be preserved when (13.2) is approximated by *finite differences* or *finite elements.*

Example: On a 2-D example we shall exhibit some of the properties of the linear system approximating the Stokes problem.

We take $\Omega =]0, 1[^2$ and (13.2) is discretized by *finite differences.* Let M be a positive integer and let $h = 1/M$. On $\bar{\Omega}$ we define the nets (see Figure 13.1)

$$\mathscr{U}_h = \{M_{ij} \mid M_{ij} = \{ih, jh\}, \quad 0 \leqslant i, j \leqslant M\}$$

$$\mathscr{U}_h^0 = \{M_{ij} \mid M_{ij} \in \mathscr{U}_h, \quad 1 \leqslant i, j \leqslant M-1\} = \mathscr{U}_h \cap \Omega$$

$$\mathscr{P}_h = \{M_{i+1/2, j+1/2} \mid M_{i+1/2, j+1/2} = \{(i+\tfrac{1}{2})h, (j+\tfrac{1}{2})h\}, 0 \leqslant i, j \leqslant M-1\}$$

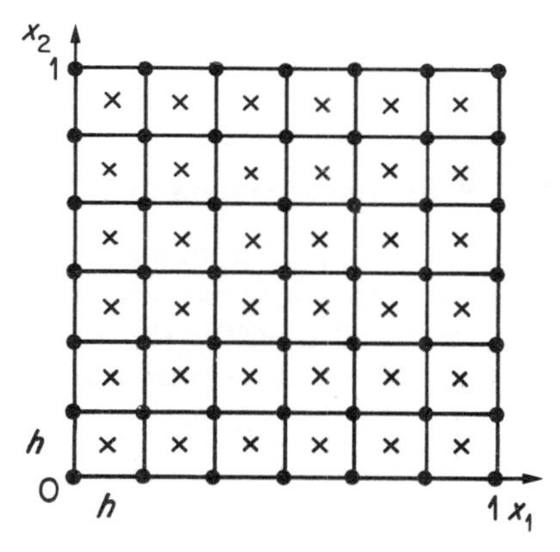

Figure 13.1 ● Nodes of $\mathscr{U}_h$; × nodes of $\mathscr{P}_h$

The *velocity* is approximated on the net $\mathcal{U}_h$ by the vector $\{\mathbf{u}_{ij}\}_{0\leq i,j\leq M}$ while the *pressure* is approximated on the net $\mathcal{P}_h$ by

$$\{p_{i+1/2,j+1/2}\}_{0\leq i\leq j\leq M-1} \text{ (do not forget that } \mathbf{u}_{ij}\in\mathbb{R}^2,\ \mathbf{u}_{ij}=\{u^1_{ij}, u^2_{ij}\})$$

Then Δ is discretized by the classical *five-point formula* and $\partial/\partial x_1$, $\partial/\partial x_2$ by *centred four-point formulae.*

Therefore the approximate Stokes problem is the *linear system:*

$$-\frac{\nu}{h^2}(u^1_{i+1j}+u^1_{i-1j}+u^1_{ij+1}+u^1_{ij-1}-4u^1_{ij})+\frac{1}{2h}(p_{i+1/2j+1/2}-p_{i-1/2j+1/2}$$
$$+p_{i+1/2j-1/2}-p_{i-1/2j-1/2})=f^1_{ij}, \qquad 1\leq i,j\leq M-1 \quad (13.4a)$$

$$-\frac{\nu}{h^2}(u^2_{i+1j}+u^2_{i-1j}+u^2_{ij+1}+u^2_{ij-1}-4u^2_{ij})+\frac{1}{2h}(p_{i+1/2j+1/2}-p_{i+1/2j-1/2}$$
$$+p_{i-1/2j+1/2}-p_{i-1/2j-1/2})=f^2_{ij}, \qquad 1\leq i,j\leq M-1 \quad (13.4b)$$

$$\frac{1}{2h}(u^1_{i+1j+1}-u^1_{ij+1}+u^1_{i+1j}-u^1_{ij})+\frac{1}{2h}(u^2_{i+1j+1}-u^2_{i+1j}+u^2_{ij+1}-u^2_{ij})=0,$$
$$0\leq i,j\leq M-1 \quad (13.5)$$

In (13.4) we assume $\mathbf{u}_{kl}=0$ if $M_{kl}\in\Gamma$.

Remark 13.3.1 Equations (13.2) (resp. (13.5)) are derived by discretizing the first equation of (13.2) (resp. the second equation of (13.2)) at the points of $\mathcal{U}_h$ (resp. $\mathcal{P}_h$).

Remark 13.3.2 If $\mathbf{f}$ is continuous one takes $\mathbf{f}_{ij}=\mathbf{f}(M_{ij})$.

Remark 13.3.3 Formulae (13.4), (13.5) can also be obtained from a *finite element* discretization with rectangles and *piecewise bilinear* approximation for $\mathbf{u}$ and *piecewise constant* pressures. Let us mention by the way that the above method is a variant of the MAC (markers and cells) method developed at Los Alamos.

13.3.1 Some properties of the linear system (13.4) and (13.5)

If the unknowns $\{u^1_{ij}\}$, $\{u^2_{ij}\}$, $\{p_{i+1/2j+1/2}\}$ are numbered properly and if (13.5) is multiplied by -1, then we obtain a linear system of type (13.3) with A positive, definite, and symmetric. It is instructive to compare some properties of this system with the system arising from the Dirichlet problem

$$\begin{aligned} -\Delta u &= f, \\ u|_\Gamma &= 0 \end{aligned} \qquad (13.6)$$

(see Table 13.1).

If (13.6) is discretized with the five-point formula we have

$$-\frac{u_{i+1j}+u_{i-1j}+u_{ij+1}+u_{ij-1}-4u_{ij}}{h^2}=f_{ij}, \qquad 1\leq i, j\leq M-1;$$

$$u_{kl}=0 \quad \text{if} \quad M_{kl}\in\Gamma \quad (13.7)$$

By inspection of Table 13.1 it appears that the numerical solution of the Stokes problem may cost much more than the one of the Dirichlet problem. This comparison is even worse in the 3-D case.

Table 13.1

Problem	Discrete Stokes		Discrete Dirichlet's
Number of unknowns	$2(N-1)^2+N^2$		$(N-1)^2$
Number of non-zero matrix elements	$2(13N-17)(N-1)$		$(5N-9)(N-1)$
Properties of the matrix	—sparse —symmetric —indefinite		—sparse —symmetric —positive definite
Bandwidth	Bandwidth Stokes	$\gg$	Bandwidth Dirichlet

Orientation. It appears from the short analysis above that two directions may be pursued for the solution of the Stokes problem:

(1) Use the general methods for *symmetric, indefinite, linear systems.* Either the recent *direct* methods of Duff *et al.*[8] which seems very interesting for sparse matrices; or use the *iterative* methods of Lanczos type like, for example, Paige and Saunders,[9] Widlund[10] (some recent tests done by Thomasset and Widlund at IRIA and at the Courant Institute, demonstrate the interesting properties of Lanczos methods for the Stokes and Navier–Stokes problems).
(2) Use specific methods based upon the particular structure of the problem.

In the sequel we shall focus on the second approach. In particular we shall break down the Stokes problem into a finite number of Dirichlet problems for $-\Delta$ (for which a very sophisticated methodology can be used either with finite differences or finite elements).

13.3.2 Gradient and conjugate gradient methods

13.3.2.1 *Generalities*

From now on Ω is bounded and Γ is regular (Lipschitz continuous). We define $H \subset L^2(\Omega)$ by

$$H = \left\{ q \in L^2(\Omega) \,\middle|\, \int_\Omega q(x)\,\mathrm{d}x = 0 \right\}$$

The iterative methods below are based upon the following result.

Theorem 13.3.1 *Let* $\mathscr{a}: L^2(\Omega) \to L^2(\Omega)$ *be defined by*

$$q \in L^2(\Omega) \tag{13.8}$$

$$\Delta \mathbf{v} = \nabla q \text{ in } \Omega, \qquad \mathbf{v} \in (H_0^1(\Omega))^N \textit{ (which implies } \mathbf{v}|_\Gamma = \mathbf{0}) \tag{13.9}$$

$$\mathscr{a} q = \nabla \cdot \mathbf{v} \tag{13.10}$$

Then $\mathscr{a}$ *is* H*-elliptic, self-adjoint, automorphic from* H *onto* H (i.e. $\exists \alpha > 0$ *such that* $(\mathscr{a} q, q)_{L^2} \geqslant \alpha \|q\|^2_{L^2} \forall q \in H$).

The proof can be found in Crouzeix.[11]

Remark 13.3.4 The *discrete* forms of $\mathscr{a}$ are in general *full* matrices.

From Theorem 13.3.1 we shall derive a family of gradient methods (steepest descent) for the solution of Stokes problem.

13.3.2.2 *Gradient methods and variant*

Let $\{\mathbf{u}, p\} \in (H_0^1(\Omega))^N \times L^2(\Omega)$ be the solution of the Stokes problem (13.2) and let $\mathbf{u}_0$ be the solution of

$$-\nu \Delta \mathbf{u}_0 = \mathbf{f} \quad \text{in} \quad \Omega, \qquad \mathbf{u}_0 \in (H_0^1(\Omega))^N \tag{13.11}$$

By subtracting (13.2) and (13.11) we have

$$\nu \Delta(\mathbf{u} - \mathbf{u}_0) = \nabla p \quad \text{in} \quad \Omega, \qquad \mathbf{u} - \mathbf{u}_0 \in (H_0^1(\Omega))^N$$

Hence $\mathscr{a} p = \nu \nabla \cdot (\mathbf{u} - \mathbf{u}_0) = -\nu \nabla \cdot \mathbf{u}_0$. In other words the pressure is the unique solution in $L^2(\Omega)/\mathbb{R}$ of

$$\mathscr{a} p = -\nu \nabla \cdot \mathbf{u}_0 \tag{13.12}$$

Owing to the properties of $\mathscr{a}$ (see Theorem 13.3.1) it is natural to solve (13.12) (and therefore (13.2)) by iterative methods such as the method of steepest descent.

Gradient method with fixed step size. For a given $\rho>0$ consider the following algorithm:

$$p^0 \in L^2(\Omega) \quad \text{given arbitrarily,} \tag{13.13}$$

for $n \geq 0$, p^n given, compute,

$$p^{n+1} = p^n - \rho(\alpha p^n + \nu\nabla \cdot \mathbf{u}_0) \tag{13.14}$$

In practice one has to replace (13.14) by

$$-\nu\Delta\mathbf{u}^n = \mathbf{f} - \nabla p^n \quad \text{in} \quad \Omega, \qquad \mathbf{u}^n \in (H_0^1(\Omega))^N \tag{13.14a}$$

$$p^{n+1} = p^n - \rho\nu\nabla \cdot \mathbf{u}^n \tag{13.14b}$$

Remark 13.3.5 To solve (13.14a) one has to solve N independent Dirichlet problems for $-\Delta$ (in practice $N=2$ or 3).

Remark 13.3.6 The previous method is close to the artificial compressibility methods of Chorin and Yanenko.

We recall the following result.

Theorem 13.3.2 If in (13.13), (13.14) *we have*

$$0<\rho<\frac{2}{N} \tag{13.15}$$

then $\forall p^0 \in L^2(\Omega)$ *we have*

$$\lim_{n\to+\infty} \{\mathbf{u}^n, p^n\} = \{\mathbf{u}, p\} \quad \textit{in} \quad (H_0^1(\Omega))^N \times L^2(\Omega), \quad \textit{strongly} \tag{13.16}$$

where $\{\mathbf{u}, p\}$ *is the solution of the Stokes problem* (13.2) *with* $\int_\Omega p \, dx = \int_\Omega p^0 \, dx$. *Moreover the rate of convergence is linear.*

We remind the reader that the $(H_0^1(\Omega))^N$-norm is

$$\|\mathbf{v}\| = \left(\int_\Omega |\nabla\mathbf{v}|^2 \, dx\right)^{1/2} = \sum_{i=1}^{N}\left(\int_\Omega |\nabla v_i|^2 \, dx\right)^{1/2}$$

Variants of (13.13) (13.14). One can find in Reference 11 variants of (13.13), (13.14) where a sequence of parameters $\{\rho_n\}_{n\geq 0}$ (cyclic in particular) is used instead of a fixed ρ. Accelerating methods of Tchebycheff type can also be found in Reference 11 for (13.13), (13.14).

Steepest descent and *minimal residual* procedures for (13.13) (13.14) can also be found in Fortin and Glowinski[12] and Fortin and Thomasset.[13] Each of these methods requires N uncoupled Dirichlet problems for $-\Delta$ to be solved at each iteration. However, these variants of (13.13), (13.14) seem less efficient than the *conjugate gradient method* of Section 13.3.2.3 which, by the way, is only slightly costlier to implement.

13.3.2.3 *A conjugate gradient method*

It follows from Daniel[14] that one may solve (13.2) via (13.12) by a *conjugate gradient method*. Referring to References 12 and 13 for more details, we shall limit ourselves to the description of the algorithm. For the sake of clarity, but without loss of generality we set $\nu = 1$. Then the conjugate gradient algorithm is as follows:

$$p^0 \in L^2(\Omega), \quad \text{given arbitrarily,} \tag{13.17}$$

$$-\Delta \mathbf{u}^0 = \mathbf{f} - \nabla p^0, \qquad \mathbf{u}^0 \in (H_0^1(\Omega))^N \tag{13.18}$$

$$g^0 = \nabla \cdot \mathbf{u}^0 \tag{13.19}$$

$$z^0 = g^0 \tag{13.20}$$

then for $n \geqslant 0$,

$$\rho_n = \frac{(z^n, g^n)_{L^2(\Omega)}}{(az^n, z^n)_{L^2(\Omega)}} = \frac{\|g^n\|^2_{L^2(\Omega)}}{(az^n, z^n)_{L^2(\Omega)}} \tag{13.21}$$

$$p^{n+1} = p^n - \rho_n z^n \tag{13.22}$$

$$g^{n+1} = g^n - \rho_n az^n \tag{13.23}$$

$$\gamma_n = \frac{\|g^{n+1}\|^2_{L^2(\Omega)}}{\|g^n\|^2_{L^2(\Omega)}} \tag{13.24}$$

$$z^{n+1} = g^{n+1} + \gamma_n z^n \tag{13.25}$$

then $n = n + 1$ and go to (13.12).

To implement (13.17)–(13.25) it is necessary to know az^n. From Theorem 13.3.1, az^n can be obtained by

$$\Delta \boldsymbol{\chi}^n = \nabla z^n, \qquad \boldsymbol{\chi}^n \in (H_0^1(\Omega))^N \tag{13.26}$$

$$az^n = \nabla \cdot \boldsymbol{\chi}^n \tag{13.27}$$

Thus each iteration costs N uncoupled Dirichlet problem for $-\Delta$. The *strong convergence* of p^n to p can be shown as in Theorem 13.3.2.

Remark 13.3.7 Owing to the H-ellipticity of a *it is not necessary to precondition* (i.e. *to scale*) the conjugate gradient algorithm above.

13.3.3 Penalty-duality methods

It is shown in References 12 and 13, for example (see also Reference 1, that the Stokes problem can be solved by a penalty-duality method (in the sense of Hestenes[15] and Powell[16]).

Therefore let $r>0$. We note that the Stokes problem (13.2) is equivalent to

$$\left.\begin{aligned} &-\Delta\mathbf{u}-r\nabla(\nabla\cdot\mathbf{u})+\nabla p=\mathbf{f} \quad \text{in} \quad \Omega \\ &\nabla\cdot\mathbf{u}=0 \quad \text{in} \quad \Omega \\ &\mathbf{u}|_\Gamma=0 \end{aligned}\right\} \tag{13.28}$$

It is then natural to generalize algorithm (13.13), (13.14) by

$$p^0\in L^2(\Omega) \quad \text{arbitrarily given,} \tag{13.29}$$

and for $n\geqslant 0$, p^n being known:

$$-\Delta\mathbf{u}^n-r\nabla(\nabla\cdot\mathbf{u}^n)=\mathbf{f}-\nabla p^n \quad \text{in} \quad \Omega, \qquad \mathbf{u}^n\in(H_0^1(\Omega))^N(\Rightarrow\mathbf{u}^n|_\Gamma=\mathbf{0}) \tag{13.30}$$

$$p^{n+1}=p^n-\rho\nabla\cdot\mathbf{u}^n, \qquad \rho>0 \tag{13.31}$$

For the convergence of (13.29)–(13.31) one shows the following.

Theorem 13.3.3 If in (13.29)–(13.31), *ρ satisfies*

$$0<\rho<2\left(r+\frac{1}{N}\right) \tag{13.32}$$

then $\forall p^0\in L^2(\Omega)$ one has

$$\lim_{n\to\infty}\{\mathbf{u}^n, p^n\}=\{\mathbf{u}, p\} \quad \textit{in} \quad (H_0^1(\Omega))^N\times L^2(\Omega) \quad \textit{strongly} \tag{13.33}$$

where $\{\mathbf{u}, p\}$ is the solution of the Stokes problem (13.2) *with $\int_\Omega p\,\mathrm{d}x=\int p^0\,\mathrm{d}x$. Moreover the convergence is linear.* ■

The above results can be made more precise by observing that

$$p^{n+1}-p=(I-\rho(rI+\mathscr{a}^{-1})^{-1}(p^n-p)$$

(where $\mathscr{a}$ is as in Theorem 13.3.1). Each operator being in $\mathscr{L}(L^2(\Omega), L^2(\Omega))$ we have

$$\|p^{n+1}-p\|_{L^2(\Omega)}\leqslant\|I-\rho(rI+\mathscr{a}^{-1})^{-1}\|\,\|p^n-p\|_{L^2(\Omega)} \tag{13.34}$$

And

$$I-\rho(rI+\mathscr{a}^{-1})^{-1}=(rI+\mathscr{a}^{-1})^{-1}((r-\rho)I+\mathscr{a}^{-1})$$

yields

$$\|I-\rho(rI+\alpha^{-1})^{-1}\| \leq \frac{1}{r}(|r-\rho|+\|\alpha^{-1}\|). \tag{13.35}$$

It follows from (13.24) (13.35) that for the classical choice (see Reference 12) $\rho=r$, we have

$$\|p^{n+1}-p\|_{L^2} \leq \frac{\|\alpha^{-1}\|}{r}\|p^n-p\|_{L^2} \tag{13.36}$$

Therefore if r is large enough the convergence ratio of algorithm (13.29)–(13.31) is of order $1/r$.

13.3.3.1 *Remarks on algorithm* (13.29)–(13.31):

Remark 13.3.8 The system (13.30) is closely related to the *linear elasticity system.* Once it is discretized by *finite differences* or *finite elements,* it can be solved using a *Cholesky's factorization* LL^t or LDL^t, done once and for all (this remark holds also for the algorithms of Section 13.3.2).

Remark 13.3.9 The method of (13.29)–(13.31) has the drawback of requiring the solution of a system of N partial differential equations coupled (if $r>0$) by $r\nabla(\nabla\cdot)$, while this is not so for algorithms of Section 13.3.2. Hence much more computer storage is required.

Remark 13.3.10 By inspection of (13.3.6) it seems that one should take $\rho=r$, and r as large as possible. However, (13.30) and its discrete forms will be *ill-conditioned* when *r is large.* In practice if (13.36) is solved by a *direct method* (Gauss, Cholesky) one should take r in the range of 10^2 to 10^5. In such cases and if $\rho=r$ the convergence of (13.29), (13.31) is extremely fast (about three iterations). Under such conditions it is not necessary to use a conjugate gradient accelerating scheme.

Remark 13.3.11 In fact, (13.29)–(13.31) is a *UZAWA* algorithm (see, for example, Reference 12 and Glowinski, Lions, and Tremolieres (Reference 17, Chapter 2)) applied to the computation of the *saddle-points* of the *augmented Lagrangian* $\mathcal{L}_r:(H_0^1(\Omega))^N\times L^2(\Omega)\to\mathbb{R}$ defined by

$$\mathcal{L}_r(\mathbf{v},q)=\frac{1}{2}\int_\Omega|\nabla\mathbf{v}|^2\,dx+\frac{r}{2}\int_\Omega(\nabla\cdot\mathbf{v})^2\,dx-\int_\Omega f\cdot\mathbf{v}\,dx-\int_\Omega q\nabla\cdot\mathbf{v}\,dx \tag{13.37}$$

This remark holds also for algorithms of Section 13.3.2 with $r=0$ in (13.37). Formula (13.37) is directly related to the fact that the *pressure* p is a

Lagrange multiplier to the *condition of incompressibility* $\nabla \cdot \mathbf{v}=0$ in the equivalent formulation of the Stokes problem:

$$\left.\begin{aligned} &\operatorname*{Min}_{\mathbf{v}\in V}\left\{\frac{1}{2}\int_\Omega |\nabla \mathbf{v}|^2\,dx - \int_\Omega \mathbf{f}\cdot\mathbf{v}\,dx\right\} \\ &V=\{v\in (H_0^1(\Omega))^N;\ \nabla\cdot\mathbf{v}=0\}\end{aligned}\right\} \tag{13.38}$$

13.4 ON A NEW METHOD FOR THE SOLUTION OF THE STOKES PROBLEM

In this section we shall describe a new class of methods, due to Glowinski and Pironneau,[18,19] for the numerical solution of the Stokes problem. Unlike the previous methods, the trace of the pressure on $\partial\Omega$ will play an important role. It leads also to the construction of a *Stokes solver* easy to implement, once in possession of a subroutine for the numerical solution of the Dirichlet problem for $-\Delta$. This method is closely related to the ideas used by the authors in Reference 20 for the *biharmonic equation.*

13.4.1 The continuous case: motivation

As before Ω is bounded and $\nu=1$. Let

$$\mathcal{H}^{1/2}(\Gamma)=\left\{\mu\in H^{1/2}(\Gamma),\ \int_\Gamma \mu\,d\Gamma=0\right\}$$

The methods below are based on the following result:

Theorem 13.4.1 *Let $\lambda\in H^{-1/2}(\Gamma)$; let $A: H^{-1/2}(\Gamma)\to H^{1/2}(\Gamma)$ be defined by*

$$\left.\begin{aligned} &\Delta p_\lambda = 0 \quad \text{in} \quad \Omega \\ &p_\lambda \in H(\Omega;\Delta)=\{q\mid q\in L^2(\Omega),\ \Delta q\in L^2(\Omega)\} \\ &p_\lambda=\lambda \quad \text{on} \quad \Gamma \end{aligned}\right\} \tag{13.39}$$

$$\Delta\mathbf{u}_\lambda=\nabla p_\lambda \quad \text{in} \quad \Omega, \qquad u_\lambda\in (H_0^1(\Omega))^N \tag{13.40}$$

$$-\Delta\psi_\lambda=\nabla\cdot u_\lambda \quad \text{in} \quad \Omega, \qquad \psi_\lambda\in H_0^1(\Omega) \tag{13.41}$$

$$A\lambda=-\left.\frac{\partial\psi_\lambda}{\partial n}\right|_\Gamma \tag{13.42}$$

Then A is an isomorphism from $H^{-1/2}(\Gamma)/R$ onto $\mathcal{H}^{1/2}(\Gamma)$. Moreover the bilinear form $a(\cdot,\cdot)$ defined by

$$a(\lambda,\mu)=\langle A\lambda,\mu\rangle$$

where $\langle \cdot, \cdot \rangle$ *denotes the duality pairing between* $H^{1/2}(\Gamma)$ *and* $H^{-1/2}(\Gamma)$, *is continuous, symmetric and* $H^{-1/2}(\Gamma)/\mathbb{R}$*-elliptic.*

The reader is referred to Reference 21 for the proof.

Application of Theorem 13.4.1 to the solution of the Stokes problem. Assume that $\mathbf{f} \in L^2(\Omega))^N$, *and define* $p_0, \mathbf{u}_0, \psi_0$ by

$$\Delta p_0 = \nabla \cdot \mathbf{f} \quad \text{in} \quad \Omega, \qquad p_0 \in H_0^1(\Omega) \tag{13.43}$$

$$-\Delta \mathbf{u}_0 = \mathbf{f} - \nabla p_0 \quad \text{in} \quad \Omega, \qquad \mathbf{u}_0 \in (H_0^1(\Omega))^N \tag{13.44}$$

$$-\Delta \psi_0 = \nabla \cdot \mathbf{u}_0 \quad \text{in} \quad \Omega, \qquad \psi_0 \in H_0^1(\Omega) \tag{13.45}$$

The following is easy to prove.

Theorem 13.4.2 *If* $\{\mathbf{u}, p\}$ *is the solution of the Stokes problem* (13.2), *then the trace* λ *of* p *on* Γ *is the unique solution of the linear variational equation:*

$$(E) \qquad \begin{cases} \lambda \in H^{-1/2}(\Gamma)/\mathbb{R}, \\ \langle A\lambda, \mu \rangle = \left\langle \dfrac{\partial \psi_0}{\partial n}, \mu \right\rangle \qquad \forall \mu \in H^{-1/2}(\Gamma)/\mathbb{R}. \end{cases}$$ ■

Theorem 13.4.2 implies that the Stokes problem (13.2) can be broken down to a *finite number* of Dirichlet problems for $-\Delta$ ($N+2$ for ψ_0, $N+1$ for $\{\mathbf{u}, p\}$ once λ is known) plus the problem (E) on $\partial\Omega$; the main difficulty being that *A is not known explicitly.*

Remark 13.4.1 If μ is sufficiently regular, Green's formula yields

$$\left\langle \frac{\partial \psi_0}{\partial n}, \mu \right\rangle = \int_\Omega \nabla \psi_0 \cdot \nabla \tilde{\mu} \, dx - \int_\Omega \nabla \cdot \mathbf{u}_0 \tilde{\mu} \, dx = \int_\Omega (\nabla \psi_0 + \mathbf{u}_0) \cdot \nabla \tilde{\mu} \, dx \tag{13.46}$$

where $\tilde{\mu}$ is a regular extention of μ in Ω. Note that in (13.46) $\partial \psi_0 / \partial n$ does not appear explicitly. We shall use this remark to approximate (E). ■

To approximate (E) will require us to introduce a new variational formulation of the Stokes problem, discretized in turn by mixed finite elements (see Section 13.4.3).

13.4.2 A new variational formulation of the Stokes problem

Let

$$W_0 = \left\{ \{\mathbf{v}, \phi\} \in (H_0^1(\Omega))^{N+1}, \int_\Omega \nabla \phi \cdot \nabla w \, dx = \int_\Omega \nabla \cdot \mathbf{v} w \, dx \qquad \forall w \in H^1(\Omega) \right\}$$

Proposition 13.4.1 *If* $\{\mathbf{u}, \phi\} \in W_0$ *then* $-\Delta\phi = \nabla \cdot \mathbf{v}$ *in* Ω *and* $\phi = \partial\phi/\partial n = 0$ *on* Γ.

As above for the sake of clarity we assume that $\mathbf{f} \in (L^2(\Omega))^N$. Consider the following problem

$$(P) \qquad \begin{cases} \text{Find } \{\mathbf{u}, \psi\} \in W_0 \text{ such that} \\ \int_\Omega \nabla\mathbf{u} \cdot \nabla\mathbf{v}\, dx = \int_\Omega \mathbf{f} \cdot (\mathbf{v} + \nabla\phi)\, dx \qquad \forall \{\mathbf{v}, \phi\} \in W_0 \end{cases}$$

Then we have

Theorem 13.4.3 *(P) has a unique solution* $\{\mathbf{u}, \psi\}$ *where* $\psi = 0$ *and* $\mathbf{u}$ *is the solution of the Stokes problem* (13.2).

Remark 13.4.3 The formulation (P) can be interpreted as follows: if $\mathbf{v} \in (H_0^1(\Omega))^N$ and $\partial\Omega$ is sufficiently smooth, there exists $\phi \in H^2(\Omega) \cap H_0^1(\Omega)$ and $\boldsymbol{\omega} \in (H^1(\Omega))^N$ with $\nabla \cdot \boldsymbol{\omega} = 0$, such that

$$\mathbf{v} = -\nabla\phi + \boldsymbol{\omega} \tag{13.47}$$

and the decomposition (13.47) is *unique.*

In the formulation (P), instead of directly imposing $\nabla \cdot \mathbf{v} = 0$, we try to impose $\phi = 0$; these procedures are equivalent in the continuous case but not in the discrete case.

13.4.3 A mixed finite element approximation

In this section we proceed to define a mixed finite element approximation to the Stokes problem. We limit ourselves to the case where Ω is polygonal and bounded in $\mathbb{R}^2$, but the following extends to $\Omega \subset \mathbb{R}^3$ (see Reference 22 for computational results).

13.4.3.1 *Triangulation of Ω. Fundamental discrete spaces*

Let $\{\mathcal{T}_h\}_h$ be a family of *regular triangulations* of Ω such that $\bar{\Omega} = \bigcup_{T \in \mathcal{T}_h} T$. We set $h(T) =$ length of the greatest side of T, $h = \max_{T \in \mathcal{T}_h} h(T)$ and we assume that

$$\frac{h}{\min_{T \in \mathcal{T}_h} h(T)} \leq \beta \ \forall \mathcal{T}_h \tag{13.48}$$

Then we define the following finite-dimensional spaces:

$$\left.\begin{array}{l} H_h^1=\{\phi_h \in C^0(\bar{\Omega}), \quad \phi_h|_T \in P_1 \quad \forall T \in \mathcal{T}_h\} \\ H_{0h}^1=H_h^1 \cap H_0^1(\Omega)=\{\phi_h \in H_h^1, \phi_h|_{\partial\Omega}=0\} \end{array}\right\} \tag{13.49}$$

$$\left.\begin{array}{l} V_h=\{\mathbf{v}_h \in (C^0(\bar{\Omega}))^2, \mathbf{v}_h|_T \in (P_2)^2 \quad \forall T \in \mathcal{T}_h\} \\ V_{0h}=V_h \cap (H_0^1(\Omega))^2 \end{array}\right\} \tag{13.50}$$

We will also consider V_h defined by

$$V_h=\{\mathbf{v}_h \in C^0(\bar{\Omega})^2, \mathbf{v}_h|_T \in (P_1)^2 \qquad \forall T \in \tilde{\mathcal{T}}_h\} \qquad \text{(13.50)bis}$$

where $\tilde{\mathcal{T}}_h$ is the triangulation deduced from $\mathcal{T}_h$ by dividing each triangle $T \in \mathcal{T}_h$ into four equal triangles (by joining the mid-sides). We record that P_k denote the space of polynomial of degree $\leqslant k$. Finally we define

$$W_{0h}=\left\{\{\mathbf{v}_h, \phi_h\} \in V_{0h} \times H_{0h}^1, \int_\Omega \nabla\phi_h \cdot \nabla w_h \,\mathrm{d}x=\int_\Omega \nabla \cdot \mathbf{v}_h w_h \,\mathrm{d}x \qquad \forall w_h \in H_h^1\right\}$$

13.4.3.2 *Definition of the approximate problem; characterization of the approximate solution*

We approximate (P) (i.e. the Stokes problem) by

$$(P_h) \qquad \begin{cases} \textit{Find } \{\mathbf{u}_h, \psi_h\} \in W_{0h} \quad \textit{such that} \\ \displaystyle\int_\Omega \nabla\mathbf{u}_h \cdot \nabla\mathbf{v}_h \,\mathrm{d}x=\int_\Omega \mathbf{f} \cdot (\mathbf{v}_h+\nabla\phi_h)\,\mathrm{d}x \qquad \forall\{\mathbf{v}_h, \phi_h\} \in W_{oh} \end{cases}$$

Then the following is shown in Reference 6:

Theorem 13.4.4 (P_h) *has a unique solution and it satisfies*

$$\int_\Omega \nabla p_h \cdot \nabla w_h \,\mathrm{d}x=\int_\Omega \mathbf{f} \cdot \nabla w_h \,\mathrm{d}x \qquad \forall w_h \in H_{0h}^1, \qquad p_h \in H_h^1 \tag{13.51}$$

$$\int_\Omega \nabla\mathbf{u}_h \cdot \nabla\mathbf{v}_h \,\mathrm{d}x=\int_\Omega (-\nabla p_h+\mathbf{f}) \cdot \mathbf{v}_h \,\mathrm{d}x \qquad \forall\mathbf{v}_h \in V_{0h}, \qquad \mathbf{u}_h \in V_{0h} \tag{13.52}$$

$$\{\mathbf{u}_h, \psi_h\} \in W_{0h} \tag{13.53}$$

Remark 13.4.4 The discrete pressure p_h is the *Lagrange multiplier* of condition (13.53).

Remark 13.4.5 If in W_{0h} and (P_h) we impose $\phi_h=\psi_h=0$ (which may be, since $\psi=0$), then the scheme is identical to the one in Taylor and Hood[23] for the Stokes problem whose convergence was established by Bercovier and Pironneau.[24]

13.4.3.3 *Error estimates*

In the sequel C will denote various constants. The following lemma, proved in References 21 and 24, plays a fundamental part.

Lemma 13.4.1 *It is assumed that no $T \in \mathcal{T}_h$ has two sides or more belonging to $\partial\Omega$. Then, provided that* (13.48)–(13.50) (*resp.* (13.50)bis) *holds, there exists C independent of h such that*

$$\|\nabla q_h\|_{L^2(\Omega)} \leq C \max_{\mathbf{v}_h \in V_{0h} - \{\mathbf{0}\}} \frac{\int_\Omega \mathbf{v}_h \cdot \nabla q_h \, dx}{\|\mathbf{v}_h\|_{L^2(\Omega)}} \qquad \forall q_h \in H_h^1. \quad \blacksquare \tag{13.54}$$

It is easy to show that (13.54) implies the *uniqueness* of p_h in $H_h^1/\mathbb{R}$. From Lemma 13.4.1 and following Thomas,[25] one can show the following.

Theorem 13.4.5 *Assume that* (13.48)–(13.50), (13.54) *hold and that Ω is a convex polygonal. Then if $\{\mathbf{u}, p\}$, solution of the Stokes problem, belongs to* $(H^3(\Omega))^2 \times H^2(\Omega)$:

$$\|\mathbf{u}_h - \mathbf{u}\|_{(H^1(\Omega))^2} \leq Ch^2(\|\mathbf{u}\|_{(H^3(\Omega))^2} + \|p\|_{H^2(\Omega)/\mathbb{R}}) \tag{13.55}$$

$$\|p_h - p\|_{H^1(\Omega)/\mathbb{R}} \leq Ch(\|\mathbf{u}\|_{(H^3(\Omega))^2} + \|p\|_{H^2(\Omega)/\mathbb{R}}) \tag{13.56}$$

Remark 13.4.6 If we use V_h defined by (13.50)bis and if $\{\mathbf{u}, p\} \in (H^2(\Omega))^2 \times H^1(\Omega)/\mathbb{R}$ then,

$$\|\mathbf{u}_h - \mathbf{u}\|_{(H^1(\Omega))^2} \leq Ch(\|\mathbf{u}\|_{(H^2(\Omega))^2} + \|p\|_{H^1(\Omega)/\mathbb{R}}) \tag{13.57}$$

Remark 13.4.7 The above error estimates have an optimal order.

13.4.3.4 *Comments*

The above methods, based on Lagrangian finite triangular elements, conforming in $H^1(\Omega)$, are easier to implement that the non-conforming methods (cf. References 1, 26, and 27). They generalize naturally to the 3-D case, to quadrilateral elements as well as curved boundaries (with curved elements (see Zienkiewicz[28] isoparametric for the velocity, superparametric for the pressure). Finally let us mention that Le Tallec[29] has extended the error estimate theorems to the *stationary Navier–Stokes equations*.

13.4.4 Approximation of Problem (E)

We shall now use the finite elements of Section 13.4.3 to approximate (E) defined in Section 13.4.1.

13.4.4.1 *The space $\mathcal{M}_h$. Approximation of $a(\cdot,\cdot)$*

Let $\mathcal{M}_h$ be a complementary space of H^1_{0h} in H^1_h, i.e. $H^1_h = \mathcal{M}_h \oplus H^1_{0h}$. In practice $\mathcal{M}_h$ is defined by

$$\left.\begin{aligned} &\mathcal{M}_h \oplus H^1_{0h} = H^1_h \\ &\phi_h \in \mathcal{M}_h \Rightarrow \phi_h|_T = 0 \qquad \forall T \in \mathcal{T}_h \quad \text{such that} \quad \partial T \cap \partial\Omega = \varnothing \end{aligned}\right\} \tag{13.58}$$

Let $N_h = \dim \mathcal{M}_h$; if H^1_h, H^1_{0h} are defined by (13.49) then N_h equals the number of nodes of $\mathcal{T}_h$ which belong to $\partial\Omega$. Notice that if $\phi_h \in \mathcal{M}_h$, $\text{supp.}\,(\phi_h) \subset \bar{\Omega}_{\gamma_h} = \bigcup_{T \cap \partial\Omega \neq \varnothing} T$ and that, $\lim_{h\to 0} \text{meas}\,(\bar{\Omega}_{\gamma_h}) = 0$.

Approximation of $a(\cdot,\cdot)$. With the notation of Section 13.4.1, if μ is sufficiently regular, Green's formula yields

$$\left.\begin{aligned} a(\lambda,\mu) &= -\int_\Gamma \frac{\partial\psi_\lambda}{\partial n}\,\mu\,d\Gamma = -\int_\Omega \nabla\psi_\lambda\cdot\nabla\tilde{\mu}\,dx - \int_\Omega \Delta\psi_\lambda\tilde{\mu}\,dx \\ &= -\int_\Omega \nabla\psi_\lambda\cdot\nabla\tilde{\mu}\,dx + \int_\Omega \nabla\cdot\mathbf{u}_\lambda\tilde{\mu}\,dx = -\int_\Omega (\nabla\psi_\lambda + \mathbf{u}_\lambda)\cdot\nabla\tilde{\mu}\,dx \end{aligned}\right\} \tag{13.59}$$

where $\tilde{\mu}$ is a *regular extension* of μ in Ω. Now let $\lambda_h, \mu_h \in \mathcal{M}_h$ and define $a_h(\cdot,\cdot): \mathcal{M}_h \times \mathcal{M}_h \to \mathbb{R}$ by

$$\int_\Omega \nabla p_h\cdot\nabla q_h\,dx = 0 \qquad \forall q_h \in H^1_{0h}, \qquad p_h - \lambda_h \in H^1_{0h} \tag{13.60}$$

$$\int_\Omega \nabla\mathbf{u}_h\cdot\nabla\mathbf{v}_h\,dx = -\int_\Omega \nabla p_h\cdot\mathbf{v}_h\,dx \qquad \forall\mathbf{v}_h \in V_{0h}, \qquad \mathbf{u}_h \in V_{0h} \tag{13.61}$$

$$\int_\Omega \nabla\psi_h\cdot\nabla\phi_h\,dx = \int_\Omega \nabla\cdot\mathbf{u}_h\phi_h\,dx \qquad \forall\phi_h \in H^1_{0h}, \qquad \psi_h \in H^1_{0h} \tag{13.62}$$

$$a_h(\lambda_h,\mu_h) = -\int_\Omega (\nabla\psi_h + \mathbf{u}_h)\cdot\nabla\mu_h\,dx \tag{13.63}$$

Then the following holds (see Reference 21).

Lemma 13.4.2 If (13.54) holds, the bilinear form $a_h(\cdot,\cdot)$ is symmetric, positive definite on $(\mathcal{M}_h/\mathbb{R}_h)^2$ where $\mathbb{R}_h = \{\mu_h \in \mathcal{M}_h,\ \mu_h = \text{constant on } \partial\Omega\}$.

13.4.4.2 *Transformation of* (P_h) *into a variational problem in* $\mathcal{M}_h$

In (13.51)–(13.53) of Section 13.4.3, an *approximate pressure* p_h was found unique in $H_h^1/\mathbb{R}$ once (13.54) holds. Therefore we can now state the discrete analogue of Theorem 13.4.2 (see Reference 21):

Theorem 13.4.6 Let p_h *be the discrete pressure. If* (13.54) *holds the component* λ_h *of* p_h *in* $\mathcal{M}_h$ *is the unique solution of*

$$(E_h) \qquad \begin{cases} \lambda_h \in \mathcal{M}_h/\mathbb{R}_h, \\ a_h(\lambda_h, \mu_h) = \int_\Omega (\nabla \psi_{0h} + \mathbf{u}_{0h}) \cdot \nabla \mu_h \, dx \qquad \forall \mu_h \in \mathcal{M}_h/\mathbb{R}_h \end{cases}$$

where p_{0h}, $\mathbf{u}_{0h}$, ψ_{0h} *are respectively the solutions of*

$$\int_\Omega \nabla p_{0h} \cdot \nabla q_h \, dx = \int_\Omega \mathbf{f} \cdot \nabla q_h \, dx \qquad \forall q_h \in H_{0h}^1, \qquad p_{0h} \in H_{0h}^1 \tag{13.64}$$

$$\int_\Omega \nabla \mathbf{u}_{0h} \cdot \nabla \mathbf{v}_h \, dx = \int_\Omega (\mathbf{f} - \nabla p_{0h}) \cdot \mathbf{v}_h \, dx \qquad \forall \mathbf{v}_h \in V_{0h}, \qquad \mathbf{u}_{0h} \in V_{0h} \tag{13.65}$$

$$\int_\Omega \nabla \psi_{0h} \cdot \nabla \phi_h \, dx = \int_\Omega \nabla \cdot \mathbf{u}_{0h} \phi_h \, dx \qquad \forall \phi_h \in H_{0h}^1, \qquad \psi_{0h} \in H_{0h}^1. \ \blacksquare \tag{13.66}$$

Remark 13.4.8 The reader will recognize that (13.64)–(13.66) are the discrete analogue of (13.43)–(13.45).

Remark 13.4.9 To compute the right-hand side of (E_h) it is necessary to solve the four (five if $\Omega \subset \mathbb{R}^3$) approximate Dirichlet problems (13.64), (13.65). Similarly if λ_h is known, to compute the approximate solution $\{\mathbf{u}_h, p_h\}$ of the Stokes problem (13.2) it is necessary to solve

$$\int_\Omega \nabla p_h \cdot \nabla q_h \, dx = \int_\Omega \mathbf{f} \cdot \nabla q_h \, dx \qquad \forall q_h \in H_{0h}^1, \qquad p_h - \lambda_h \in H_{0h}^1 \tag{13.67}$$

and (13.52); i.e. three approximate Dirichlet problems (four in $\mathbb{R}^3$).

Remark 13.4.10 On account of the choice (13.58) for the space $\mathcal{M}_h$, the integrals in the definition of $a_h(\cdot, \cdot)$ (see (13.63)) and of the right-hand side of (E_h), involve functions whose supports are in the neighbourhood of $\partial\Omega$ only.

13.4.5 Solution of (E_h) by a direct method

13.4.5.1 *Construction of the linear system equivalent to* (E_h)

Generalities. As before $\mathcal{M}_h$ is defined by (13.58); let $\mathcal{B}_h = \{w_i\}_{i=1}^{N_h}$ be a basis of $\mathcal{M}_h$. Then $\forall \mu_h \in \mathcal{M}_h$

$$\mu_h = \sum_{i=1}^{N_h} \mu_i w_i \tag{13.68}$$

and from now on we shall write

$$r_h \mu_h = \{\mu_1, \ldots, \mu_{N_h}\} \in \mathbb{R}^{N_h} \tag{13.69}$$

In practice $\mathcal{B}_h$ is defined by

$$\mathcal{B}_h = \{w_i\}_{i=1}^{N_h} \tag{13.70}$$

and (see Figure 13.2)

$$\left.\begin{aligned} &\forall i = 1, \ldots, N_h \\ &w_i(P_i) = 1 \\ &w_i(Q) = 0 \qquad \forall Q \ \text{ vertex of } \ \mathcal{T}_h, \qquad Q \neq P_i \end{aligned}\right\} \tag{13.71}$$

where we assumed implicitly (but in practice it is not necessary) that the boundary nodes are numbered first.

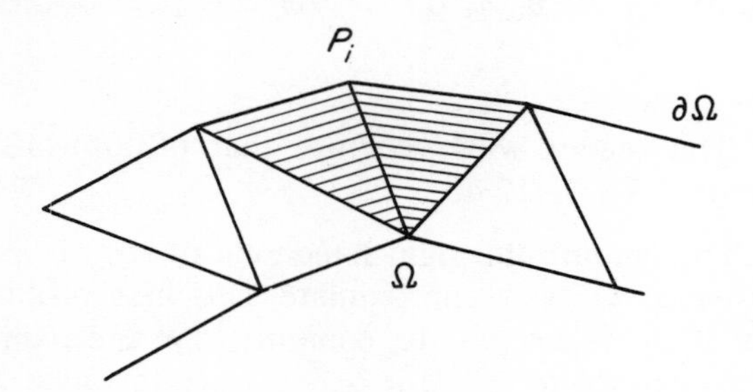

Figure 13.2 (The support of w_i is shown)

With this choice for $\mathcal{B}_h$, $\mu_i = \mu_h(P_i)$ in (13.68). Then problem (E_h) is equivalent to the *linear system*

$$\sum_{j=1}^{N_h} a_h(w_j, w_i)\lambda_j = \int_\Omega (\nabla\psi_{0h} + \mathbf{u}_{0h}) \cdot \nabla w_i \, dx, \qquad 1 \leq i \leq N_h \tag{13.72}$$

Let $a_{ij} = a_h(w_j, w_i)$, $A_h = (a_{ij})_{1\leq i,j\leq N_h}$, $b_i = \int_\Omega (\nabla\psi_{0h} + \mathbf{u}_{0h}) \cdot \nabla w_i \, dx$, $b_h = \{b_i\}_{i=1}^{N_h}$. The matrix A_h is *full* and *symmetric, positive, semi-definite.* If (13.54) is verified, then 0 is a *single eigenvalue* of A_h; furthermore if $\mathcal{B}_h$ is defined by (13.71) then

$$\text{Ker}\,(A_h)=\{\mathbf{y}\in\mathbb{R}^{N_h},\ y_1=y_2=\ldots=y_{N_h}\} \tag{13.73}$$

As to the conditioning of A_h restricted to $R(A_h)$ $(=\mathbb{R}^{N_h}-\text{Ker}\,(A_h))$, it can be shown that the ratio $\nu(A_h)$ of the largest eigenvalue to the smallest is of order h^{-2}, if (13.54) holds. In fact, by analogy with Reference 20, Section 4 it is reasonable to conjecture that $\nu(A_h)=0(1/h)$ but we were not able to obtain this estimate.

Construction of A_h · A_h is constructed column by column according to the relation $a_{ij}=a_h(w_j, w_i)$. To compute the jth column of A_h we solve (13.60)–(13.62) with $\lambda_h=w_j$ and compute a_{ij} from (13.63). Thus four Dirichlet problems must be solved for each column (five in $\mathbb{R}^3$). The matrix A_h being *symmetric* one may restrict i to be greater or equal to j. By the way, Remark 13.4.10 applies for the computation of the b_i and a_{ij}.

13.4.5.2 *Solution of (13.72) by the Cholesky method*

Assume that (13.54), (13.71) hold. Then one shows from (13.73) (see Reference 21) that the submatrix $\tilde{\mathbf{A}}_h=(a_{ij})_{1\leqslant i,j\leqslant N_h-1}$ is symmetric and positive definite. Therefore one may proceed as follows.

Take $\lambda_{N_h}=0$ and solve

$$\tilde{A}_h\tilde{r}_h\lambda_h=\tilde{b}_h \tag{13.74}$$

(where $\tilde{r}_h\lambda_h=\{\lambda_1\ldots\lambda_{N_{h-1}}\}$, $\tilde{b}_h=\{b_1,\ldots,b_{N_h-1}\}$) by the Cholesky method via a factorization:

$$\tilde{A}_h=\tilde{L}_h\tilde{L}_h^t\ (\text{or } \tilde{A}_h=\tilde{L}_h\tilde{D}_h\tilde{L}_h^t) \tag{13.75}$$

where $\tilde{L}_h$ is lower triangular non-singular (and $\tilde{D}_h$ is diagonal).

Let us review the sub-problems arising in the computation of $\{\mathbf{u}_h, p_h\}$ via (E_h) if the Cholesky method is used:

—The four approximate Dirichlet problems (13.64)–(13.66) to compute p_{0h}, $\mathbf{u}_{0h}$, ψ_{0h} and $\tilde{b}_h$ (five if $\Omega\in\mathbb{R}^3$)
—four(N_h-1) approximate Dirichlet problems to construct $\tilde{A}_h$ (five(N_h-1) if $\Omega\subset\mathbb{R}^3$)
—five triangular systems to compute $\lambda_h:\tilde{L}_h\tilde{y}_h=\tilde{b}_h$, $\tilde{L}_h^t\tilde{r}_h\lambda_h=\tilde{y}_h$
—three approximate Dirichlet problems to obtain p_h and $\mathbf{u}_h$ from λ_h (four if $\Omega\subset\mathbb{R}^3$).

Hence if $\Omega\subset\mathbb{R}^N$ $(N=2, 3)$ *it is necessary to solve* $(N+2)(N_h+1)-1$ *approximate Dirichlet problems.*

In practice the matrices of the approximate Dirichlet problem should be factorized once and for all (there are two symmetric positive matrices, one for the affine elements, one for the quadratic elements (or affine on $\tilde{\mathscr{T}}_h$ if (13.50)bis is used)).

13.4.6 Solution of (E_h) by the conjugate gradient method

We may also solve (E_h) (and therefore (P_h)) by a *conjugate gradient method, which does not require the knowledge of* A_h but requires four approximate Dirichlet problems to be solved at each iteration (five if $\Omega \subset \mathbb{R}^3$):

$$\lambda_h^0 \in \mathcal{M}_h, \quad \text{arbitrarily given} \tag{13.76}$$

$$g_h^0 = A_h r_h \lambda_h^0 - b_h \tag{13.77}$$

$$z_h^0 = g_h^0 \tag{13.78}$$

and for $n \geq 0$

$$\rho_n = \frac{(z_h, g_h^n)_h}{(A_h z_h^n, z_h^n)_h} \left(\text{or } \frac{\|g_h^n\|_h^2}{(A_h z_h^n, z_h^n)_h} \right) \tag{13.79}$$

$$r_h \lambda_h^{n+1} = r_h \lambda_h^n - \rho_n z_h^n \tag{13.80}$$

$$g_h^{n+1} = g_h^n - \rho_n A_h z_h^n \tag{13.81}$$

$$\gamma_n = \frac{\|g_h^{n+1}\|_h^2}{\|g_h^n\|_h^2} \tag{13.82}$$

$$z_h^{n+1} = z_h^n + \gamma_n z_h^n \tag{13.83}$$

In (13.76)–(13.83), $(\cdot, \cdot)_h$ stands for the standard Euclidian scalar product of $\mathbb{R}^{N_h}$ (but one could use a conjugate gradient method with preconditioning in the sense of Reference 30).

The matrix A_h being *symmetric, positive semi-definite,* one can show that $\{\lambda_h^n\}_{n\geq 0}$ converges to λ_h, solution of (E_h); the component of λ_h in $\mathbb{R}_h$ is that of λ_h^0. Implementing (13.76)–(13.83) requires the solution of four Dirichlet problems at each iteration (five if $\Omega \subset \mathbb{R}^3$) to compute $A_h z_h^n$ from

$$a_h(\lambda_h, \mu_h) = (A_h r_h \lambda_h, r_h \mu_h)_h \qquad \forall \lambda_h, \mu_h \in \mathcal{M}_h \tag{13.84}$$

Here also one should factorize the matrices of the approximate Dirichlet problem.

13.4.7 Comments

In Section 13.4 a new mixed finite element method was described for the Stokes problem (13.2). The direct method described in Section 13.4.4 has been used in 2-D and 3-D cases for the computation of *unsteady incompressible viscous* flows. We recommend the method if the Stokes problem has to be solved many times on a given domain. On the other hand if the Stokes problem is to be solved once only or if N_h, the number of boundary nodes, is large, we recommend the conjugate gradient method of Section 13.4.6. The ideas of Section 13.4 will be developed in References 6 and 21 where the proofs will be included together with most of the results shown here.

13.5 FURTHER REFERENCES AND CONCLUSION

To conclude with we would like to mention the works of Bercovier,[31] Argyris and Dunne,[32] and Johnson[33] on Stokes and Navier–Stokes equations, and incompressible media. These appear to us connected with some of the ideas developed in this chapter.

REFERENCES

1. R. Temam, *Theory and Numerical Analysis of the Navier–Stokes equations*, North-Holland, Amsterdam, 1977.
2. R. A. Adams, *Sobolev spaces*, Academic Press, New York, 1976.
3. J. L. Lions, E. Magenes, *Non-Homogeneous Boundary Value Problems and Applications*, Vol. 1, Springer-Verlag, New York, 1972.
4. J. Necas, *Les Méthodes directes en théorie des équations elliptiques*, Masson, Paris, 1967.
5. J. T. Oden and J. N. Reddy, *An Introduction to the Mathematical Theory of Finite Elements*, Wiley, New York, 1976.
6. R. Glowinski and O. Pironneau, 'On a mixed finite element approximation for the Stokes problem. (II) Solution of the approximate problems' (to appear).
7. O. A. Ladyshenskaya, *The Mathematical Theory of Viscous Incompressible Flow*, Gordon and Breach, 1969.
8. I. S. Duff, N. Munksgaard, H. B. Nielsen, and J. K. Reid, 'Direct solution of sets of linear equations whose matrix is sparse, symmetric and indefinite'. *Harwell report, C.S.S. Division, A.E.R.E.*, Harwell, January 1977.
9. C. C. Paige and M. A. Saunders, 'Solution of sparse indefinite systems of linear equations', *SIAM J. Num. Anal.*, **12,** 617–629 (1975).
10. O. Widlund, 'A Lanczos method for a class of non-symmetric systems of linear equations' (to appear).
11. M. Crouzeix, 'Etude d'une méthode de linéarisation. Résolution numérique des équations de Stokes stationnaires. Application aux équations de Navier–Stokes stationnaires', in *Approximation et méthodes itératives de résolution d'inéquations variationnelles et de problèmes non linéaires*, Cahier de l'Iria, N°. 12, May 1974, pp. 139–224.
12. M. Fortin and R. Glowinski, 'Augmented Lagragian in quadratic programming', Ch. 1 of *Numerical Solution of Boundary Value Problems by Augmented Lagragians* (M. Fortin and R. Glowinski, Eds.), (to appear).
13. M. Fortin and F. Thomasset, 'Application to Stokes and Navier–Stokes equations', Ch. 2. of *Numerical Solution of Boundary Value Problems by Augmented Lagrangian* (M. Fortin and R. Glowinski, Eds.), (to appear).

14. J. W. Daniel, *The Approximate Minimization of Functionals*, Prentice-Hall, Englewood Cliffs, 1970.
15. M. R. Hestenes, 'Multiplier and gradient methods', *J.O.T.A.*, **4,** 5, 303–320 (1969).
16. M. J. D. Powell, 'A method for non linear optimization in minimization problems', in *Optimization* (R. Fletcher, Ed.), Academic Press, London, 1969.
17. R. Glowinski, J. L. Lions, and R. Tremolieres, *Analyse numérique des inéquations variationnelles*, Vol. 1, Dunod-Bordas, Paris, 1976.
18. R. Glowinski and O. Pironneau, 'Approximation par éléments finis mixtes du problème de Stokes en formulation vitesse-pression. Convergence des solutions approchées'. *C.R.A.S. Paris*, T.**268A,** 181–183 (1978).
19. R. Glowinski and O. Pironneau, 'Approximation par éléments finis mixtes du problème de Stokes en formulation vitesse-pression. Résolution des problèmes approchés'. *C.R.A.S. Paris*, T.**268A,** 225–228 (1978).
20. R. Glowinski and O. Pironneau, 'Numerical methods for the first biharmonic equation and for the two-dimensional Stokes problem'. *Comp. Science Dpt., Report STAN-CS-77-615*, Stanford University, 1977, and *SIAM Review* (to appear).
21. R. Glowinski and O. Pironneau, 'On a mixed finite element approximation for the Stokes problem. (I) Convergence of the approximate solutions' (to appear).
22. M. O. Bristeau, R. Glowinski, O. Pironneau, J. Periaux, P. Perier, and G. Poirier, 'Application of Optimal Control Methods to the Calculation of Transonic Flows and Incompressible Viscous Flows', in *Numerical Methods in Applied Fluid Dynamics* (B. Hunt, Ed.), Academic Press, London (to appear).
23. C. Taylor and P. Hood, 'A numerical solution of the Navier–Stokes equations using the finite element technique', *Comp. and Fluids*, **1,** 73–100 (1973).
24. M. Bercovier and O. Pironneau, 'Estimations d'erreurs pour la résolution du problème de Stokes en éléments finis conformes de Lagrange', *C.R.A.S. Paris*, T.**285A,** 1085–1087 (1977).
25. J. M. Thomas, *Sur l'Analyse numérique des méthodes d'éléments finis hybrides et mixtes*, Thesis, Université Paris VI, 1977.
26. M. Crouzeix and P. A. Raviart, 'Conforming and non conforming finite element methods for solving the stationary Stokes equations', *R.A.I.R.O.*, **R-3,** 33–76 (1973).
27. P. A. Raviart, 'Finite element methods and Navier–Stokes equations'. *Proceedings of the Third Iria Symposium on Numerical Methods in Engineering and Applied Sciences* (to appear).
28. O. C. Zienkiewicz, *The Finite Element Method in Engineering Sciences*, McGraw-Hill, London 1978.
29. P. Le Tallec, Thesis (to appear).
30. O. Axelsson, 'A class of iterative methods for finite element equations', *Comp. Methods Applied Mech. Eng.*, **9,** 2, 123–138 (1976).
31. M. Bercovier, *Régularisation duale des problèmes variationnels mixtes et extension à quelques problèmes non linéaires*, Thesis, Université de Rouen, 1976.
32. J. H. Argyris and P. C. Dunne, 'Improved displacement finite elements for incompressible materials', Chapter 12 of this book.
33. C. Johnson, 'A mixed finite element method for the Navier–Stokes equations', *Research Report 77-IAR*, Department of Computer Sciences, Chalmers University of Technology and the University of Göteborg, 1977.

Chapter 14

Use of the Current Stiffness Parameter in Solution of Nonlinear Problems

P. G. Bergan, I. Holand, and T. H. Söreide

14.1 INTRODUCTION

The use of finite elements for solving nonlinear structural problems results in a set of nonlinear algebraic equations. What is the 'best' technique for solving these equations depends on the physical behaviour of the structure under consideration. While in linear analysis the solution always is unique, this is often not the case in nonlinear problems. Geometric instabilities, plasticity and non-conservative loading make the solution path or history dependent. Therefore, insight into the nature of structural problems as well as good knowledge of numerical analysis is essential. In this way the solution algorithm can be set up according to the characteristics of the problem at hand. However, a good solution algorithm should also, at least to some extent, be capable of adjusting itself automatically according to the degree and type of nonlinearity experienced along the solution path.

Many discussions of solution techniques have been presented, see for instance Reference 1–4. When some energy functional is available the solution is given by its stationary points. The most widely used formulation of this type is the principle of stationary potential energy. A lower level of formulation uses the equilibrium equations directly, stating that the internal forces should balance the external loads.

Most solution techniques utilize the incremental form of the equilibrium equations. In a finite element framework this leads to an incremental relationship between the external forces and the nodal degrees of freedom. The solution of such a set of differential equations is normally carried out by applying the external loading in steps. Often the simple forward integration scheme of Euler–Cauchy is used.[5] Unfortunately, such a crude solution scheme results in considerable drift-off from the true solution path. This may be rectified to some extent by applying unbalanced force corrections in addition to the external loads. Better still is to use equilibrium iterations at each level of loading. Since the tangential stiffness is available, Newton–Raphson iteration[6] is easily incorporated. 'Quasi–Newton' methods are also commonly used; the computation of incremental stiffness and triangularization of stiffness matrix is then performed only once or a few times at each level of loading. An efficient solution algorithm requires that there is a good balance between incrementation and iteration.

The present paper discusses techniques for optimizing the process of load incrementation. A scheme for automatic generation of load steps is presented. The basic idea behind this method is that the change in incremental stiffness should be nearly the same for all steps. Special effort is made in order to obtain a reliable algorithm for problems involving instabilities.

14.2 CURRENT STIFFNESS PARAMETER

14.2.1 Definition of the current stiffness parameter

When using an incremental solution scheme some simple way of characterizing the structural stiffness during load application may be useful. For this purpose a quantity called the 'current stiffness parameter' is proposed.

Considering load step number i, the incremental load vector $\Delta\mathbf{R}^i$ produces an incremental displacement $\Delta\mathbf{r}^i$. Letting $\|\Delta\mathbf{R}_i\|$ denote some norm of $\Delta\mathbf{R}^i$, a normalized incremental load vector $\Delta\mathbf{R}_n^i$ is defined by

$$\Delta\mathbf{R}_n^i = \frac{\Delta\mathbf{R}^i}{\|\Delta\mathbf{R}^i\|} \tag{14.1}$$

The corresponding scaled incremental displacement vector reads

$$\Delta\mathbf{r}_n^i = \frac{\Delta\mathbf{r}^i}{\|\Delta\mathbf{R}^i\|} \tag{14.2}$$

In the current state a unit of the incremental load $\Delta\mathbf{R}_n^i$ produces the incremental displacements $\Delta\mathbf{r}_n^i$. The 'work' that results from this load application is $\Delta w_n^i = \Delta\mathbf{r}_n^{i\mathrm{T}}\Delta\mathbf{R}_n^i$. A stiff structural system produces small displacements, hence the incremental work would be small. A measure of the current stiffness of the structure may be expressed in terms of the inverse quantity

$$S_p^{i*} = \frac{1}{\Delta w_n^i} = \frac{\|\Delta\mathbf{R}^i\|^2}{\Delta\mathbf{r}^{i\mathrm{T}}\Delta\mathbf{R}^i} \tag{14.3}$$

The stiffer the system, the greater this quantity will be. Note that the scaling by the norm of the load increment has made this S_p^* independent of the actual size of $\Delta\mathbf{R}$.

It is convenient to scale the stiffness parameter S_p^{i*} by the corresponding value for the initially 'linear' system (undeformed configuration). The expression for the current stiffness parameter at load level number i then becomes

$$S_p^i = \frac{\Delta \mathbf{r}^{1T}\,\Delta \mathbf{R}^1}{\Delta \mathbf{r}^{iT}\,\Delta \mathbf{R}^i}\frac{\|\Delta \mathbf{R}^i\|^2}{\|\Delta \mathbf{R}^1\|^2} \tag{14.4}$$

Index 1 indicates the first linear step. On the form (14.4) S_p always has the initial value 1.0. For softening systems S_p decreases and for stiffening systems S_p increases.

An average measure of the external loading is obtained by using the Euclidean norm of $\Delta\mathbf{R}^i$ in (14.1)–(14.4). It is recommended that the modified Euclidean norm suggested in Reference 11 is used. The forces and moments are then scaled by some reference quantities in order to make them non-dimensional.

For proportional loading an arbitrary load level may be expressed in terms of a reference load $\mathbf{R}_{\text{ref}}$:

$$\mathbf{R} = p\mathbf{R}_{\text{ref}} \tag{14.5}$$

In (14.5), p is a loading parameter. Expression (14.4) now turns into

$$S_p^i = \frac{\Delta p^i}{\Delta p^1}\frac{\Delta \mathbf{r}^{1T}\mathbf{R}_{\text{ref}}}{\Delta \mathbf{r}^{iT}\mathbf{R}_{\text{ref}}} \tag{14.6}$$

Introducing the incremental stiffness K_{I}, the current stiffness parameter S_p for proportional loading may also be written

$$S_p^i = \left(\frac{\Delta p^i}{\Delta p^1}\right)^2 \frac{\Delta \mathbf{r}^{1T} K_{\text{I}}^1 \Delta \mathbf{r}^1}{\Delta \mathbf{r}^{iT} K_{\text{I}}^i\, \Delta \mathbf{r}^i} \tag{14.7}$$

(14.7) expresses the ratio between scaled quadratic forms of the incremental stiffness in the initial and the current step, respectively. This ratio gives a measure of the incremental stiffness of the system when moving in the direction $\Delta\mathbf{r}^i$ in the solution space.

In case of non-proportional loading a characterizing reference load $\tilde{\mathbf{R}}$ may be defined. The magnitude of external forces may be given by a loading parameter p which is some norm of the external forces or some time-dependent reference parameter for load application. Letting the displacements be written as functions of p in the way

$$\mathbf{r} = \mathbf{r}(p) \tag{14.8}$$

and using differentiation instead of finite increments, an alternative expression for S_p reads

$$S_p^i = \frac{\dot{\mathbf{r}}^{1T}\tilde{\mathbf{R}}}{\dot{\mathbf{r}}^{iT}\tilde{\mathbf{R}}} \tag{14.9}$$

In (14.9) 'dot' denotes differentiation with respect to the time or loading parameter p. The analogy with (14.6) for proportional loading is clear.

14.2.2 Behaviour of current stiffness parameter

The variation of the current stiffness parameter during load application will now be discussed in more detail. Especially for instability problems valuable information about the structural behaviour may be gained from S_p. In Figure 14.1 this is illustrated for a typical snap-through problem.

Fig. 14.1(a) shows the load parameter p versus some norm of the displacement vector $\|\mathbf{r}\|$. The associated curve for S_p as function of $\|\mathbf{r}\|$ is traced in Figure 14.1(b). It is seen that at the extremum points of the load–displacement curve S_p has the value zero. In this situation the incremental stiffness is singular, and even a small load increment produces infinite incremental displacements; thus, according to (14.7), S_p^i becomes zero. S_p is positive for the stable branches of the load–displacement curve.

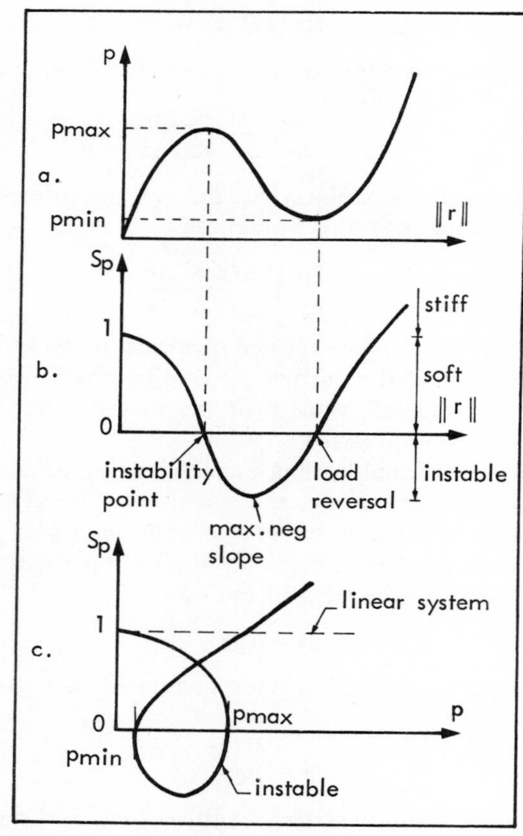

Figure 14.1 Behaviour of current stiffness parameter for instability problem

This is because the quadratic form of $\Delta\mathbf{r}$ and K_I expresses the second variation of the potential energy of the system. Disregarding some exceptions, this implies that the system is stable. The unstable configurations are characterized by negative values of S_p. In Figure 14.1(c) the relationship between current stiffness parameter S_p and loading parameter p is illustrated.

Although there may be some exceptions to the above rules for interpretation of S_p, a wide range of uses of S_p for practical problems have all proved to fit into this pattern. S_p may be useful for estimating the critical load. S_p also gives a good indication of the structural behaviour in the post-buckling range of deformation. Thus, in addition to guiding the load incrementation, the current stiffness parameter may also be used to characterize the type of buckling: the more abrupt change of S_p at an instability point, the more violent type of buckling. The applicability of the current stiffness parameter for instability analysis will be demonstrated in the following sections.

14.3 USE OF CURRENT STIFFNESS PARAMETER TO GUIDE SOLUTION PROCESS

14.3.1 Procedure for automatic load incrementation

A previously proposed scheme for automatic load incrementation[4] was based upon control of truncation errors due to linearization. A new method using the current stiffness parameter for control will now be described. The basic idea behind this method is that the change in S_p should be close to the same for all load steps. This implies that the incremental stiffness should be allowed to change by a prescribed magnitude for each new step. Figure 14.2 gives an illustration of the process. The anticipated change in S_p per step is denoted $\Delta\tilde{S}_p$. This quantity (positive) is given as input to the computer program. Since S_p as defined by (14.4) is normalized so that the initial value becomes unity, $\Delta\tilde{S}_p$ may easily be chosen (e.g. in the range 0.05–0.10). The smaller $\Delta\tilde{S}_p$, the smaller the step will be. Referring to Figure 14.2 the scheme for automatic load incrementation is as follows:

(a) Prescribe the first step Δp^1. After equilibrium iteration, point 1 on the S_p–p diagram is obtained.
(b) Extrapolate through points 0 and 1 and calculate the next load increment Δp^2 that tentatively would change the current stiffness parameter by $\Delta\tilde{S}_p$. Δp^2 is then given by the expression (for Δp^2: $i=1$, $i+1=2$):

$$\Delta p^{i+1} = \Delta p^i \frac{\Delta\tilde{S}_p}{|\Delta S_p^i|}, \qquad \Delta S_p^i = S_p^i - S_p^{i-1} \tag{14.10}$$

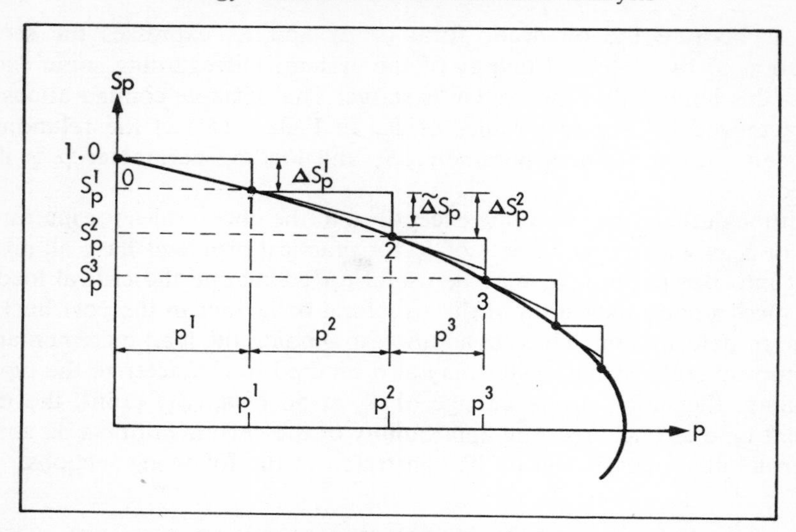

Figure 14.2 Automatic load generation

(c) Perform equilibrium iteration at load level p^2 until convergence is reached and point 2 on the S_p–p curve is found.

Procedure (b)–(c) is repeated for subsequent increments. Generally, load increment Δp^{i+1} is found by extrapolating through points $i-1$ and i. In connection with point (b), it is necessary to use a trial load increment to compute an incremental displacement vector and S_p^i. The increment Δp^{i+1} can then be found from (14.10) and the corresponding Δr^{i+1} by scaling the trial step.

It is seen from (14.10) that the process yields extremely large load increments in regions where the current stiffness parameter is nearly constant (ΔS_p^i close to zero). When the incremental stiffness becomes infinite ($S_p - p$ diagram vertical), Δp^{i+1} according to (14.10) tends towards zero. In order to eliminate these problems, upper and lower limits for the load step and incremental displacement norm should also be given.

When using the present technique an estimate of the number of increments up to maximum load ($S_p = 0$) is found as the inverse of $\Delta \tilde{S}_P$.

As an alternative to the linear extrapolation of S_p suggested here, it can be expected that a parabolic extrapolation would yield an even better algorithm. However, this possibility has not been tested yet.

14.3.2 Control of iteration around extremum points

For instability problems the incremental stiffness matrix becomes singular at extremum points. It is well known that equilibrium iteration normally fails in these regions. It will now be demonstrated how the transition zone from a step-iterative to a pure incremental procedure may conveniently be defined using the current stiffness parameter.

The basic idea is that equilibrium iteration can safely be carried out only for states that are sufficiently far away from the singularity at the displacement path. This singularity is recognized by $S_p = 0$. Moreover, such a singularity can be detected ahead of time by observing the magnitude of S_p. Thus, a threshold value $\bar{S}_p$ can be prescribed and iteration is carried out only as long as the absolute value of S_p is greater than $\bar{S}_p$. This procedure is illustrated in Figure 14.3. At point A, S_p has become less than $\bar{S}_p$, thus the solution proceeds with pure incrementation without iterations. When point B has been reached, the absolute value of S_p has become greater than $\bar{S}_p$ and equilibrium iteration is resumed. Usually $\bar{S}_p$ is chosen in the range 0.05–0.10.

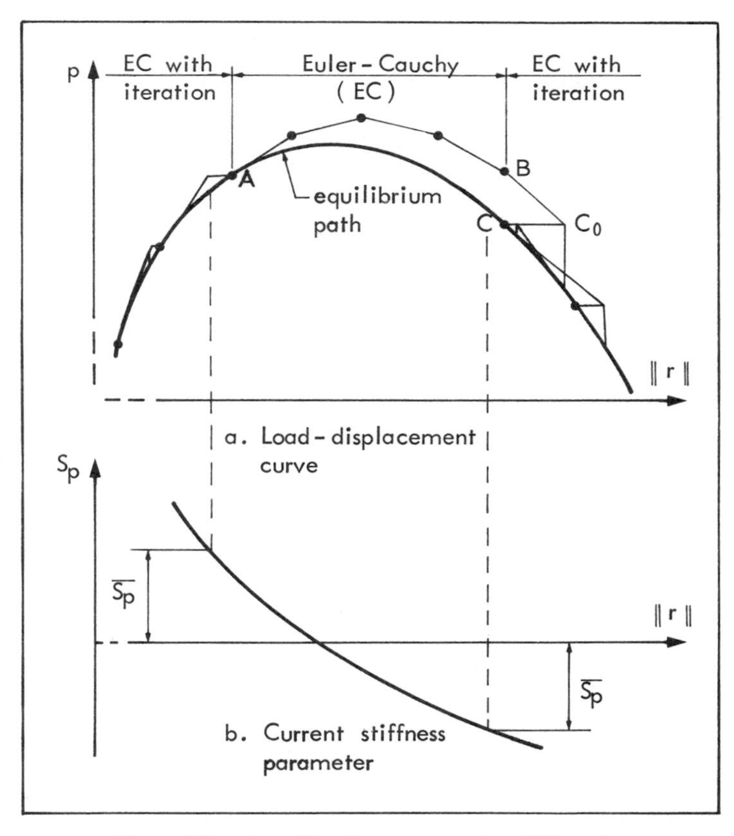

Figure 14.3 Algorithm for passing an instability point

In the pure incremental phase the maximum bounds on the incremental displacements ensure that dramatic drift-off is avoided. As shown in Figure 14.3 the load incrementation procedure is automatically reversed when the maximum point has been reached. This is done by checking for sign changes of S_p for each new step (corresponds to computing incremental work). When a sign change is detected both $\Delta \mathbf{R}$ and the computed $\Delta \mathbf{r}$ are given opposite sign.

14.3.3 Characterization of bifurcation

In the present section it will be demonstrated how the current stiffness parameter may be used to characterize a bifurcation. A 'bifurcation index' I_p is defined by the following expression

$$I_p = \frac{S_p^- \cdot S_p^+}{S_p^- - S_p^+} \tag{14.11}$$

where S_p^- = current stiffness parameter prior to bifurcation point; S_p^+ = current stiffness parameter immediately after bifurcation point.

The bifurcation index is designed to express the change in direction between the two branches. A change from an ascending to a descending and unstable branch is always reflected in a negative index.

Values of I_p for some types of bifurcation are given in Figure 14.4(a). The two cases shown in Figure 14.4(b) give the largest negative value of I_p to the

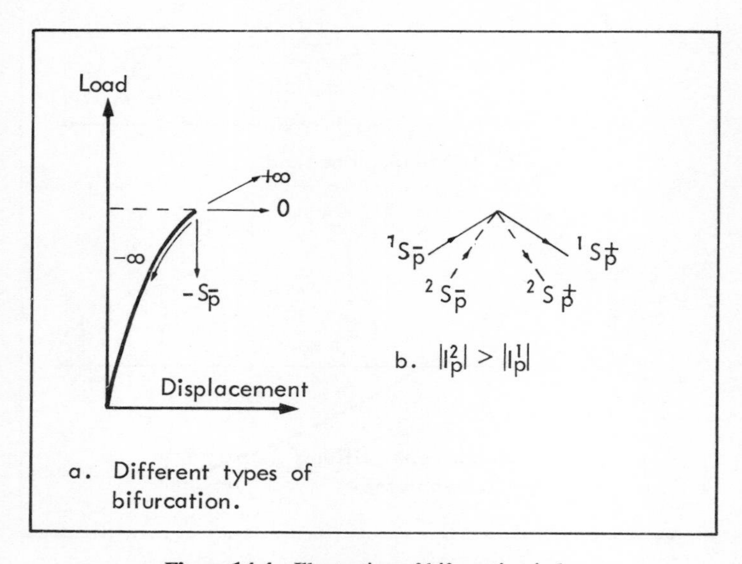

Figure 14.4 Illustration of bifurcation index

most peaked situation. When the second branch follows neutral equilibrium, the bifurcation index is zero. When a linear system suddenly loses all its carrying capacity (vertical second branch), I_p is equal to minus unity. Infinite I_p simply means that there is no bifurcation.

Since I_p is based on the non-dimensional quantity, I_p becomes an absolute measure that can be used to characterize and compare different types of structures. A general classification of bifurcations is suggested in Table 14.1.

Table 14.1 Characterization of bifurcation problems using the bifurcation index

I_p	Type of bifurcation
∞ (very high)	No bifurcation
$\infty > I_p > 0$	Bifurcation into initially stable branch
0	Bifurcation into neutral equilibrium
$0 > I_p > -0.5$	Unstable branch. Moderately sensitive system
$-0.5 \geq I_p > -1.0$	Unstable branch. Highly sensitive system
$-1 \geq I_p > -\infty$	Unstable branch. Violently sensitive system

The numerical values that this classification is based on, are chosen on intuition and may be subjects of discussion. It should also be noted that the table serves as a guideline and there may be exceptions, e.g. a system may be unstable even when $I_p = 0$.

In practical cases the bifurcation will not appear as a mathematically sharp directional change of the displacement path; imperfections and impurities tend to make the bend rounded. From a computational point of view it is also advantageous to introduce such disturbances when using an incremental-iterative technique. In such situations where the change in S_p appears more gradually, S_p^- and S_p^+ are taken as the values closest to the transition zone. Extrapolation of S_p for the two branches to the instability point can also be used.

14.4 NUMERICAL STUDIES

14.4.1 Computer program

In the following sections some instability analyses of arches and stiffened plates will be presented. The calculations are carried out using a nonlinear

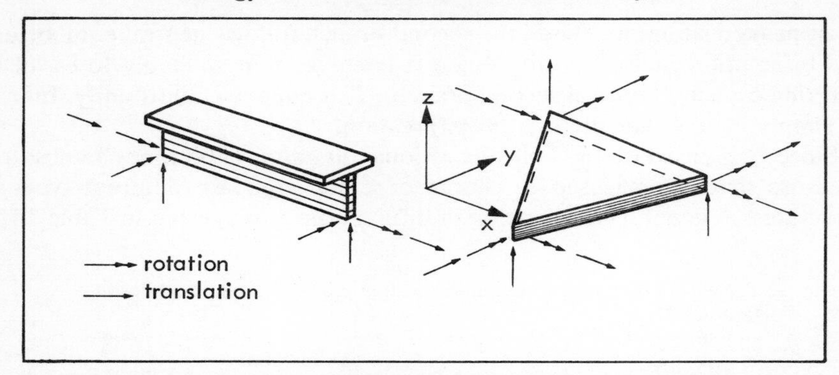

Figure 14.5 Division of stiffener height and plate thickness into layers

finite element program that originally was developed to study collapse of stiffened plates. The program accounts for both material and geometric nonlinearities. The large displacement effects are taken care of by using the updated Lagrangian description of motion.[7] The mathematical modelling of the material is based upon the flow theory of plasticity.[8]

The present calculations only make use of the stiffener elements in the program. In these spar elements[9,10] the integration of stiffness is performed by dividing the height into layers which may have independent elastoplastic material properties (see Figure 14.5).

In order to account for flanges, bar elements may be coupled to the spar elements.

A detailed description of the computer program is given in Reference 11.

14.4.2 Shallow sinusoidal arch under gravity loading

The unstable behaviour of an elastic shallow arch of sinusoidal shape subjected to gravity loading will now be studied. The arch is hinged and horizontally restrained at the ends. With dimensions as given in Figure 14.6 the span/thickness ratio becomes 50; this is in the practical range for steel structures.

Only symmetric buckling is considered and one half of the arch is modelled by five stiffener elements. The scheme for automatic load generation described in Section 14.3.1 is used. Newton–Raphson iteration is performed at each level of loading until the modified Euclidean norm of the incremental displacements[12] becomes less than 0.005. Two cases of automatic load incrementation are considered, namely those corresponding to $\Delta\tilde{S}_p = 0.10$ and $\Delta\tilde{S}_p = 0.05$. For both alternatives the limit value of the current stiffness parameter defining transition from step-iterative to pure

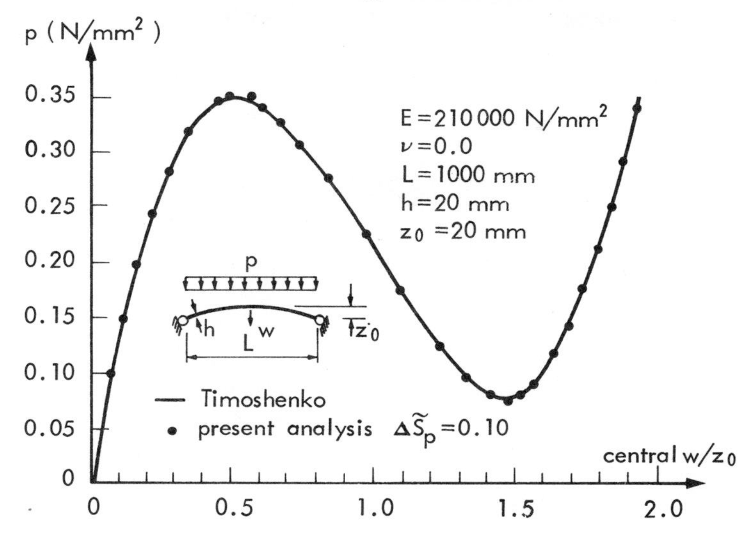

Figure 14.6 Load-deflection curve for shallow arch. Anticipated change $\Delta\tilde{S}_p = 0.10$

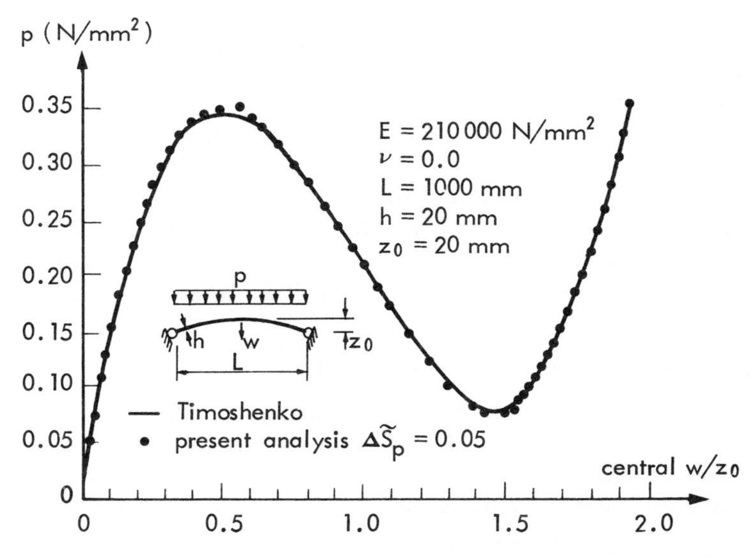

Figure 14.7 Load-deflection curve for shallow arch. Anticipated change $\Delta\tilde{S}_p = 0.05$

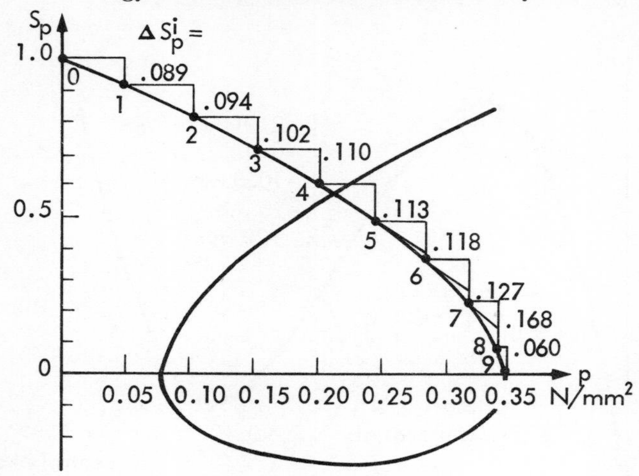

Figure 14.8 Variation of S_p for shallow arch

incremental calculation is $\bar{S}_p = 0.10$. Figure 14.6 and 14. 7 show the results obtained for $\Delta\tilde{S}_p = 0.10$ and 0.05, respectively. It is seen that the finite element solution agrees very well with the analytical solution by Timoshenko and Gere.[13] The points along the load–displacement curve have automatically been distributed according to the degree of nonlinearity of the system; 10 and 19 steps are required to reach the first instability for $\Delta\tilde{S}_p = 0.10$ and 0.05, respectively. This verifies that the inverse of $\Delta\tilde{S}_p$ gives a good estimate of the number of increments from the initial state up to the first extremum point.

The variation of the current stiffness parameter as function of loading p is shown in Figure 14.8. For the case $\Delta\tilde{S}_p = 0.10$ the points obtained after equilibrium iterations are indicated for the first nine load levels. It is seen that the actual changes in S_p are close to the prescribed value 0.10 except for the steps near the instability point where the algorithm is governed by the bounds on the loading and displacement increments.

14.4.3 Beam-column failure of stiffened plates

The present example deals with buckling of stiffened plates. The beam–column failure mode of a panel with close stiffeners may be studied by considering a column with T-type cross-section. In the present investigation one-span and two-span columns are analysed. The cross-sections have equal areas for web and flange and are of the same type used by Dowling.[14] The width/thickness ratios are 10 and 20 for web and flange, respectively.

Three slenderness ratios are considered, namely $l/r = 40$, 70, and 100 (l is the span length, r denotes radius of inertia of the cross-section). The beam-columns have initial deflections in the form of a single half sine wave over each span. The amplitude of imperfection is $l/750$. Imperfections with flange on convex side and with flange on concave side are studied.

The beam–columns are free to rotate at the supports and subjected to a concentrated axial load P through the centroid of the cross-section. Young's modulus is 210 000 N/mm^2 and Poisson's ratio 0.3. The material is elastic–ideally plastic with yield stress $\sigma_Y = 240$ N/mm^2.

Figure 14.9 shows the relationship between nominal axial stress ratio σ/σ_Y and relative axial displacement u/l for the one-span cases. The load steps generated by the automatic process are indicated for $l/r = 100$ with flange on convex side. The effect of reversing the mode of imperfection is clearly demonstrated in Figure 14.9.

In Figure 14.10 the load–displacement curves obtained for the two-span cases are traced. For $l/r = 70$ and 100 the unloading paths are even more dramatic than the one-span curves in Figure 14.9. This is due to the sudden snap from a symmetric mode of deformation to an antisymmetric mode, see Figure 14.11.

The variation of current stiffness parameter S_p as function of load level is shown in Figures 14.12 and 14.13 for the two-span beam–columns with

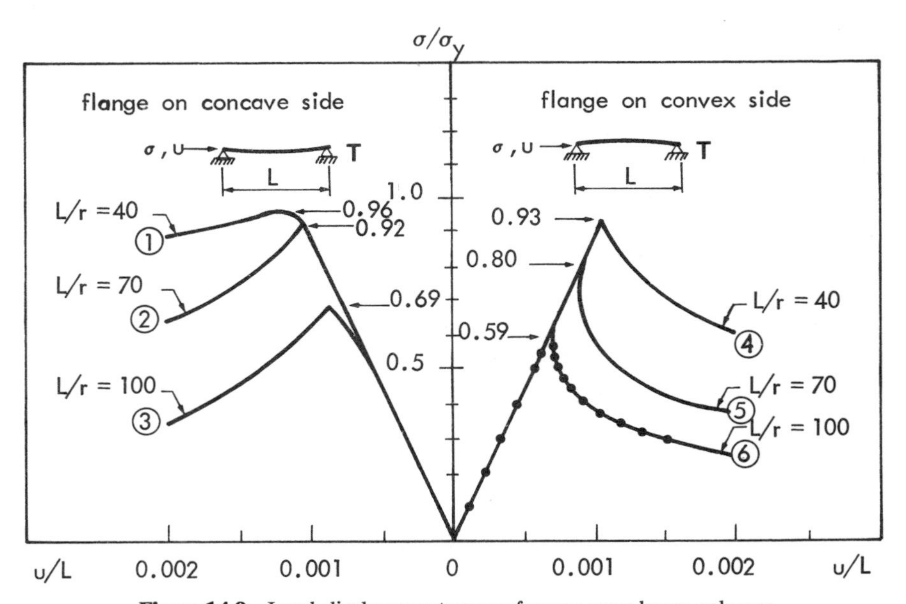

Figure 14.9 Load-displacement curves for one-span beam-columns

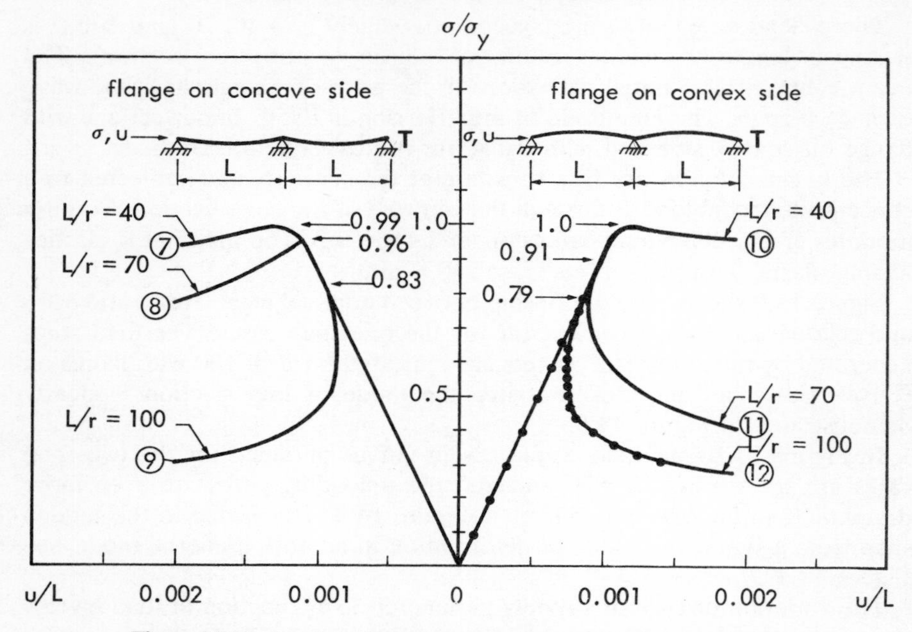

Figure 14.10 Load-displacement curves for two-span beam-columns

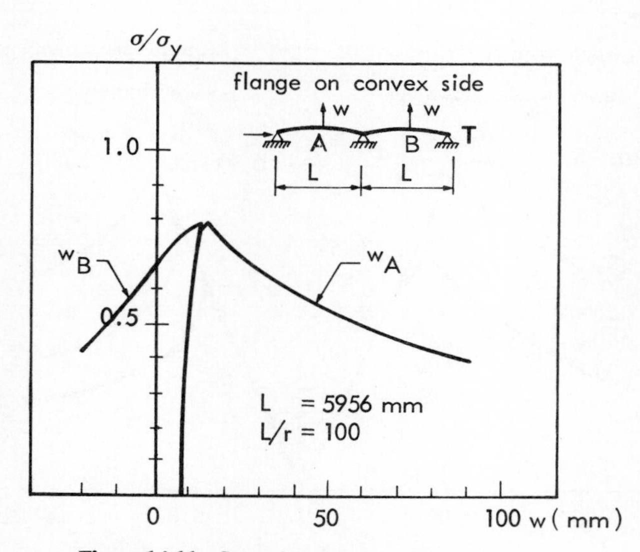

Figure 14.11 Snap-through of beam-column

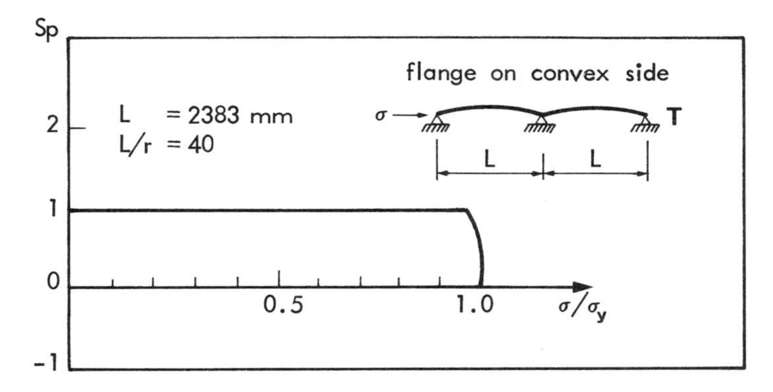

Figure 14.12 Variation of current stiffness parameter for two-span beam-column with $1/r = 40$

slenderness ratios $l/r = 40$ and 100, respectively. Both cases have imperfections with flange on concave side. For $l/r = 40$, S_p is nearly constant up to first yield at $\sigma/\sigma_Y = 0.95$. Thereafter, the stiffness decreases rapidly towards zero as the nominal stress ratio approaches its ultimate value. Beyond the maximum load the incremental stiffness is almost zero.

The violent post-buckling behaviour for $l/r = 100$ is clearly demonstrated by the variation of S_p in Figure 14.13. Up to $\sigma/\sigma_Y = 0.79$ the current stiffness parameter is constant and equal to unity. This region is marked ⓐ in Figure 14.13. For $\sigma/\sigma_Y = 0.79$ first yield and collapse occur simultaneously. The curve for S_p is discontinuous at this level of loading. In the first region of unloading S_p is positive (marked ⓑ in Figure 14.13). The curve for S_p has a new discontinuity for $\sigma/\sigma_Y = 0.55$. At this stage the load–displacement curve in Figure 14.10 has vertical slope. By further unloading S_p is negative (marked ⓒ in Figure 14.13).

In Table 14.2 comparison is made between the results from present analyses and those obtained by Dowling.[14] Values for the bifurcation index I_p defined in Section 14.3.3 are also listed. It is seen that I_p gives a good characterization of the type of bifurcation. According to the classification in Table 14.1, the cases ①, ③, ⑦, ⑧, and ⑩ are moderately sensitive, cases ② and ④ are highly sensitive and ⑤, ⑥, ⑨, ⑪, and ⑫ are violently sensitive systems.

The two-span case $l/r = 100$ with flange on concave side has also been analysed by Crisfield.[15] The ultimate stress ratio that Crisfield obtained was $\sigma/\sigma_Y = 0.83$ which coincides with the present result.

Examples of use of the current stiffness parameter are also given in Reference 16 and 17.

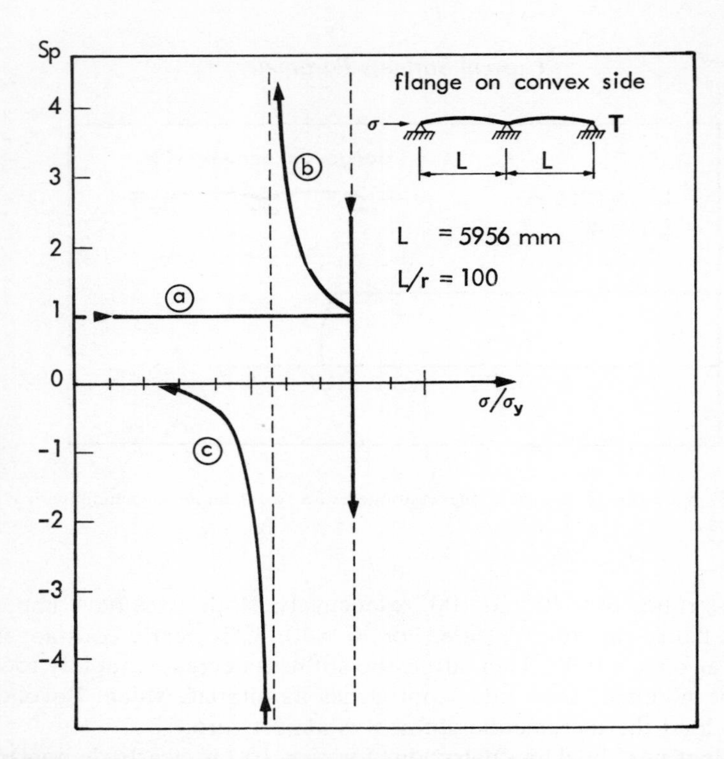

Figure 14.13 Variation of current stiffness parameter for two-span beam-column with $l/r = 100$

Table 14.2 Values of ultimate stress ratios and of bifurcation index

Imperfection	Case no.	l/r	σ/σ_Y first yield	σ/σ_Y collapse Present	σ/σ_Y collapse Dowling	I_p
Flange on concave side	①	40	0.95	0.96	0.95	−0.12
	②	70	0.88	0.92	0.86	−0.50
	③	100	0.69	0.69	0.64	−0.39
Flange on convex side	④	40	0.89	0.93	0.91	−0.52
	⑤	70	0.77	0.80	0.76	−1.22
	⑥	100	0.59	0.59	0.54	−1.18
Flange on concave side	⑦	40	0.92	0.99	0.97	−0.06
	⑧	70	0.83	0.96	0.91	−0.25
	⑨	100	0.75	0.83	0.79	−4.33
Flange on convex side	⑩	40	0.95	1.00		−0.06
	⑪	70	0.90	0.91		−3.50
	⑫	100	0.79	0.79		−11.00

14.5 CONCLUDING REMARKS

A new quantity for characterizing the nonlinear behaviour of structures has been presented. This 'current stiffness parameter' is a normalized measure of the incremental stiffness in the direction of motion.

It has been demonstrated how this parameter may be used to guide the incremental solution process. An algorithm for automatic load incrementation has been proposed together with a scheme for control of equilibrium iterations in the vicinity of extremum loads. Since the current stiffness parameter is normalized with initial value 1.0, it is relatively easy to specify the limits to be used in the automatic process.

The examples verify that the current stiffness parameter represents a valuable tool for nonlinear analyses. Especially for passing extremum points the new technique for automatic loading seems to be very efficient. The method also makes it easy to adjust the solution algorithm according to a desired number of load steps up to instability. Experience has shown that in order to make the automatic process generally reliable, the steps generated should be checked against upper and lower limits for load and displacement increments.

It seems that the bifurcation index may be a useful tool for characterization of what happens with a structure at a bifurcation point. However, experience with several more types of structures is necessary in order to explore the usefulness of this concept.

REFERENCES

1. Oden, J. T., 'Finite element applications in nonlinear structural analysis', presented at *ASCE Symposium on Application of Finite Element Methods in Civil Engineering*, Nashville, 1969.
2. Gallagher, R. H., 'Finite element analysis of geometrically nonlinear problems', in *Theory and Practice in Finite Element Structural Analysis* (Y. Yamada and R. H. Gallagher, Eds.), University of Tokyo Press, Tokyo, 1973.
3. Tillerson, J. R., Stricklin, J. A., and Haisler, W. E., 'Numerical methods for the solution of nonlinear problems in structural analysis', in *Numerical Solution of Nonlinear Structural Problems*, presented at the ASME Winter Annual Meeting, Detroit, 1973, pp. 67–101.
4. Bergan, P. G., and Söreide, T., 'A comparative study of different numerical solution techniques as applied to a nonlinear structural problem', *Computer Methods in Applied Mechanics and Engineering*, **2,** 185–201 1973.
5. Isaacson, E., and Keller, H. B., *Analysis of Numerical Methods*, Wiley, New York, 1966.
6. Ortega, J. M., and Rheinboldt, W. C., *Iterative Solution of Nonlinear Equations in Several Variables*, Academic Press, New York, 1970.
7. Murray, D. W., and Wilson, E. L., 'Finite element large deflection analysis of plates', *Journal of the Engineering Mechanics Division, ASCE*, **95,** EM1, (1969).

8. Hill, R., *The Mathematical Theory of Plasticity*, Oxford University Press, London, 1950.
9. Willam, K., '*Finite element analysis of cellular structures*', Doctoral Dissertation, University of California, Berkeley, 1969.
10. Bergan, P. G., Clough, R. W., and Mojtahedi, S., 'Analysis of stiffened plates using the finite element method', *Report No. UCSESM* 70–1, *University of California*, Berkeley, 1970.
11. Söreide, T. H., 'Collapse behavior of stiffened plates using alternative finite element formulations', *Report No.* 77–3, *Division of Structural Mechanics, The Norwegian Institute of Technology*, The University of Trondheim, Norway, June 1977.
12. Bergan, P. G., and Clough, R. W., 'Convergence criteria for iterative processes', *AIAA Journal*, **10,** 8, 1107–1108 (1972).
13. Timoshenko, S. P., and Gere, J. M., *Theory of Elastic Stability*, McGraw-Hill, New York, 1961.
14. Dowling, P. J., 'Some approaches to the nonlinear analysis of plated structures', presented at *Symposium on Nonlinear Techniques and Behaviour in Structural Analysis, Transport and Road Research Laboratory*, Crowthorne, U.K., 1974.
15. Crisfield, M. A., 'Large deflection elasto-plastic buckling analysis of eccentrically stiffened plates using finite elements, *Report No.* 725, *Transport and Road Research Laboratory*, Crowthorne, U. K., 1967.
16. Bergan, P. G., and Söreide, T. H., 'Solution of large displacement and instability problems using the current stiffness parameter', *Proceedings of International Conference on Finite Elements in Nonlinear Solid and Structural Mechanics*, Geilo, Norway, August 29–September 1, 1977.
17. Remseth, S. N., Holthe, K. H., Bergan, P. G., and Holand, I., 'Tube buckling analysis by the finite element method', *Proceedings of International Conference on Finite Elements in Nonlinear Solid and Structural Mechanics*, Geilo, Norway, August 29–September 1, 1977.

Chapter 15

Total and Updated Lagrangian Descriptions in Nonlinear Structural Analysis: A Unified Approach

S. Cescotto, F. Frey, and G. Fonder

15.1 INTRODUCTION

Total Lagrangian description (TLD) has successfully been used for about ten years since the fundamental work done by Oden,[1] Yaghmai,[2] and Marcal[3] among others. Both geometrical and material nonlinearities have been taken into account, together with unrestricted large displacements and deformations.

Updated Lagrangian description (ULD) has been less successful, that is until these very last years. This type of nonlinear analysis, however, began about twenty years ago with an intuitive formulation[4,5,6] which has since been forsaken. It is now recognized that this formulation often presents more simplicity than the TLD, especially for 'large displacement–small strain' problems.

Basic equations of both formulations may be derived by means of virtual work or energy principles. The former is more general and directly enables to consider history dependent constitutive laws. It will therefore be used here.

Nonlinear equations are generally written in an incremental way and solved as linear equations. The incremental forms may be obtained, either by subtracting the equilibrium equations of two neighbouring configurations and then linearizing the result, or by direct linearization of the nonlinear equations. The former method has the advantage of clearly showing the neglected nonlinear terms; however it will not be used here (see, for instance, references 2, 7, and 12). The latter is more straightforward and may be interpreted in several equivalent ways: second variation, total differential, linear incrementation, Taylor expansion limited to linear terms, direct consequence of Newton–Raphson method for solution of nonlinear equations, etc.

In the motion of the body, three successive configurations can be considered (Figure 15.1): the initial (unstressed, undeformed) configuration γ^0,

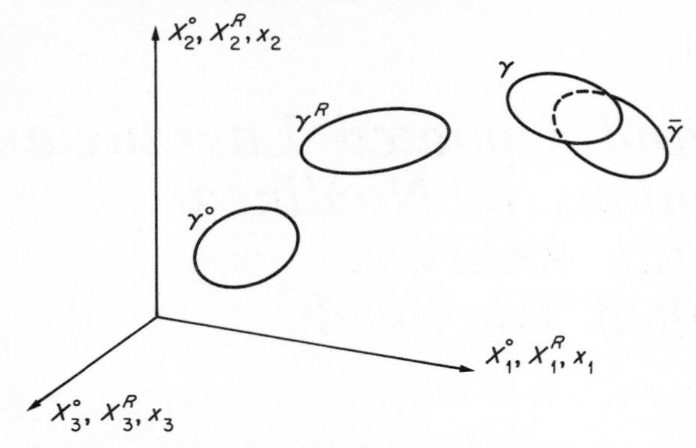

Figure 15.1 Successive configurations of the body

the current configuration γ and the incremented configuration $\bar{\gamma}$. The Lagrangian description requires a known reference configuration to formulate the equilibrium equations: the TLD uses the initial unstressed configuration of the body γ^0, and the ULD uses the current deformed configuration γ (incidentally, it may be mentioned that an Eulerian approach would use $\bar{\gamma}$ as its reference configuration). However, any intermediate configuration may be used as a Lagrangian reference. The first objective of this paper is to present a generalized Lagrangian description (GLD) where the equilibrium equations are referred to an arbitrary known deformed configuration γ^R.

In the two limit cases where γ^R coincides with γ^0 or γ, the TLD or ULD respectively will be recovered as special applications of this generalized formulation, so that both descriptions are now unified.

Then, the linearized or incremental equation of the GLD will be derived; it enables us to find linearized increments of displacements from the configuration γ to the neighbouring (incremented) configuration $\bar{\gamma}$ (Figure 15.1). Incremental equations of the TLD and ULD will again be derived as special applications by letting γ^R coincide with γ^0 and γ. Until now, these incremental equations were generally obtained by subtraction; incorrect use of this method in TLD has sometimes led to an incomplete incremental equation as shown in Reference 7. Basic knowledge of the nonlinear theory of continuum mechanics is supposed to be familiar.[8,11] The virtual displacement principle in rectangular Cartesian co-ordinates and displacement models of the finite element method are used throughout. 'Dead' forces, unaffected by the change in configuration, as well as 'following' forces, which depend on displacements (for instance, pressure loads), are considered; they are supposed to act statically for the sake of simplicity (dynamic loading may be easily introduced via the D'Alembert principle).

15.2 DERIVATION OF THE GENERALIZED LAGRANGIAN DESCRIPTION

15.2.1 Overall equilibrium of a body

The equilibrium of a body in a deformed configuration γ may be expressed by means of the virtual work principle as follows

$$\int_V \sigma_{ij}\,\delta\varepsilon_{ij}\,dv = \int_V \rho f_i\,\delta u_i\,dv + \int_S t_i\,\delta u_i\,ds \tag{15.1}$$

where v, dv and s, ds are current (deformed) volumes and areas of the loaded surface, respectively; σ_{ij} are the Cauchy (Euler, true) stresses; ρ is the current mass density; f_i the body force per unit mass; t_i the surface force per unit deformed area; δu_i is a compatible field of virtual displacements; and $\delta\varepsilon_{ij}$ the corresponding field of compatible virtual strains

$$2\,\delta\varepsilon_{ij} = \frac{\partial(\delta u_j)}{\partial x_i} + \frac{\partial(\delta u_i)}{\partial x_j} \tag{15.2}$$

x_i are the current Cartesian co-ordinates of material points of the body in the deformed configuration.

It is possible to express equation (15.1) with respect to any known reference configuration γ^R of the body. Therefore, to be general, it is only required that γ^R be an equilibrium configuration located between γ^0 and γ.

The transformation of equation (15.1) follows the same scheme as for the TLD (see Reference 8, for example). The result is

$$\int_{V^R} S_{ij}^R\,\delta E_{ij}^R\,dV^R = \int_{V^R} \rho^R f_i\,\delta u_i^R\,dV^R + \int_{S^R} T_i^R\,\delta u_i^R\,ds^R \tag{15.3}$$

where V^R, dV^R and S^R, dS^R are the volumes and areas of the body in γ^R; ρ^R is the mass density in γ^R, T_i^R the surface force acting in γ but measured per unit area in γ^R; S_{ij}^R are the second Piola–Kirchhoff stresses, acting in γ but measured in γ^R, with

$$S_{ij}^R = \frac{\rho^R}{\rho}\frac{\partial X_i^R}{\partial x_n}\sigma_{mn}\frac{\partial X_j^R}{\partial x_m} \tag{15.4}$$

E_{ij}^R are the Green strains describing the deformation of the body between γ^R and γ

$$2E_{ij}^R = \frac{\partial x_k}{\partial X_i^R}\frac{\partial x_k}{\partial X_j^R} - \delta_{ij} \tag{15.5}$$

X_i^R are the Cartesian co-ordinates of material points of the body in γ^R; u_i^R

are the relative displacements of material points between γ^R and γ

$$u_i^R = x_i - X_i^R \tag{15.6}$$

and δu_i^R is any compatible variation of the displacement field u_i^R.

Generally the body force f_i may be written as[3]

$$f_i = \lambda \Phi_i(x_j) \tag{15.7}$$

where λ is a load parameter and $\Phi_i(x_j)$ is a function of the deformed geometry that should be defined for each particular case of body force.

On the other hand, if the deformed surface in γ has the following parametric equations

$$x_i = x_i(\xi_1, \xi_2), \qquad i = 1, 2, 3 \tag{15.8}$$

a local base may be defined by

$$\mathbf{e}_1 = \mathbf{a}_1/\sqrt{a_{11}}; \qquad \mathbf{e}_2 = \mathbf{a}_2/\sqrt{a_{22}}, \qquad \mathbf{e}_3 = \mathbf{a}_1 \times \mathbf{a}_2/\|\mathbf{a}_1 \times \mathbf{a}_2\| \tag{15.9}$$

with

$$\mathbf{a}_1 = \frac{\partial x_i}{\partial \xi_1}\mathbf{i}_i; \qquad \mathbf{a}_2 = \frac{\partial x_i}{\partial \xi_2}\mathbf{i}_i, \qquad \sqrt{a_{11}} = \|\mathbf{a}_1\|; \qquad \sqrt{a_{22}} = \|\mathbf{a}_2\| \tag{15.10}$$

where $\mathbf{i}_1$, $\mathbf{i}_2$, $\mathbf{i}_3$ are the unit base vectors of the global Cartesian co-ordinates.

The components of the surface force t_i in the local base are

$$\mathbf{t} = t_i \mathbf{i}_i = \tau^r e_{ri} \mathbf{i}_i \tag{15.11}$$

where $e_{ri} = \mathbf{e}_r \cdot \mathbf{i}_i$, and the element of deformed area is

$$\mathrm{d}s = \sqrt{a}\, \mathrm{d}\xi_1\, \mathrm{d}\xi_2 \tag{15.12}$$

Similarly, if the parametric equations of the surface of the body in γ^R are $X_i^R = X_i^R(\xi_1, \xi_2)$, one has

$$\mathrm{d}S^R = \sqrt{A^R}\, \mathrm{d}\xi_1\, \mathrm{d}\xi_2 \tag{15.13}$$

where $\sqrt{A^R}$ is defined in the same way as $\sqrt{a}$.

Finally, using (15.11), (15.12), and (15.13)

$$T_i^R = \tau^r e_{ri} \sqrt{a}/\sqrt{A^R} \tag{15.14}$$

and the equilibrium equation of the body in the deformed configuration γ is expressed using γ^R as reference by

$$\int_{V^R} S_{ij}^R \,\delta E_{ij}^R \,\mathrm{d}V^R = \int_{V^R} \rho^R \lambda \Phi_i \,\delta u_i^R \,\mathrm{d}V^R + \int_{S^R} \tau^r e_{ri} \frac{\sqrt{a}}{\sqrt{A^R}} \,\delta u_i^R \,\mathrm{d}S^R \tag{15.15}$$

Formally, this equation is similar to equation (13) in Reference 7; however, it is more general because the reference configuration γ^R has not been defined yet. So, equation (15.15) potentially contains both the TLD and ULD as will be seen hereafter.

15.2.2 Discretization of the equilibrium equation

The large displacement field

$$u_i^R = x_i - X_i^R$$

may be discretized as follows

$$u_i^R = g_i(p_\alpha, X_j^R) \tag{15.16}$$

where $p_\alpha (\alpha = 1, 2, \ldots, n)$ are the discretization parameters (unknown nodal displacements, for example).

The functions g_i are defined with respect to the configuration γ^R. As the relative displacements u_i^R can be large, they are not assumed to be linear functions of the parameters p_α. It is assumed that they are at least piecewise continuous and that their first and second derivatives exist.

From (15.16)

$$\delta u_i^R = H_{i\alpha} \,\delta p_\alpha \tag{15.17}$$

with

$$H_{i\alpha} = \frac{\partial g_i}{\partial p_a} = H_{i\alpha}(p_\alpha, X_j^R) \tag{15.18}$$

From (15.15), the general equilibrium equation may be written as:

$$\int_{V^R} S_{ij}^R \left(\delta_{ik} + \frac{\partial g_k}{\partial X_i^R} \right) \frac{\partial H_{k\alpha}}{\partial X_j^R} \,\mathrm{d}V^R = \int_{V^R} \rho^R \lambda \Phi_i H_{i\alpha} \,\mathrm{d}V^R + \int_{S^R} \tau^r e_{ri} \frac{\sqrt{a}}{\sqrt{A^R}} H_{i\alpha} \,\mathrm{d}S^R \tag{15.19}$$

15.2.3 Incremental equilibrium equation

The incremental equilibrium equation is obtained by differentiating both members of (15.19). As the integrals are taken over constant volumes and surfaces, the differentiation may be carried out under the integral signs.

Define

$$(K^R_{LD})_{\alpha\beta} = \int_{V^R} \left(\delta_{ir} + \frac{\partial g_r}{\partial X^R_i}\right)\frac{\partial H_{r\alpha}}{\partial X^R_j} C^R_{ijkl}\left(\delta_{ks} + \frac{\partial g_s}{\partial X^R_k}\right)\frac{\partial H_{s\beta}}{\partial X^R_l}\,\mathrm{d}V^R \tag{15.20}$$

$$(K^R_S)_{\alpha\beta} = \int_{V^R} \frac{\partial H_{k\alpha}}{\partial X^R_i} S^R_{ij} \frac{\partial H_{k\beta}}{\partial X^R_j}\,\mathrm{d}V^R \tag{15.21}$$

$$(K^R_{SS})_{\alpha\beta} = \int_{V^R} S^R_{ij}\left(\delta_{ik} + \frac{\partial g_k}{\partial X^R_i}\right)\frac{\partial}{\partial X^R_j}\left(\frac{\partial^2 g_k}{\partial p_\alpha\,\partial p_\beta}\right)\mathrm{d}V^R \tag{15.22}$$

$$(K^R_{NC})_{\alpha\beta} = \int_{V^R} \rho^R\lambda\frac{\partial \Phi_i}{\partial x_l} H_{l\beta} H_{l\alpha}\,\mathrm{d}V^R + \int_{S^R} \tau^r M_{ri\beta} H_{i\alpha}\,\mathrm{d}S^R \tag{15.23}$$

$$(K^R_{IL})_{\alpha\beta} = \int_{V^R} \rho^R\lambda\Phi_i\frac{\partial^2 g_i}{\partial p_\alpha\,\partial p_\beta}\,\mathrm{d}V^R + \int_{S^R} \tau^r e_{ri}\frac{\sqrt{a}}{\sqrt{A^R}}\frac{\partial^2 g_i}{\partial p_\alpha\,\partial p_\beta}\,\mathrm{d}S^R \tag{15.24}$$

$$\mathrm{d}P^R_\alpha = \int_{V^R} \rho^R\,\mathrm{d}\lambda\,\Phi_i H_{i\alpha}\,\mathrm{d}V^R + \int_{S^R} \mathrm{d}\tau^r e_{ri}\frac{\sqrt{a}}{\sqrt{A^R}} H_{i\alpha}\,\mathrm{d}S^R \tag{15.25}$$

where the incremental constitutive law has been assumed to be

$$\mathrm{d}S^R_{ij} = C^R_{ijkl}\,\mathrm{d}E^R_{kl} \tag{15.26}$$

and where

$$\mathrm{d}\left(e_{ri}\frac{\sqrt{a}}{\sqrt{A^R}}\right) = M_{ri\beta}\,\mathrm{d}p_\beta \tag{15.27}$$

Then the result of the differentiation, that is the *general incremental equilibrium equation* is

$$(K^R_{LD} + K^R_S + K^R_{SS} - K^R_{NC} - K^R_{IL})_{\alpha\beta}\,\mathrm{d}p_\beta = (K^R_T)_{\alpha\beta}\,\mathrm{d}p_\beta = \mathrm{d}P^R_\alpha \tag{15.28}$$

or, in matrix form (underlined symbols denote matrices and curly brackets vectors)

$$\underline{K}^R_T\{\mathrm{d}p\} = \{\mathrm{d}P^R\}$$

This equation defines the tangent stiffness matrix $\underline{K}^R_T$ as the sum of

$\underline{K}^R_{LD}$: the symmetrical 'large displacement' matrix

$\underline{K}^R_S$: the symmetrical 'initial stress' matrix

$\underline{K}^R_{SS}$: a symmetrical 'additional initial stress' matrix

$\underline{K}^R_{NC}$: the non-symmetrical 'displacement dependent load' matrix

$\underline{K}^R_{IL}$: a symmetrical 'initial load' matrix

$\{\mathrm{d}P^R\}$ is the incremental load vector and $\{\mathrm{d}p\}$ is the incremental unknown displacement vector.

As recently mentioned in Reference 7 for the TLD several differences with respect to the more traditional approach[1,2,3,9,10,] arise because of the nonlinearity of the assumed displacement field:[16]

(a) two new symmetrical stiffness matrices ($\underline{K}^R_{SS}$ and $\underline{K}^R_{IL}$) appear; indeed the sum $\underline{K}^R_S + \underline{K}^R_{SS}$ is the real 'initial stress' matrix;
(b) the incremental load vector $\{dP^R)$ is not constant but depends on the total displacements p_α;
(c) equation (15.19) states the equilibrium between internal forces $\{Q^R\}$ and external forces $\{P^R\}$, where $\{Q^R\}$ and $\{P^R\}$ are vectors represented by the left- and right-hand side of this equation respectively; in the residual forces $\{R^R\}$, defined as

$$\{R^R\} = \{P^R\} - \{Q^R\} \tag{15.29}$$

the 'load level' $\{P^R\}$ is not constant, as it also depends on the displacements p_α.

15.3 TOTAL LAGRANGIAN DESCRIPTION

If the configuration γ^R is made to coincide with the undeformed, unstressed initial configuration γ^0 (which is assumed to exist), one can easily verify that the formulation of the classical TLD is recovered.

Indeed, in this case

$X^R_i = X^0_i$: initial co-ordinates of material points

$S^R_{ij} = S^0_{ij}$: Piola–Kirchhoff stresses measured in the initial configuration γ^0

$E^R_{ij} = E^0_{ij}$: Green strains describing the deformation of the body from its initial undeformed state

$C^R_{ijkl} = C^0_{ijkl}$: elasticity coefficients refered to the initial configuration γ^0

Then, the *discretized equilibrium equation* of the TLD is, from (15.19),

$$\int_{V^0} S^0_{ij}\left(\delta_{ik} + \frac{\partial g_k}{\partial X^0_i}\right)\frac{\partial H_{k\alpha}}{\partial X^0_j}\,dV^0 = \int_{V^0} \rho^0 \lambda \Phi_i H_{i\alpha}\,dV^0 + \int_{S^0} \tau^r e_{ri}\frac{\sqrt{a}}{\sqrt{A^0}} H_{i\alpha}\,dS^0 \tag{15.30}$$

On the other hand, the *incremental equilibrium equation* writes, from (15.28),

$$(\underline{K}^0_{LD} + \underline{K}^0_S + \underline{K}^0_{SS} - \underline{K}^0_{NC} - \underline{K}^0_{IL})\{dp\} = \underline{K}^0_T\{dp\} = \{dP^0\} \tag{15.31}$$

where $\underline{K}^0_{LD}$, $\underline{K}^0_S$, $\underline{K}^0_{SS}$, $\underline{K}^0_{NC}$, $\underline{K}^0_{IL}$, $\{dP^0\}$ are the stiffness matrices and incremental load vector given by equations (15.20) to (15.27) in which the superscript R must be replaced by the superscript 0. It is easily seen that these equations are identical to equations (17) to (23) of Reference 7.

15.4 UPDATED LAGRANGIAN DESCRIPTION

Let the reference configuration γ^R coincide with the deformed configuration γ; still, one has to keep in mind that a Lagrangian description is used: this means that one observes the motion of material particles. This motion is studied only in these instants immediately following the passing of particles through the configuration γ. So, very small displacement increments are dealt with that have to be added to the displacements of the body in γ, Now, the co-ordinates X_i^R are equal to the current co-ordinates of material points x_i. However, in order to avoid possible confusion with an Eulerian description, a formal distinction is introduced by retaining capital letters in the notation

$$\begin{aligned} X_i^R &= X_i \equiv x_i \\ V^R &= V \equiv v \\ S^R &= S \equiv s \end{aligned} \tag{15.32}$$

But the functions discretizing the small displacements measured from γ

$$u_i = g_i(p_\alpha, X_k) \tag{15.33}$$

must be such that

$$[g_i(p_\alpha, X_k)]_{p_\alpha=0} = 0 \tag{15.34}$$

$$\left[\frac{\partial g_i(p_\alpha, X_k)}{\partial X_j}\right]_{p_\alpha=0} = 0 \tag{15.35}$$

$$\left[\frac{\partial^2 g_i(p_\alpha, X_k)}{\partial X_j\, \partial X_m}\right]_{p_\alpha=0} = 0 \tag{15.36}$$

This means that the function g_i as well as their material derivatives must vanish when the particles are in configuration γ. But the partial derivatives of g_i with respect to the parameters p_α

$$\frac{\partial g_i}{\partial p_\alpha}; \qquad \frac{\partial^2 g_i}{\partial p_\alpha\, \partial p_\beta}; \qquad \frac{\partial^2 g_i}{\partial p_\alpha\, \partial X_j} \cdots$$

must not vanish for $p_\alpha = 0$ because they describe the motion of the body.

Furthermore, as only small displacements are discretized, the functions g_i may be expanded in Taylor series with only the first order terms retained

$$u_i = [g_i(p_\alpha, Xk)]_{p_\alpha=0} + \left[\frac{\partial g_i(p_\alpha, X_k)}{\partial X_l}\right]_{p_\alpha=0} \cdot \mathrm{d}X_l + \left[\frac{\partial g_i(p_\alpha, X_k)}{\partial p_\alpha}\right]_{p_\alpha=0} \cdot \mathrm{d}p_\alpha \tag{15.37}$$

With (15.34), (15.35) and (15.18), one obtains

$$u_i = h_{i\alpha}(X_k) \cdot q_\alpha \tag{15.38}$$

with

$$h_{i\alpha}(X_k) = [H_{i\alpha}(p_\alpha, X_k)]_{p_\alpha=0} \tag{15.39}$$

and

$$q_\alpha = \mathrm{d}p_\alpha \tag{15.40}$$

where it is understood that the q_α are very small (incremental) quantities.

This shows that the displacement field discretized by (15.38) is a linear function of the discretization parameters q_α (incremental field). Therefore, in ULD,

$$(K_{SS})_{\alpha\beta} = 0 \tag{15.41}$$

and

$$(K_{IL})_{\alpha\beta} = 0 \tag{15.42}$$

On the other hand, the S^R_{ij} stresses become the Cauchy (true) stresses σ_{ij} in γ (see equation 15.4). So, the matrices $\underline{K}^R_{LD}$ and $\underline{K}^R_S$ respectively transform into

$$(K)_{\alpha\beta} = \int_V \frac{\partial h_{i\alpha}}{\partial X_j} C_{ijkl} \frac{\partial h_{k\beta}}{\partial X_l} \mathrm{d}V \tag{15.43}$$

$$(K_\sigma)_{\alpha\beta} = \int_V \frac{\partial h_{k\alpha}}{\partial X_i} \sigma_{ij} \frac{\partial h_{k\beta}}{\partial X_j} \mathrm{d}V \tag{15.44}$$

where C_{ijkl} are the elasticity coefficients related to the current configuration γ as they appear in the constitutive equation

$$\mathrm{d}\sigma_{ij} = C_{ijkl} \cdot \mathrm{d}\varepsilon_{kl} \tag{15.45}$$

deduced from (15.26).

Let us now examine what the 'displacement dependent load' matrix $\underline{K}^R_{NC}$ becomes in ULD. In equation (15.23), the volume integral is the contribution of the body forces f_i, the change of which is given by

$$\mathrm{d}f_i = \mathrm{d}\lambda\, \Phi_i(X_j) + \lambda \frac{\partial \Phi_i}{\partial X_j} \mathrm{d}X_j \tag{15.46}$$

In the general derivation of Section 15.2, the first term of this equation produced the volume integral in the incremental load vector $\mathrm{d}P^R_\alpha$ (equation 15.25) while the second term was responsible for the volume integral of $\underline{K}^R_{NC}$ (equation 15.23). But here

$$\mathrm{d}X_j = u_j = h_{j\alpha} q_\alpha \tag{15.47}$$

Therefore (15.46) becomes

$$\mathrm{d}f_i = \mathrm{d}\lambda\,\Phi_i(X_j) + \lambda \frac{\partial \Phi_i}{\partial X_j} h_{j\alpha} q_\alpha \tag{15.48}$$

Consequently, the volume integral in the displacement dependent lood matrix $\underset{\sim}{K}_{NC}$ writes

$$\int_V \rho\lambda \frac{\partial \Phi_i}{\partial X_j} h_{j\beta} h_{i\alpha}\, \mathrm{d}V$$

It is seen that this term can also be obtained from the corresponding one in (15.23) simply by putting

$$V^R = V, \qquad \rho^R = \rho, \qquad H_{i\alpha} = h_{i\alpha}$$

Using a similar transformation, it can be show that the surface integral of (15.23) becomes

$$\int_S \tau^r N_{ri\beta} h_{i\alpha}\, \mathrm{d}S$$

where $N_{ri\beta}$ is defined by

$$\frac{\partial(e_{ri}\sqrt{a})}{\partial X_j} \bar{u}_j = N_{ri\beta} q_\beta$$

and is equivalent to $M_{ri\beta}$ when $A^R = A$.

So, in ULD, the displacement dependent load matrix is

$$(K_{NC})_{\alpha\beta} = \int_V \rho\lambda \frac{\partial \Phi_i}{\partial X_j} h_{j\beta} h_{i\alpha}\, \mathrm{d}V + \int_S \tau^r N_{ri\beta} h_{i\alpha}\, \mathrm{d}S \tag{15.49}$$

This shows that

$$(K_{NC})_{\alpha\beta} = \lim_{\gamma^R \to \gamma} (K_{NC}^R)_{\alpha\beta} \tag{15.50}$$

We see that $(K_{NC})_{\alpha\beta}$ does not vanish in ULD. The fact has already been mentioned in Reference 10.

Finally, the incremental load vector becomes

$$\mathrm{d}P_\alpha = \int_V \rho\, \mathrm{d}\lambda\, \Phi_i h_{i\alpha}\, \mathrm{d}V + \int_S \mathrm{d}\tau^r e_{ri} h_{i\alpha}\, \mathrm{d}S$$

or

$$\mathrm{d}P_\alpha = \int_V \rho\, \mathrm{d}f_i h_{i\alpha}\, \mathrm{d}V + \int_S \mathrm{d}t_i h_{i\alpha}\, \mathrm{d}S \tag{15.51}$$

Thus, the *incremental equilibrium equation in ULD* is

$$(\underset{\sim}{K}+\underset{\sim}{K}_{\sigma}+\underset{\sim}{K}_{NC})\{dp\}=\{dP\} \tag{15.52}$$

with $\underset{\sim}{K}$, $\underset{\sim}{K}_{\sigma}$, $\underset{\sim}{K}_{NC}$ and $\{dp\}$ given respectively by (15.43), (15.44), (15.50), and (15.51). $\underset{\sim}{K}$ is sometimes called 'small displacement' matrix, and $\underset{\sim}{K}_{\sigma}$ is the initial stress matrix; both are symmetrical.

Furthermore, it is easily seen that the *discretized equilibrium equation* becomes here, from (15.19),

$$\int_V \sigma_{ij}\frac{\partial h_{i\alpha}}{\partial X_j}\,dV=\int_V \rho f_i h_{i\alpha}\,dV+\int_S t_i h_{i\alpha}\,dS \tag{15.53}$$

Equations (15.52) and (15.53) represent the classical equations of the ULD: here, this formulation has been completely deduced from the generalized formulation developed in Section 15.2.

15.5 CONCLUSIONS

Let us recall and comment on the basic equations obtained for the three Lagrangian formulations.

15.5.1 Equilibrium equations

GLD:
$$\int_{V^R} S^R_{ij}\left(\delta_{ik}+\frac{\partial g_k}{\partial X^R_i}\right)\frac{\partial H_{k\alpha}}{\partial X^R_J}\,dV^R=\int_{V^R}\rho^R\lambda\Phi_i H_{i\alpha}\,dV^R+\int_{S^R}\tau^r e_{ri}\frac{\sqrt{a}}{\sqrt{A^R}}H_{i\alpha}\,dS^R$$

TLD:
$$\int_{V^0} S^0_{ij}\left(\delta_{ik}+\frac{\partial g_k}{\partial X^0_i}\right)\frac{\partial H_{k\alpha}}{\partial X^0_j}\,dV^0=\int_{V^0}\rho^0\lambda\Phi_i H_{i\alpha}\,dV^0+\int_{S^0}\tau^r e_{ri}\frac{\sqrt{a}}{\sqrt{A^0}}H_{i\alpha}\,dS^0$$

ULD:
$$\int_V \sigma_{ij}\frac{\partial h_{i\alpha}}{\partial X_j}\,dV=\int_V \rho\lambda\Phi_i h_{i\alpha}\,dV+\int_S \tau^r e_{ri}h_{i\alpha}\,dS$$

15.5.2 Incremental equilibrium equations

GLD: $(\underset{\sim}{K}^R_{LD}+\underset{\sim}{K}^R_S+\underset{\sim}{K}^R_{SS}-\underset{\sim}{K}^R_{NC}-\underset{\sim}{K}^R_{IL})\{dp\}=\{dP^R\}$

TLD: $(\underset{\sim}{K}^0_{LD}+\underset{\sim}{K}^0_S+\underset{\sim}{K}^0_{SS}-\underset{\sim}{K}^0_{NC}-\underset{\sim}{K}^0_{IL})\{dp\}=\{dP^0\}$

ULD: $(\underset{\sim}{K}+\underset{\sim}{K}_{\sigma}+\underset{\sim}{K}_{NC})\{dp\}=\{dP\}$

15.5.3 GLD and TLD

These descriptions produce very similar formulae and the equilibrium equations are rather complicated; in the general case, the incremental equilibrium equations contain five different stiffness matrices and the incremental load vectors $\{dP^R\}$ and $\{dP^0\}$ depend on the displacements.
These properties are typical of so-called 'oriented bodies' i.e. defined by means of a middle surface or axis, subjected to large displacements and large rotations, when forces do not act on this middle surface or axis (this allows us to consider eccentric loading without a fictitious element between the force and the middle surface or axis; see Reference 7, for instance). However, if the discretized displacement field is assumed to be linear with respect to the discretization parameters, the matrices $\underline{K}^R_{SS}$, $\underline{K}^0_{SS}$, $\underline{K}^R_{IL}$, $\underline{K}^0_{IL}$ vanish, and the incremental load vectors are constant. This occurs for any type of finite element when displacements are only moderate, but also in case of large displacements when the elements are described with the help of an intrinsic theory (such as isoparametric elements). But even in this simplified case, the expressions of $\underline{K}^0_{LD}$ and $\underline{K}^R_{LD}$ remain complex, and their computation may be time consuming.

The main advantages of the GLD and TLD are the use of a unique reference configuration and, consequently, of a unique type of stresses, i.e. the Piola–Kirchhoff stresses. For elastic (or hyperelastic) materials, these stresses can be computed directly, once the total displacements are known. Then, true stresses may be deduced from (15.4)

$$\sigma_{ij} = \frac{\rho}{\rho^R} \frac{\partial x_n}{\partial X_i^R} S_{mn}^R \frac{\partial x_m}{\partial X_j^R} \tag{15.54}$$

15.5.4 ULD

Obviously, the equations of the ULD are much simpler than in GLD or TLD. Physically, this can be understood as a result of the choice of a reference configuration for which all the nonlinear contributions of the discretized displacement field vanish. On the other hand, the reference configuration is variable and may be complicated to use. For instance, in finite strips models analytical integration of series of sines and cosines is no longer possible because the reference configuration itself already contains such terms.[13] Similar problems arise with oriented bodies but not with isoparametric elements because numerical integration is used. In practice, problems of large displacements accompanied by small strains are very common. In this particular case, simple and approximate formulations of the ULD may be developed, which make it very efficient and clearly superior to the TLD. This interpretation justifies the intuitive use of the ULD mentioned in the introduction.[12] As for the stresses, their computation is slightly more complicated than in TLD because, in a step-by-step procedure, the load increments are not infinitely small. Consequently, the stress increments from γ to $\bar{\gamma}$ are Piola–Kirchhoff stresses referred to γ, which must be transformed into Cauchy stresses by means of a formula similar to (15.54).

15.5.5 Constitutive relations

This aspect of the problem has been carefully examined elsewhere[10] and therefore will not be discussed here. However, it is worth recalling that the elasticity coefficients C^R_{ijkl}, C^0_{ijkl} and C_{ijkl} are related by

$$C_{ijkl} = \frac{\rho}{\rho^0} \frac{\partial X_i}{\partial X^0_m} \frac{\partial X_j}{\partial X^0_n} C^0_{mnpq} \frac{\partial X_k}{\partial X^0_p} \frac{\partial X_l}{\partial X^0_q} \tag{15.55}$$

$$C^R_{ijkl} = \frac{\rho^R}{\rho^0} \frac{\partial X^R_i}{\partial X^0_m} \frac{\partial X^R_j}{\partial X^0_n} C^0_{mnpq} \frac{\partial X^R_k}{\partial X^0_p} \frac{\partial X^R_l}{\partial X^0_q} \tag{15.56}$$

If these transformations are performed, the three formulations GLD, TLD, and ULD give the same results.[10]

REFERENCES

1. Oden, J. T., *Finite Elements of Non linear Continua.* McGraw-Hill, New York, 1972.
2. Yaghmai, S., 'Incremental analysis of large deformations in mechanics of solids with applications to axisymmetric shells of revolution'. *Report SESM 68-17*, University of California, Berkeley, December 1968.
3. Hibbit, H. D., Marcal, P. V., and Rice, J. R., 'A finite element formulation for problems of large strain and large displacement', *Int. Jl. Solids Struct.*, **6,** 8, 1069–1086 (1970).
4. Turner, M. J., Martin, H. C., and Weikel, B. C., 'Further development and applications of the stiffness method', *Matrix Methods in Structural Analysis*, AGARDOGRAPH 72 (edited by B. Fraeijs de Veubeke), Pergamon Press, Oxford, 1964, pp. 203–266.
5. Argyris, J. H., Kelsey, S., and Kamel, H., 'Matrix methods of structural analysis—a précis of recent developments'. *Matrix Methods in Structural Analysis*, AGARDOGRAPH 72 (edited by B. Fraeijs de Veubeke), Pergamon Press, Oxford, 1964, pp. 1–164.
6. Murray, D. W., 'Large deflection analysis of plates', *Report n° SESM 67-44*, University of California, Berkeley, September 1967.
7. Frey, F., and Cescotto, S., 'Some new aspects of the incremental total Lagrangian description in nonlinear analysis', in Finite Elements in Nonlinear Mechanics, (edited by Pal Bergan *et al.*), (Proc. of the Int. Conf. Finite Elements in Nonlinear Solids and Structural Mechanics, Geilo, 1977) Tapir Publishers, University of Trondheim, 1978.

8. Malvern, L. E., *Introduction to the Mechanics of a Continuous Medium*, Prentice-Hall, Englewood Cliffs, 1969.
9. Larsen, P. K., 'Large displacement analysis of shells of revolution including creep, plasticity and viscoelasticity', *Report No. UC SESM 71-22*, University of California, Berkeley, December 1971.
10. Bathe, K. J., Ramm, E., and Wilson, E.L., 'Finite element formulations for large deformation dynamic analysis', *Int. Jl. Num. Methods Eng.*, **9**, 2, 353–386 (1975).
11. Fraeijs de Veubeke, B., Sander, G., and Beckers, P., 'Dual analysis by finite elements—linear and nonlinear applications', in *Nonlinear Theories by Fraeijs de Veubeke, B.*, Report SA-22, LTAS, University of Liège, 1970.
12. Frey, F., 'L'analyse statique nonlinéaire des structures par la méthode des éléments finis et son application à la construction métallique', Doctoral thesis, University of Liège, 1978.
13. Godinas, A., 'Problèmes de grands déplacements dans les bandes finies', (to be published).

Chapter 16

Hyper-Beams, Generalized Splines, and Practical Curve Fitting

B. M. Irons

16.1 INTRODUCTION AND MOTIVATION

The capabilities of most curve fitting packages are woefully inadequate. We should expect good performance in circles and ellipses and other analytic curves, even if the points are given in wildly irregular spacing, for that may be how the user receives the data. Further, it should be possible to plot a curve with a cusp. For example, an involute gear profile possesses a cusp at the base circle, where the curvature becomes infinite. Such properties can be inferred from a consecutive list of co-ordinates given a sufficiently intelligent algorithm. This paper is an interdisciplinary exercise, which aims to satisfy a practical geometric needs by exploiting some rather unusual beam theory. The finite element formulation is straightforward, but the nonlinear equations have unfortunate properties. It is a curious observation that the finite element technique, which goes from geometry to field problems, encounters unexpected difficulties when it returns to geometry.

16.2 LINEAR SPLINES

Let us first consider the simplest physical analogy to the generalized spline. A veteran car travels along the roads for which it was built. The going is so rough that passenger discomfort is a function of speed alone. The problem is how to pass through the given points A to Z in alphabetical order, in some alloted time, with the least total discomfort:

$$\phi = \int x^2 + y^2 \, \mathrm{d}t = \text{minimum} \qquad (16.1)$$

The linear spline is the exact solution. We simply draw the straight line segments AB, BC, $\cdots$ and we travel along them at constant speed v. The time of arrival at B is $t_B = AB/v$, at C is $t_C = (AB+BC)/v$, etc.

In the sequel, another physical model will be more useful. Figure 16.1 shows the curve ABCDEFZ being generated by linear splines. The given points are represented by rigid smooth needles, erected normal to the drawing plane, i.e. in the direction t. A flexible string is constrained to pass

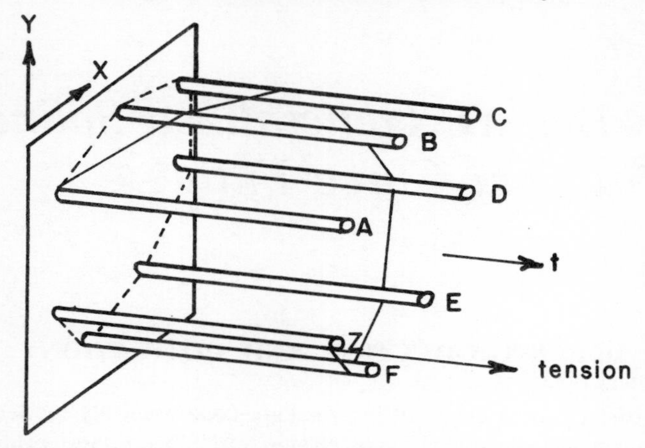

Figure 16.1 Mechanical analogue for the linear spline in 2D

through one end of needle A and the other end of needle Z. When the string is tensioned and its image is projected on the drawing plane, the dotted polygonal curve results. An experienced draughtsman would regard this construction in an even simpler way. He would imagine a sheet of paper wrapped around the needles, forming a polyhedral prism. On unwrapping the paper, i.e. taking a 'developed' view, the locus of the string becomes a single straight line from A to Z. Again, the direction cosine between the string and the t-axis is constant; for this is merely another way of saying that the 'speed' is constant.

16.3 CUBIC SPLINES

In fact $C^{(0)}$ is not good enough. To produce a more useful algorithm, we now consider a modern car, for which a better measure of discomfort is vector acceleration:

$$\phi = \int \left(\frac{\ddot{x}}{2}\right)^2 + \left(\frac{\ddot{y}}{2}\right)^2 \mathrm{d}t = \text{minimum} \tag{16.2}$$

For example, the driver anticipates a sharp bend by reducing speed, thus subjecting his passenger to increased fore–aft acceleration in order to reduce the subsequent sideways acceleration. In this case, the exact solution makes x and y cubics in t in each segment. The computer must find the velocity $(\dot{x}, \dot{y})$ at each nodal point, A, B, · · · , Z, and also the times of arrival at the intermediate nodes.

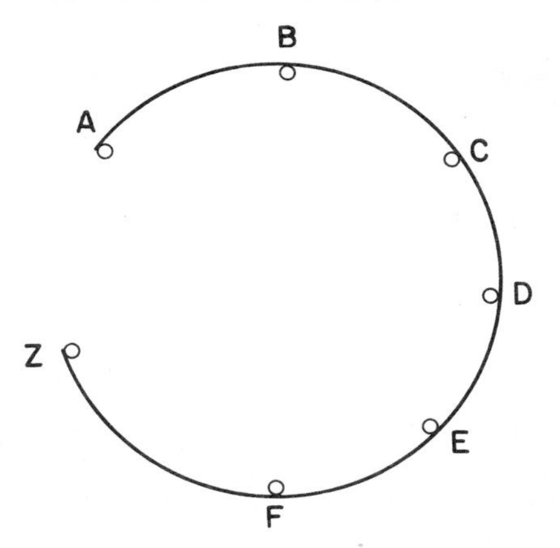

Figure 16.2 View along the needles of Figure 16.1 for a cubic spline

Again, an analogy from statics will be useful. Figure 16.2 should be interpreted as a view in the t-direction of a thin isotropic uniform beam, wrapped around the smooth needles of Figure 16.1. This model begs the question of negative reactions: the (x, y) values at the needles are prescribed, but the t values are not, except at A and Z. Thus x and y considered separately are cubic splines in t and have $C^{(2)}$ continuity.

We might be tempted to assert that the final x, y curve must be $C^{(2)}$ everywhere, but this is not universally true. At the cusp of Figure 16.3, the car stops momentarily, and reverses along a different curve. We probably have $C^{(1)}$ even here, and elsewhere we have $C^{(2)}$. Evidently, this would be useful algorithm, if it could be implemented. We do not expect to see cusps

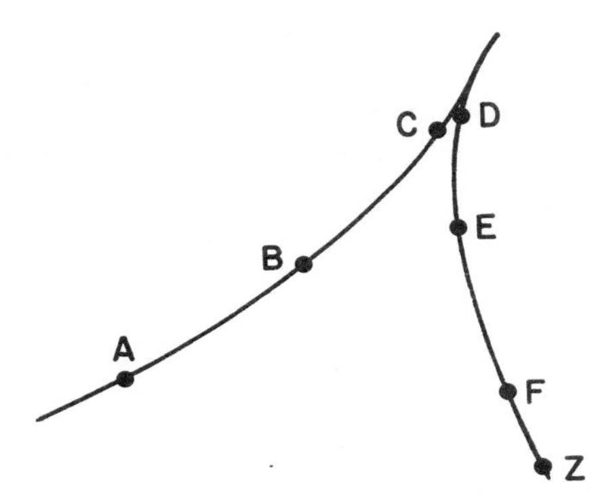

Figure 16.3 A cusp as discussed in the text

frequently, but curves with small loops and with local regions of flurried activity are not uncommon.

16.4 QUINTIC AND HIGHER SPLINES

A more sophisticated measure of passenger discomfort is the rate of change of acceleration, or 'jerk':

$$\phi = \int \left(\frac{\dddot{x}}{6}\right)^2 + \left(\frac{\dddot{y}}{6}\right)^2 \mathrm{d}t = \text{minimum} \tag{16.3}$$

(We must point out that, strictly, this model is in error. The jerk as sensed in the passenger seat is a much more complicated expression. For our purposes the simpler expression is almost certainly better.) The exact solution of (16.3) is a piecewise quintic, $C^{(4)}$ in t, and we seem to have a very attractive algorithm. The computer must now find $\dot{x}_1, \ddot{x}_1, \dot{y}_1, \ddot{y}_1$ at each node. But high-order processes usually exact some kind of revenge on the over-confident user.

The quintic spline will cost more in arithmetic. Increasingly we can discount such objections, as central processors become faster and cheaper. More important questions prevail: the reliability of the equation solver, whether the package is temperamental, and whether new difficulties appear in the quintic case. For example, St. Venant's principle is less valid in beams than in truss structures. A cantilever beam gives notoriously illconditioned equations.

Let us consider more quantifiable bad behaviour. Two given points are insufficient to define x as a quintic spline in t: the matrix is singular. With three distinct given points, an interesting case arises, for the three points define x as a quadratic in t. Thus $\dddot{x}$ is zero and the functional ϕ is minimized. So a quintic spline through three points gives the parabola. There is no harm in this, but suppose in our enthusiasm we generated a fifteenth order spline. With eight given points, this would generate the unique heptic through these points. This is dangerous; for the curve probably misbehaves near the end-points, where the interpolated values are abnormally sensitive to the given values near the centre of the range covered. The dimensionless multipliers in the Lagrange interpolation formula tend to be large. (Perhaps this is a case of St. Venant's principle not working very well?)

In the x–y–t case with three points, we can show that the quintic spline *always* gives a semi-definite coefficient matrix. Let us assume that t_A, t_B, and t_C are fixed, so that x and y are quadratics in t. Thus $\phi = 0$, again, because $\dddot{x} = 0$ and $\dddot{y} = 0$. However, if we repeated the calculation with some other t_B^* replacing t_B, ϕ would still be zero. If ϕ does not change with t_B (provided

$\dot{x}_i$, $\dot{y}_i$, etc. also change appropriately), then its Hessian matrix is singular. Admittedly the program can recognize this case, and default to a cubic spline.

A more troublesome case arises when we plot a straight line through N points. Evidently the cubic spline is stable, since the beam would be elastically stable. For the quintic, the same solution, $t_i = x_i - x_A$ or $y_i - y_A$ (one or the other could be zero for all i, but not both) gives $\phi = 0$. But in the case, $x_i - x_A$ or $y_i - y_A = t_i^* + Kt_i^* (t_Z - t_i^*)$ gives $t_A^* = 0$ and $t_Z^* = t_Z$ as before: it may be solved, at least for fairly small values of K. Because x and y are now quadratics in t, ϕ remains zero. This time the programmer cannot remove the danger by logic.

It would be foolhardy indeed to pretend that an ad hoc human search has revealed every degenerate case. Even if this list is complete, we have little hope of finding them all, in say the heptic spline. Evidently, the higher splines may yield near-analytic curves, but the penalties in practical terms would be severe.

16.5 MISBEHAVIOUR OF THE CUBIC SPLINE

The equations are exceptionally difficult to solve iteratively, even in the cubic case. Although the functionals have every appearance of positive-definiteness or at least of semi-definiteness, the Hessians have a strong tendency in practice to be indefinite if the t_i are incorrect, whether or not the x_i and y_i are consistent with the t_i.

This anti-theorem is particularly distressing to the structural intuition: one does not expect to find beam structures, without large end-loads, which are physically unstable. Yet the example we shall describe, and the argument we shall use, invoke the beam model. The demonstration curve is degenerate, a straight line which Figure 16.4 takes to lie on the x-axis. In the x–t plane, we have given points ABDE: observe that C is absent and that A and B,

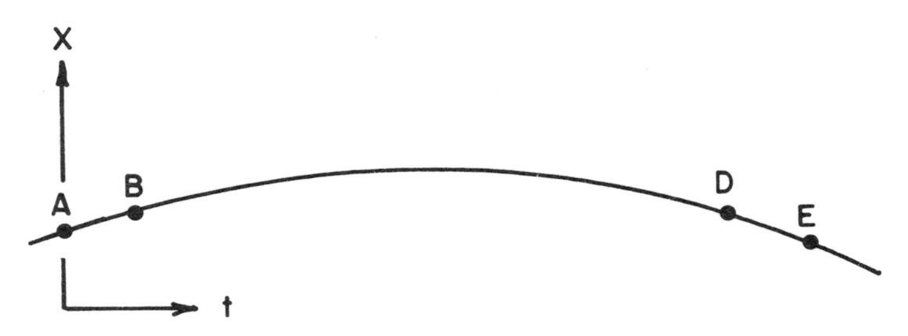

Figure 16.4 A parabola as defined by two pairs of closely spaced points

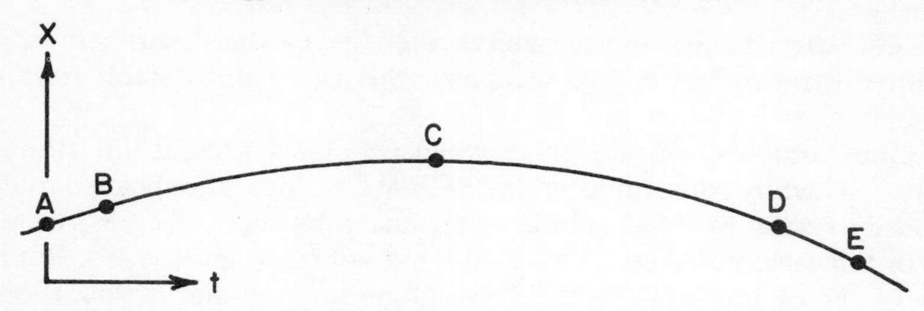

Figure 16.5 The parabola of Figure 16.4, perturbated by an extra point

and D and E are very closely spaced. Furthermore, we choose four points that lie on a parabola. Figure 16.5 introduces a fifth point C, which lies on the maximum of the parabola. Introducing C would make no difference to the plotted curve.

Let us now perturbate C downwards, so that x_C is reduced and t_C remains the same. The curvature is no longer sensibly constant over the length of the beam. Therefore, ϕ must increase, because the total change in slope between A and E is defined by the pairs of points AB and DE. If with the reduced x_C we either increase or reduce t_C so that point C again lies on the parabola, ϕ reverts to its original value. Therefore with reduced x_C and unchanged t_C we must have had a saddle point in ϕ.

In practice, if we attempt Newton iteration for solving the equations, negative pivots occur very frequently. One curious case arises when x_C lies exactly on the maximum of the parabola: ϕ in this case is indefinite because altering t_C to $t_C+\Delta t$ is equivalent to lowering C by $0\ (\Delta t^2)$ giving an energy perturbation of $0\ (\Delta t^4)$. We saw that quintic splines can give a singular matrix: evidently so can a cubic spline, if a cusp happens to occur precisely at a node, and if the reaction at the node is precisely zero.

16.6 THE EQUATION-SOLVING ALGORITHM

Clearly we must aim to avoid negative or zero pivots, and if they appear or threaten to appear the program must have the means to control them.

(1) A special technique is needed to control those cases in which the terminal Hessian is singular. We use Newtonian iteration, with the functional, e.g.

$$\phi = \int \left(\frac{\ddot{x}}{2}\right)^2 + \left(\frac{\ddot{y}}{2}\right)^2 + S_0[\dot{x}^2 + \dot{y}^2]\,dt$$

where S_0 takes a small value, say 0.0001. This gives a non-singular matrix. When the residuals reach a certain low level, we switch to modified Newton

iteration. That is, we retain the matrix but henceforth we compute the residuals with $S_0 = 0$. We have also considered giving S_0 some positive value if, for purely aesthetic reasons, we want to round off regions of high curvature, e.g. the ends of ellipses.
(2) This is an unusual nonlinear problem, in that if the t_i are fixed it becomes linear. We have observed much more severe negative pivots when the nodal $\dot{x}_i$ and $\dot{y}_i$ are not matched to the existing t_i. Therefore, we tend to run alternating iterations: in every other iteration the t_i are fixed, to ensure that whenever the t_i are free they are coupled to appropriate $\dot{x}_i$ and $\dot{y}_i$.
(3) It is still possible to encounter negative pivots, and even negative diagonals. If t_i changes to $t_i + \delta t_i$, we restrain the largest proportionate change in any interval, i.e. $(\delta t_{i+1} - \delta t_i)/(t_{i+1} - t_i)$. We try to achieve this by adding a 'spring' of constant flexibility per unit t to the stiffness matrix. If this fails, and a negative pivot appears, the limiting proportionate change is reduced.
(4) The linear spline is trivial and solutions are $t_A = 0$, $t_B = \mathrm{AB}$, $t_C = \mathrm{AB} + \mathrm{BC}$, etc. It is possible to proceed in small steps towards the cubic and thence to the quintic, as a factor R goes, say, to values 0.2, 0.4, 0.6, 0.8, 1. The basic stiffness matrix is:

$$RK^{(N)} + (1 - R)Q^{(N)}$$

where $K^{(N)}$ is the matrix for generating an ordinary spline of order N, and $Q^{(N)}$ is an artificial matrix. For a quintic segment of unit length, $Q^{(5)}\ \{x\}$ would be:

$$\begin{bmatrix} 3 & 1\tfrac{1}{2} & 0 & -3 & 1\tfrac{1}{2} & 0 \\ 1\tfrac{1}{2} & 1 & 0 & -1\tfrac{1}{2} & \tfrac{1}{2} & 0 \\ 0 & 0 & \tfrac{3}{35} & 0 & 0 & \tfrac{1}{70} \\ -3 & -1\tfrac{1}{2} & 0 & 3 & -1\tfrac{1}{2} & 0 \\ 1\tfrac{1}{2} & \tfrac{1}{2} & 0 & -1\tfrac{1}{2} & 1 & 0 \\ 0 & 0 & \tfrac{1}{70} & 0 & 0 & \tfrac{3}{35} \end{bmatrix} \begin{Bmatrix} x_0 \\ \dot{x}_0 \\ \ddot{x}_0 \\ x_1 \\ \dot{x}_1 \\ \ddot{x}_1 \end{Bmatrix}$$

The 4×4 occupying rows and columns 1, 2, 4, and 5 is $K^{(3)}$ so that if $R = 0$ and $\ddot{x}_0 = \ddot{x}_1 = 0$ the cubic solution is exactly reproduced. The terms 3/35 and 1/70 are taken from the 6×6 combining the functional for the cubic with the interplant for the quintic. (The rationale is as follows: we could take this 6×6, part-invert in rows 1, 2, 4, and 5, delete the cross-terms in the third and sixth rows and columns, then part-invert again.)

Each value of R gives a vector $t_i(R)$ whose terms are analytic functions of R. Hence with $t_i(0)$, $t_i(0.2)$, $t_i(0.4)$ for example we can use Lagrange extrapolation to predict the $t_i(0.6)$. The equation solver corrects these values, before moving to $R = 0.8$.

(5) R has a parallel task. In going from linear to cubic, the curve is progressively rolled up, starting with the segments in a straight line, and ending with the points as given. It is important to associate each segment with its accumulated angle. (In the case of a spiral, the angles could greatly exceed 2π.) As the curve is rolled up, these angles are multiplied by R. The values of t_i remain analytic in R.

16.7 UNIQUENESS OF THE SOLUTION

Our practical trials have been confined to the cubic spline. We felt that a positive definite functional as simple as this should be positive-definite in the other sense also that a stationary case should imply an optimum case. It was not possible to prove this. At the same time, numerical trials showed repeatedly that solutions were not optimal, i.e. Hessians were not positive-definite without artificial additional springs. However, the attempt revealed some interesting phenomena. We argue in terms of classical engineers' theory of thin beams.

Let us first examine the case in x–t only. By considering the virtual work when t_i is perturbated, we show that the force T_i corresponding to t_i equals $(\text{reaction})_i \times (\text{slope})_i$. Since T_i is zero, it follows either that the reaction is zero, or that the slope is zero. Thus in Figure 16.6 the reactions at C, E, and F are zero and the slope at B and D is zero. Evidently this follows because x_B and x_D are extrema in the sequence of x_i; and x_B and x_D are also extrema in the resulting curve. The solution is unique and simple enough for an undergraduate exercise. Further, in view of the sense of the reaction and of the curvature, it is evident that the energy varies positive-definitely with the t_i.

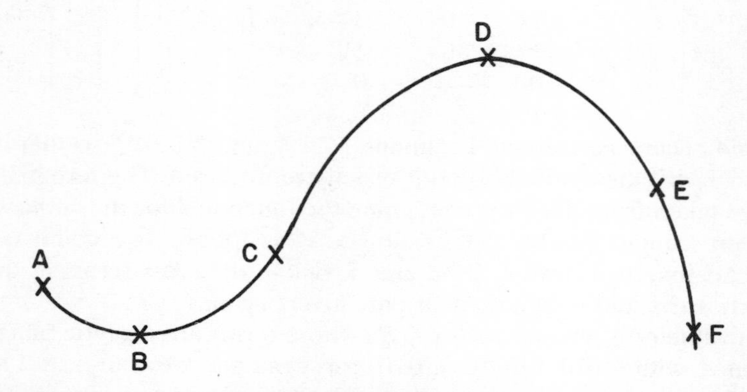

Figure 16.6 Example in 1D to show that the reaction at non-extremal points is zero and the slope at extremal points is zero

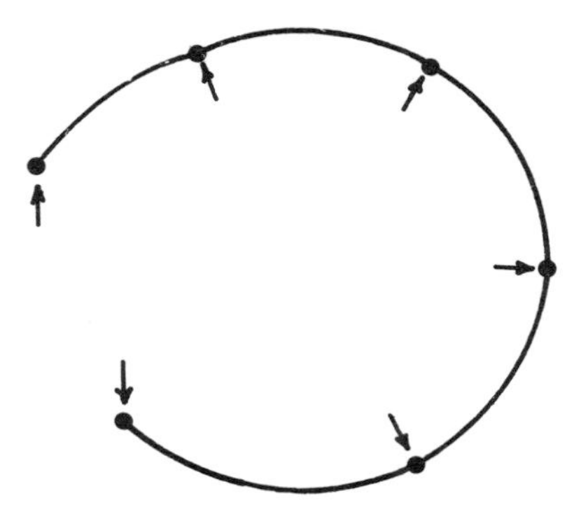

Figure 16.7 2D equivalent of Figure 16.6, showing the directions of the reactions

These trivial conclusions should carry over to the much more difficult case, in x–y–t. For example, if we choose the x–y axes, normal and tangent to the projected curve at node i, we see that the reaction must now be normal to the curve, as in Figure 16.7. Again, it seems obvious, yet it is evidently not true in general, that the reaction and the curvatures are in consistent directions. Indeed, it has been impossible to prove anything further than this.

If it had been possible to prove positive-definiteness, uniqueness would have followed, since ϕ is analytic provided $t_A < t_B < \cdots < t_Z$.

16.8 PRACTICAL ASPECTS, EXAMPLES, AND CONCLUSIONS

With cubic splines, the curvature at the ends is zero. Despite the arguments against the quintic spline, it remains attractive because the curves terminate more naturally. Perhaps a compromise is appropriate. It might be relatively safe, easy, and cheap to calculate the t_i as for a cubic spline and then, keeping the t_i constant, to interpolate x_i and y_i with quintic splines, for example. There is no obvious penalty: the curve has $C^{(4)}$ except at cusps.

Numerical experience has been unfortunate. However, we now consider that it was foolish to introduce the artificial springs, a move that encouraged the algorithm to find non-optimal stationary points. It would have been better to fix a t_i where the pivot becomes negative, simply by giving the pivot a large value. To discourage one pivot from being large throughout, randomly chosen pivots could also be made large. There are few remaining options: a crude numerical search is another, but very expensive. Therefore, we suggest that curves with no local violence would probably respond well to linear t_i followed by cubic or quintic interpolation. Figure 16.8 shows the

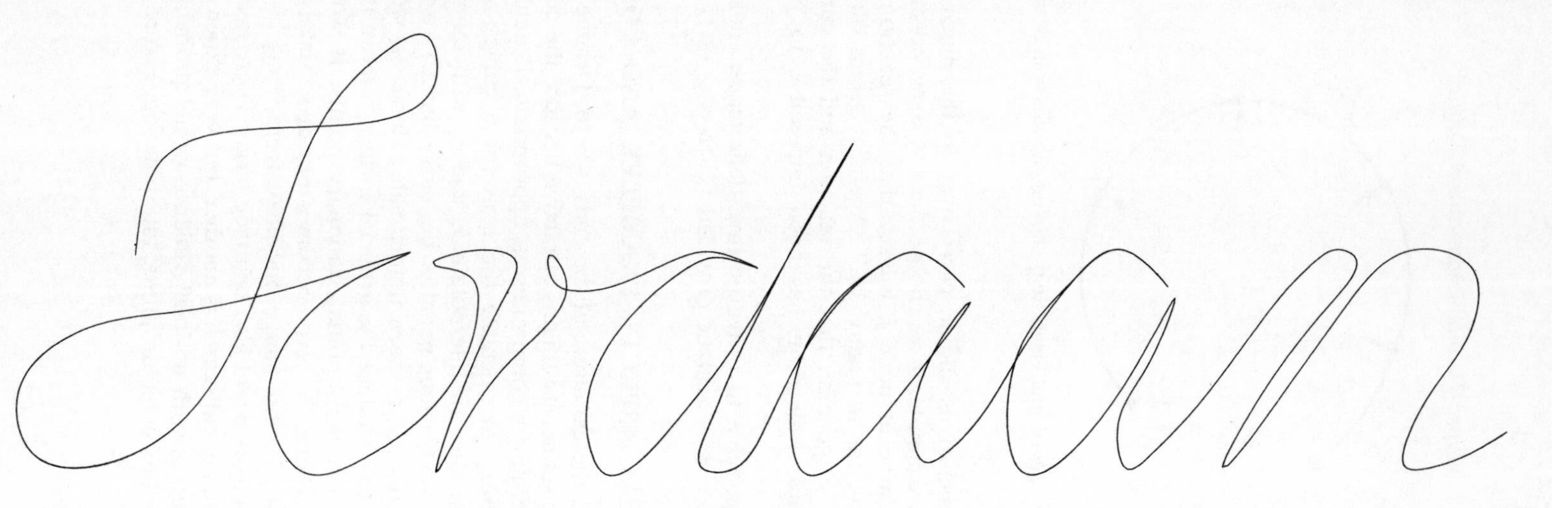

Figure 16.8 An example of cursive handwriting

sort of thing we have in mind. Some computer users would value the option of printing output data in the form of synthetic cursive handwriting. It is not yet feasible, but it seems worth the effort; not only for its own sake, but also to demonstrate that the package will not baulk at interpreting difficult curves.

The artistic guidance of Mr Leonard Jordaan is gratefully acknowledged, and will be called upon again if some way out of the present *impasse* is discovered. In particular, the help of Mr Saad Zaghlool in the final attempt to solve the equations was invaluable. I am grateful to CDC for making their plotting facilities in Calgary available to us.

Chapter 17

The Saddle Point of a Differential Program

H. Matthies, G. Strang, and E. Christiansen

17.1 INTRODUCTION

This paper is devoted to duality, a subject which Fraeijs de Veubeke must have enjoyed very much. He had a deep understanding of the relation between complementary pairs of variational principles, and that led him to new ideas for approximation of the classical problems of mechanics. His ideas will have permanent value.

We want to discuss one particular example, the duality between the static and kinematic theorems of limit analysis. That is a problem in infinite dimensional linear programming, or more generally in the optimization of a linear function subject to convex constraints. In the static case, the constraints include the differential equations of equilibrium: we maximize the scalar λ subject to two requirements,

$$\sum \frac{\partial \sigma_{ij}}{\partial x_j} + \lambda F_i = 0 \quad \text{and} \quad \sigma \in B, \tag{17.1}$$

B is the set of stresses which nowhere exceed the yield limits of perfect plasticity. It seems right to describe this as a *differential program*, and its object is to find the collapse multiplier $\bar{\lambda}$, the largest multiple of the given loads F_i which the structure can bear.

The theory of convex analysis provides a 'sup–inf' duality theorem: the supremum of all attainable λ equals the infimum in the dual problem, where the unknowns are the velocities u_i. This theorem was anticipated by the engineers who created limit analysis, and Koiter's survey article[1] gives an extremely clear history of the whole subject. It has since been re-examined by a number of authors, and used as the starting point for finite element algorithms; we thank Moreau, Nayroles, Debordes, Mercier, and Frémond for articles and conversations which have been extremely valuable. We hope that limit analysis can eventually become an efficient tool computationally, as attempted in a recent conference,[2] when it is not the whole history of the stresses which is needed but only information about the collapse state. It should be possible to get that special information at a reasonable cost.

The problem studied in this note is a much more theoretical one: to go beyond the sup–inf theorem, and to show that the maximum and minimum are actually attained. In other words, we try to choose admissible sets for the stresses and velocities within which a genuine saddle point $\bar{\sigma}$, $\bar{u}$ can be found. For the finite dimensional approximations this presents no difficulty, but in the continuous case the right choice of function spaces is still not completely evident; we have been forced to modify our original choice by the discovery that Korn's inequality fails to hold in the L_1 norm (see below). Furthermore the existence of $\bar{\sigma}$ and $\bar{u}$ can be studied in different ways; we start within the framework of the third author's MIT thesis and then change to an alternative path. (His approach is through sup–inf duality in the classical Sobolev spaces, and will be published separately.) Even though both methods aim at the same basic theorem of limit analysis, there are important questions still waiting for a complete answer.

This note will be an outline of our results. We discuss only the case in which there are no boundary conditions (surface tractions) prescribed for σ; a subsequent paper will be more general. Here we hope to go as directly as possible to the non-reflexive function spaces which seem fundamental to limit analysis.

17.2 THE CHOICE OF SPACES

There are four fundamental quantities in the problem—the forces f, the stresses σ, the strains ε, and the velocities u. For each of the four we shall have to choose an appropriate 'admissible space'; then for a given f we try to find in the other three spaces the collapse states $\bar{\sigma}$, $\bar{\varepsilon}$, and $\bar{u}$ of limit analysis. Looking ahead, we anticipate that the same spaces may be correct also for deformation theory, and even for incremental theory, in perfect plasticity. The choice is constrained by the following laws of mechanics:

(1) $\varepsilon = Du$, where D is the deformation operator given by

$$\varepsilon_{ij} = (\text{def } u)_{ij} = \frac{1}{2}\left(\frac{\partial u_i}{\partial x_j} + \frac{\partial u_j}{\partial x_i}\right)$$

(2) $D^t\sigma = f$, the equilibrium equation in (17.1) above
(3) the plastically admissible set B is defined by *pointwise* bounds on the stress components σ_{ij}
(4) collapse represents a minimax of the expression

$$(\sigma, \varepsilon) = \sum_{i,j} \int \sigma_{ij}\varepsilon_{ij}\, \mathrm{d}V \tag{17.2}$$

These laws suggest a first approximation to the choice of admissible spaces: the stresses should be bounded measurable functions ($\sigma \in L_\infty$, and $\sigma_{ij} = \sigma_{ji}$) because of the pointwise condition (3), and the strain space should be L_1 (the integrable functions) because of its pairing with the stress in the expression (σ, ε). Then the forces are derivatives of bounded functions, $f \in W^{-1,\infty}$, and the velocities are those functions which vanish on the boundary and have first derivatives in L_1, in other words, $u \in W_0^{1,1}$. These choices ought to be nearly right, but several difficulties remain.

First, we want to be able to solve $D^t\sigma = f$; the map D^t should be surjective, or 'onto', which requires that D have closed range. In analytical terms we would need

$$\|\varepsilon\|_{L_1} \geqslant c\|u\|_{W_0^{1,1}}$$

This is Korn's inequality in the L_1 norm. It asserts that the particular combinations ε_{ij} of first derivatives are enough to dominate every individual derivative,

$$\sum_{i,j} \int \frac{1}{2}\left|\frac{\partial u_i}{\partial x_j} + \frac{\partial u_j}{\partial x_i}\right| \mathrm{d}V \geqslant c \sum_{i,j} \int \left|\frac{\partial u_i}{\partial x_j}\right| \mathrm{d}V$$

With an exponent $p > 1$ in both integrands, the inequality is true; with $p = 2$, it is the foundation of elasticity theory. However, all proofs of this inequality fail in the case $p = 1$, and the inequality is actually false. We found a counter-example in Ornstein's paper,[3] which was concerned with scalar functions f rather than vector functions u; he constructs for any c an infinitely differentiable function $f(x, y)$, vanishing outside the unit square and depending on c, such that

$$\int |f_{xx}| + |f_{yy}| < c \int |f_{xy}|$$

Now we set $u = f_x$ and $v = -f_y$, and find

$$\int |u_x| + |u_y + v_x| + |v_y| < c \int |u_y|$$

Therefore Korn's inequality cannot hold.

This suggests a change in the spaces: the norm of u will be matched with the norm of the corresponding strain, so that $\|u\|$ in the velocity space equals $\|\varepsilon\|$ in the strain space. This means that the space of admissible velocities is slightly larger than before; only the particular combinations ε_{ij} are examined, and not all first derivatives.

We turn to the second difficulty: the strain $\bar{\varepsilon}$ at the moment of collapse may not lie in L_1, and the velocity $\bar{u}$ may not vanish on the boundary. These possibilities were illustrated in Reference 4 by the example of anti-plane shear, and we believe they are completely typical. In the example, the force

f has only a single non-zero component f_3, and that component depends only on $x = x_1$ and $y = x_2$; as a result, the same is true of the velocity, $u = (0, 0, u_3(x, y))$. The underlying domain is the square pipe $0 \leq x, y \leq 1$, $-\infty < z < \infty$. In this case the collapse mechanism $\bar{u} = (0, 0, \bar{u}_3)$ is found to be the characteristic function of a certain set E: $\bar{u}_3 = 1$ in E, $\bar{u}_3 = 0$ elsewhere, and E touches a substantial part of the boundary of the pipe. Thus $\bar{u}$ is not zero on the boundary, and furthermore the strains $\varepsilon_{13} = \partial\bar{u}_3/\partial x$ and $\varepsilon_{23} = \partial\bar{u}_3/\partial y$ involve δ-functions and lie outside L_1.

This last fact indicates that ultimately the admissible space for the strains should contain ε_{ij} which are measures; the integral of each $|\varepsilon_{ij}|$ should be finite but slip lines (lines of discontinuity in u) must be allowed to give a finite contribution to the integral from a set of zero measure. The corresponding set of velocities will be called the space of *functions bounded deformation.* In the example of anti-plane shear, where u is effectively a scalar and ε is its gradient, BD reduces to the space BV of *functions of bounded variation.* To compensate for the non-vanishing of u on part of the boundary, the norm must take account of the boundary integral of $|u|$; in other words, we set $u = 0$ *outside* $\bar{\Omega}$, and admit a line of δ-functions along $\partial\Omega$ as contributing to the norm (which is the total measure of ε). Roger Temam has shown how this corresponds exactly to the 'relaxed problems' of convex analysis and allows the existence theory to proceed as in the case of minimal surfaces.[5] We hope to analyse this question more fully in a joint paper.

The third difficulty is more profound, and harder to overcome: none of our spaces is reflexive. When we chose L_1 we know its dual to be L_∞—but then the dual of L_∞, the bidual of L_1, is a space in which we had hoped never to work. The same would be true of $W_0^{1,1}$ and $W^{-1,\infty}$. And for BD and BV, even their duals seem to have no satisfactory interpretation for an application to mechanics. These spaces do, however, allow the possibility of going backwards; they themselves are dual spaces.

That will be a valuable result. It is connected to the classical theorem of Riesz, that the measures $M(\bar{\Omega})$ are dual to the continuous functions $C(\bar{\Omega})$. For functions of one variable the connection is extremely close; every measure is, roughly speaking, the derivative of some u in BV. In several variables, this will not be true; the partial derivatives of u are linked by a compatibility condition $\partial/\partial x_j(\partial u/\partial x_i) = \partial/\partial x_i(\partial u/\partial x_j)$. In that case we are far from identifying every vector measure as a gradient, and we must anticipate that BV is the dual of some *subspace* of C.

First, we define BV. A better notation would be BV_0, since for smooth functions the original boundary condition $u = 0$ on $\partial\Omega$ will be literally true. The space is an enlargement of $W_0^{1,1}$, which was seen to be unsatisfactory in the example of anti-plane shear. Suppose Ω lies inside a ball $Q \subset R^n$, say with dist $(\partial\Omega, \partial Q) > 1$. Then BV contains those L_1 functions on Q which vanish outside $\bar{\Omega}$ and have gradients which are regular measures:

$$BV = \{u \in L_1(Q) \mid u = 0 \quad \text{in} \quad Q - \bar{\Omega}, \text{grad } u \in M\}$$

Since grad u will have its support in $\bar{\Omega}$, we can think of it as a (vector) measure on $\bar{\Omega}$ and thereby as a continuous linear functional on $C(\bar{\Omega})^n$. This is certainly justified if Ω is reasonable, since we can extend any continuous s from $\bar{\Omega}$ to the rest of Q, and De Giorgi's definition

$$(s, \text{grad } u) = \lim_{h \to 0} \int_Q s \cdot \text{grad}\,(u * b_h)\,dV, \qquad b_h \approx e^{-x^2/h^2}$$

will be independent of the extension.

In the case $n = 1$, when the interval Ω lies inside a larger interval Q, the result will be zero whenever the function s is contant—the integral of du/dx vanishes because $u = 0$ at both ends of Q. Thus the boundary conditon '$u = 0$ outside Ω' in this case means that BV annihilates the space N of constant functions, and is the dual of C/N. In more dimensions, the space N is larger but the pattern remains the same.

Let N be the closed subspace of $C = C(\bar{\Omega})^n$ generated by smooth functions ϕ with div $\phi = 0$. *Then BV can be identified with the subspace $N^\perp$ in $M(\bar{\Omega})$, and hence with the dual of C/N.*

For any divergence-free ϕ in $C_0^\infty(Q)$, and any u in BV,

$$O = -\int_Q u \,\text{div}\, \phi = \int_Q \phi \cdot \text{grad } u = \int_{\bar{\Omega}} \phi \cdot \text{grad } u$$

so that $\phi|_{\bar{\Omega}}$ lies in N. On the other hand, if m is in $N^\perp$ we extend it by zero on $Q - \bar{\Omega}$, and

$$\int_{\bar{Q}} \phi \cdot m = \int_{\bar{\Omega}} \phi \cdot m = 0$$

The main step is now to apply a theorem of Fleming[6] and Krickeberg:[7] Every measure which annihilates all divergence-free ϕ in $C_0^\infty(Q)$ is the gradient of some function u of bounded variation on Q. Outside $\bar{\Omega}$, $m = 0$ and thus $u =$ constant (we assume for simplicity that $Q - \bar{\Omega}$ is connected), and we may take $u = 0$. Thus $u \in BV$. The representation is unique, because if also grad $u = 0$ in $\bar{\Omega}$ we find $u \equiv 0$. The map grad: $BV \to M(\bar{\Omega})$ is injective with closed range, and with $\|u\| = \|\text{grad } u\|_M$, BV is congruent to $N^\perp$ and hence to $(C/N)^*$.

Alternatively, the predual of BV is the subspace $G = \text{div}\, C \subset W^{-1,\infty}$, i.e. the space of all divergences $g = \text{div}\, s$ of continuous functions: $\langle g, u\rangle = (s, \text{grad } u)$. The norm of g is the infimum of $\|s\|_\infty$, $g = \text{div}\, s$.

An equivalent norm on BV seems to be

$$|||u||| = \|u\|_{L_1} + \int_{\Omega} |\text{grad } u|,$$

since BV is a Banach space in this norm, and we can show that both terms are dominated by the original norm $\|\text{grad } u\|_M$.

We turn to the corresponding questions for BD. In this case u is vector-valued; but because Korn's inequality failed, BD is not just n copies of BV. It is defined by

$$BD = \{u \in (L_1(Q))^n \mid u = 0 \quad \text{in} \quad Q - \bar{\Omega}, \text{def } u \in M\}$$

where the deformation operator maps the velocity u to the strain tensor $\frac{1}{2}(\partial u_u/\partial x_i + \partial u_j/\partial x_i)$. The elements $m \in M$ are now symmetric n by n matrices whose entries are measures. Similarly the functions $s \in C$ are symmetric n by n matrices whose entries are continuous functions, and by the Riesz theorem $M = C^*$. The pairing of s and def u is given by the integral of the trace:

$$(s, \text{def } u) = \int_{\bar{\Omega}} \text{tr}(s \text{ def } u) = \int_{\bar{\Omega}} \sum s_{ij}\varepsilon_{ij} \, dV$$

The divergence of s has n components, one from each column of the matrix; it is a function in $(W^{-1,\infty})^n$.

With these changes, the predual of BD matches that of BV:

Let N be the closed subspace of C generated by smooth symmetric matrix functions ϕ with $\text{div } \phi = 0$. *Then BD can be identified with $N^{\perp}$ and hence with the dual of C/N.*

The analogue of the Fleming–Krickeberg theorem is proved by Moreau[8] even for multiply-connected domains: Every measure which annihilates all divergence-free symmetric matrices ϕ in $C_0^{\infty}(Q)$ is the deformation of some vector function u. Moreau finds u as a distribution on Q; in our case it should be of bounded deformation on $\bar{\Omega}$, and for that we introduce mollifiers a_h as in Miranda:[9]

$$\|\text{def } u\|_M \geqslant \|\text{def } u * a_h\|_{L_1(Q)} = \|\text{def } (u * a_h)\|_{L_1} \geqslant c\|u * a_h\|_{L_p}$$

Here we may take $1 < p < n/n-1$. The sequence $u * a_h$ is now weakly compact in L_p, and it converges weak-$*$ as a distribution to u. Thus $u \in L_p \subset L_1$. Then since def $u = 0$ in $Q - \bar{\Omega}$, u is a rigid body motion there (Moreau, Reference 8, Prop. 3. c) which we may choose to be zero. Therefore $u \in BD$. These properties also ensure that $\|u\|_{BD} = \|\text{def } u\|_M$ is a norm on BD, which becomes the dual of C/N.

Alternatively, the predual of BD is the space $G = \text{div}\, C$ of all divergences $g = \text{div}\, s$ of continuous symmetric matrix functions. It is paired to BD by

$$\langle g, u \rangle = (s, \text{def}\, u) = \int_{\bar{\Omega}} \text{tr}\,(s\, \text{def}\, u)$$

In this case, rather than giving G the $W^{-1,\infty}$ norm, the failure of Korn's inequality places it in a smaller space X^*; X is the closure of $(C_0^\infty(\Omega))^n$ with the norm $\|\ \|_{BD}$.

It is in this space X that we begin to look for the velocity $\bar{u}$ at the moment of collapse. It contains all velocities which literally satisfy the boundary condition $u = 0$, and if Korn's inequality had held it would be just our original guess $W_0^{1,1}$. Instead it is a little larger, but not as large as BD itself—where we hope the minimax theorem will finally locate $\bar{u}$.

17.3 THE MINIMAX THEOREM

This is the theorem which, in finite dimensions, produces a saddle point for a two-person game. In infinite dimensions the theorem has to be adapted to the function spaces involved; the hypotheses vary with the intended application. In our case the spaces are not reflexive, but nevertheless we want the minimum and maximum to be attained; this takes us into the bidual.

Minimax theorem. Suppose X is a Banach space with dual X^* and bidual X^{**}. Let K be a convex subset of X^* containing a ball centred at zero. Let $H \subset X^{**}$ be the hyperplane $\langle f, x^{**} \rangle = 1$, where f is given in X^*. Then

$$\sup_{x^* \in K} \inf_{x^{**} \in H} \langle x^*, x^{**} \rangle = \min_{x^{**} \in H} \sup_{x^* \in K} \langle x^*, x^{**} \rangle \tag{17.3}$$

If in addition K is weak-$*$ compact, then also the maximum is attained:

$$\max_{x^* \in K} \inf_{x^{**} \in H} \langle x^*, x^{**} \rangle = \min_{x^{**} \in H} \sup_{x^* \in K} \langle x^*, x^{**} \rangle \tag{17.4}$$

In other words, there is a saddle point in $K \times H$. The maximizing element is $\bar{x}^* = \bar{\lambda} f$, where $\bar{\lambda}$ is the value assumed by both sides of (17.4).

This is a variant of the duality theorems in Ekeland–Temam,[5] and its proof can go into the appendix. We plan to use it in the following (not necessarily optimal) way. We start with the velocity space X, normed so that the deformation map $D: X \to L_1$ is injective with closed range; as described earlier, the norm on X is $\|u\| = \|\varepsilon\|_{L_1}$. The load f is given in the dual space X^*, and then a collapse state $\bar{u} = x^{**}$ will be found in the larger velocity space X^{**}. The hyperplane H contains those velocities with unit dissipation rate:

$$H = \{u \in X^{**} \mid \langle f, u \rangle = 1\}$$

The same operator D is the link between the bilinear form of equation (17.2) and the pairing $\langle f, u\rangle$ of force and velocity:

$$(\sigma, \varepsilon)=(\sigma, Du)=\langle D^t\sigma, u\rangle$$

If the hypotheses of the theorem can be verified then we have a saddle point, but not in a desirable pair of spaces. Therefore there will be a final step, in which the force is restricted to a smaller space (the predual of BD), and a collapse velocity is found in BD.

First we discuss, without full detail, the hypotheses of the minimax theorem. Recall that B is the set of stresses in L_∞ which satisfy the yield condition. This set is certainly convex, with zero in its interior (in the sup norm). Therefore $K=D^t(B)$ has the same properties, because Dt is an open mapping (it is surjective since D is injective with closed range). There is more difficulty with the weak-$*$ compactness of K, because the set B is not bounded. This is a well-known situation in mechanics: pure 'hydrostatic pressure' has no effect on collapse (we assume that the force f is not a pure pressure, since otherwise the limit multiplier $\bar{\lambda}$ would be infinite) and any multiple of the identity can be added to an admissible σ without leaving B. Thus the yield conditions produce an unbounded admissible set of the special form

$$B=B_1+\Sigma_2=\{\phi I \mid \phi\in L_\infty(\Omega)\}$$

The set B_1 is bounded, and lies within the complementary subset Σ_1:

$$\Sigma_1=\{\sigma \mid \text{trace}\ \sigma=0 \text{ a.e. in } \Omega\}$$

Furthermore the set B is weak-$*$ closed; Nayroles[10] has shown how this follows from a theorem of Rockafellar.

The difficulty created by Σ_2 seems to be only technical, and we briefly outline a remedy: as space of forces consider temporarily the quotient of our present X^* with $D^t(\Sigma_2)$, and as velocity space the annihilator of $D^t(\Sigma_2)$. In other words, all velocities u are to satisfy $\langle D^t\sigma, u\rangle=0$ for σ in Σ_2. With these changes the unbounded part of the stresses is removed and K becomes weak-$*$ compact; the theorem produces a minimax value $\bar{\lambda}$ and a saddle point $\bar{\sigma}_1, \bar{u}_1$ with $\bar{\sigma}_1$ in B_1. Then when we return to the real problem and solve $D^t\bar{\sigma}=\bar{\lambda}f$ with the same $\bar{\lambda}$, the Σ_1 component of this $\bar{\sigma}$ will be the same $\bar{\sigma}_1$. Keeping $\bar{u}=\bar{u}_1$, we deduce that

$$\langle D^t\sigma, \bar{u}\rangle\leqslant\langle D^t\bar{\sigma}, \bar{u}\rangle=\bar{\lambda}\leqslant\langle D^t\bar{\sigma}, u\rangle \tag{17.5}$$

for all σ in B and u in H, from the fact that it holds for all σ in B_1. Now $\bar{\sigma}, \bar{u}$ is the saddle point we wanted.

The last step is to move the velocity to the more natural space BD. To do so, we have to ensure that the expression $\langle D^t\sigma, \bar{u}\rangle$ will be defined; the admissible stresses will include only those for which $D^t\sigma = \operatorname{div} \sigma$ lies in the space $G = \operatorname{div} C$. And we have to suppose in particular that the $\bar{\sigma}$ we found earlier is still admissible. Thus $D^t\bar{\sigma}$, which is a multiple of f, also lies in G; *our force must be in equilibrium with some continuous stress*. Then, since BD contains all continuous linear functionals on G, there is a $\bar{v}$ in BD which agrees (on G) with the given $\bar{u}$ in X^{**}:

$$\langle D^t\sigma, \bar{v}\rangle = \langle D^t\sigma, \bar{u}\rangle \quad \text{whenever} \quad D^t\sigma \in G$$

Now $\bar{\sigma}$, $\bar{v}$ is the saddle point we want:

$$\langle D^t\sigma, \bar{v}\rangle \leqslant \langle D^t\bar{\sigma}, \bar{v}\rangle = \bar{\lambda} \leqslant \langle D^t\bar{\sigma}, v\rangle$$

for all stresses $\sigma \in B$ with $\operatorname{div} \sigma \in G$, and all velocities $v \in BD$ with $\langle f, v\rangle = 1$.

ACKNOWLEDGEMENT

We are very grateful to Terry Rockafellar and David Schaeffer for their help. The first author has been supported at MIT by a grant from the DAAD in West Germany and the second by the National Science Foundation (MCS 76–22289).

We are also very grateful to Pierre Suquet for a manuscript on trace theorems and Green's formula in the space BD; it is the beginning of a valuable contribution to plasticity theory.

APPENDIX: PROOF OF THE MINIMAX THEOREM

The result we want is equation (17.3), $\sup \phi(x^*) = \min \psi(x^{**})$, where

$$\phi(x^*) = \inf_H \langle x^*, x^{**}\rangle, \ \psi(x^{**}) = \sup_K \langle x^*, x^{**}\rangle$$

As always, it is automatic that $\sup \phi \leqslant \inf \psi$; the problem is to prove the opposite inequality, and to show that the minimum is attained. Writing $\sup \phi(x^*) = \mu$, we have to construct an $\bar{x}^{**}$ in U for which $\psi(\bar{x}^{**}) \leqslant \mu$. Since $\mu = -\infty$ is impossible because 0 is in K, we assume μ to be finite and define the convex sets

$$S_1 = \{(x^*, r) \mid \phi(x^*) - \mu \geqslant r\}$$
$$S_2 = \{(x^*, r) \mid x^* \in K, r \geqslant 0\}$$

Then S_2 has $(0, 1)$ in its interior, and that interior is disjoint from S_1. The Hahn–Banach theorem yields a non-zero pair $(\bar{x}^{**}, \bar{r})$ in the dual, such that

(A) $\quad \langle x^*, \bar{x}^{**}\rangle + r\bar{r} \geqslant c \quad$ for (x^*, r) in S_1

(B) $\quad \langle x^*, \bar{x}^{**}\rangle + r\bar{r} \leqslant c \quad$ for (x^*, r) in S_2

If $\bar{r}>0$ then (B) would be impossible; take $x^*=0$, $r\to\infty$. Suppose $\bar{r}=0$. Since $(0,-\mu)\in S_1$, (A) forces $c\leqslant 0$; but then (B) is impossible since K contains a ball around zero. Thus $\bar{r}<0$, and from now on we rescale $\bar{x}^{**}$, $\bar{r}$, c to make $\bar{r}=-1$.

Since the pair $(0,-\mu)$ is in S_1, (A) gives $\mu\geqslant c$. On the other hand, the distance between S_1 and S_2 is zero so there cannot be any slack in (B):

$$\sup_K \langle x^*, \bar{x}^{**}\rangle = c$$

But this means $\psi(\bar{x}^{**})=c\leqslant\mu$ as we wished. To show that $\bar{x}^{**}$ lies in the hyperplane H, consider the pair (x^*, r) with $x^*=(r+\mu)f$. This pair lies in S_1, since $\phi(x^*)=r+\mu$ by definition of ϕ and H. Therefore (A) requires that

$$(r+\mu)\langle f, \bar{x}^{**}\rangle - r \geqslant c$$

This can hold for $r\to\pm\infty$ only if $\langle f, \bar{x}^{**}\rangle=1$, so we have found the required $\bar{x}^{**}$ in H. This proves the first half of the theorem, stated by equation (17.4).

The second part goes quickly because H is a hyperplane: $\phi(x^*)\geqslant a$ only if $x^*=\lambda f$, $\lambda\geqslant a$. Since this last set is weak-$*$ closed, and K is now assumed weak-$*$ compact, the maximum must be attained.

REFERENCES

1. W. Koiter, 'General theorems for elastic-plastic solids', *Progress in Solid Mechanics*, North-Holland, Amsterdam, 1960.
2. M. Z. Cohn (Ed.), *Engineering Plasticity by Mathematical Programming*, University of Waterloo, Canada, 1977.
3. D. Ornstein, 'A non-inequality for differential operators in the L_1 norm', *Archive Ratl. Mech. Anal.*, **11** 40–49 (1962).
4. G. Strang, 'A minimax problem in plasticity theory', *AMS Symposium on Functional Analysis Methods in Numerical Analysis* (M. Z. Nashed, Ed.), Springer Lecture Notes (to appear).
5. I. Ekeland and R. Temam, *Convex Analysis and Variational Problems*, North-Holland, Amsterdam (American Elsevier), 1976.
6. W. H. Fleming, 'Functions with generalized gradient and generalized surfaces', *Annali di Matematica* (4), **44,** 83–103 (1957).
7. K. Krickeberg, 'Distributionen, Funktionen beschränkter Variation und Lebesguescher Inhalt nichtparametrischer Flächen', *Annali di Matematica* (4), **44,** 105–133 (1957).
8. J. J. Moreau, 'Champs et distributions de tenseurs déformation sur un ouvert de connexité quelconque', *Seminaire d'Analyse Convexe*, Montpellier, 1976.
9. M. Miranda, 'Distribuzioni aventi derivate misure insiemi di perimetro localmente finito', *Ann. Scuola Norm. Sup. Pisa* (3), **18,** 27–56 (1964).
10. B. Nayroles, 'Essai de théorie fonctionelle des structures rigides plastiques parfaites', *J. de Mécanique*, **9,** 491–506 (1970).

Chapter 18

A Finite Element Method for Cyclically Loaded Creeping Structure

A. R. S. Ponter and P. Brown

18.1 INTRODUCTION

In this paper we attempt to show that the deformation of creeping structures may be understood from simple finite element calculations. The method assumes that for cyclic loading, the cycle time is sufficiently short for the residual stress field to effectively remain constant. This form of solution provides the exact solution in the limit when the cycle time reduces to zero, but forms a sufficiently accurate approximation for many of the cycle times which occur in practice.

The need for such solutions is clear. Although it is possible to solve creep problems numerically by forward integration procedures over the history of loading, the process is time consuming and often numerically unstable. Such solutions have not provided any broad insight into structural behaviour. It is hoped that the solution method described here may be sufficiently simple to allow fairly exhaustive study of problems in structural design.

We study two constitutive relationships which describe viscous and strain-hardening material behaviour. For uniaxial stress they take the form:[1]

$$\dot{v} = k\sigma^n \qquad \text{(viscous)} \qquad (18.1)$$

$$\dot{v} = mB^{1/m}\sigma^{n/m}v^{(m-1)/m} \qquad \text{(strain-hardening)} \qquad (18.2)$$

where

$$k = \frac{\dot{v}_0}{\sigma_0^n} \quad \text{and} \quad B = \dot{v}_0/\sigma_0^n m t_0^{m-1}$$

where n is a creep index, m a time index and $\dot{v}_0$ is the creep rate at time t_0 due to a constant stress σ_0 maintained from time $t = 0$. The quantities k and B are functions of temperature θ, and their form will be discussed later. The strain-hardening equation provides a good description of creep for temperature ranges where recovery effects are small. Most design problems lie within this range ($\theta < 0.5\theta_m$, where θ_m is the melting temperature) and this equation has been used extensively in design calculations (see, for example References 1–6).

We consider a body subjected to a cyclic history of temperature and load. Temperature enters into the problem through the formation of thermal expansion strains and the temperature dependence of the constitutive relationship. The solution is constructed under the assumption that the cycle time Δt is very small compared with characteristic material time scales. The derivation of the relevant equations is given elsewhere,[2] and yields a stress history of the form

$$\sigma_{ij} = \hat{\sigma}_{ij} + \bar{\rho}_{ij} \tag{18.3}$$

where $\hat{\sigma}_{ij}$, is the thermo-elastic solution (i.e. ignoring creep strains) and $\bar{\rho}_{ij}$ is a residual stress field which remains constant during the cycle.

For a viscous material the solution provides an upper bound on the energy dissipated by creep,[3] and for situations where the creep rates do not reverse in sign during a cycle it appears that the solution also provides an upper bound on displacement for any cycle time.

For a strain-hardening material two alternative solutions are possible. The simpler one relates to a particular initial residual stress field $\bar{\rho}_{ij}$ which remains constant for all $t>0$. The simple solution, as well as the viscous solution, may be generated by a finite element method. As the only unknown quantity is the residual stress field $\bar{\rho}_{ij}$ (once, of course, the elastic solution $\hat{\sigma}_{ij}$ has been evaluated) the numerical methods are of a comparable complexity to those involved in solving a constant load problem.

In Section 18.2 the rapid cycle solution is discussed and in Section 18.3, the finite element method is described. In Section 18.4 a range of solutions is computed for a single problem. A plate containing a central hole is subjected to tension and a radial temperature distribution. The behaviour of the plate for variable temperature is described and it is shown that the deformation may be related to constant load and temperature solutions.

As the finite element method is essentially an adaptation of elastic finite element methods, the methods described here may easily be included in existing finite element systems. The solutions of the plate problem indicate that the deformations which occur under variable temperature may be estimated from knowledge of the stationary state solutions for time-constant load and temperature, together with the plastic limit and shake-down loads for a perfectly plastic material. Further, the stationary state solutions may be expressed in terms of a reference stress and a reference temperature. It should be possible, using the techniques described here, to express the behaviour of creeping structures in terms of relatively few reference quantities.

An extended version of this paper has been published elsewhere.[10]

18.2 THE RAPID CYCLING SOLUTION

The total strain rate $\dot{\varepsilon}_{ij}$ is composed of three constituent parts,

$$\dot{\varepsilon}_{ij} = \dot{e}_{ij} + \dot{v}_{ij} + \dot{\Delta}_{ij} \tag{18.4}$$

where $\dot{e}_{ij}$ denotes the elastic strain rate and $\dot{\Delta}_{ij}$ the thermal expansion strain rate, which is given in terms of temperature θ by,

$$\Delta_{ij} = \tfrac{1}{3}\delta_{ij}\alpha(\theta - \theta_0) \tag{18.5}$$

where δ_{ij} denotes the Kronecker delta tensor. The coefficient of volumetric expansion is denoted by α and θ_0 is some designated temperature at which Δ_{ij} is zero. The general form of the strain-hardening relationship (18.2) is given by

$$\dot{v}_{ij}(v_{kl}, \sigma_{kl}) = mB^{1/m}\phi^{n/m}(v_{kl}v_{kl}/n_{rs}n_{rs})n_{ij} \tag{18.6}$$

where $n_{ij} = \dfrac{\partial \phi}{\partial s_{ij}}$, ϕ being a homogeneous function of degree one in the component of the deviatoric stress,

$$s_{ij} = \sigma_{ij} - \tfrac{1}{3}\delta_{ij}\sigma_{kk} \tag{18.7}$$

When s_{ij} remains constant in time and $v_{ij} = 0$ at $t = 0$, then equation (18.6) may be solved for v_{ij} to yield,

$$\dot{v}_{ij} = \dot{v}^{s}_{ij} = B\frac{\partial}{\partial_{s_{ij}}}\left(\frac{\phi^{n+1}}{n+1}\right)mt^{m-1} \tag{18.8}$$

Hence n may be recognized as the creep index and m the time index which has often been shown to be close to $m = 1/3$. The quantity B is a function of temperature. Experimental evidence indicates that for an initially isotropic metal ϕ correlates with the octahedral shear stress or J_2 theory (see Reference 1, for example):

$$\phi = \sqrt{\tfrac{3}{2}s_{ij}s_{ij}} \tag{18.9}$$

and this form was adopted in the calculation. Two cases of equation (18.6) will be discussed, when $m = 1$, (viscous) and $m = 1/3$.

Although the case $m = 1$ has no clear relationship with the behaviour of metals under variable stress and temperature, we show that the solution in the two cases may be described in terms of reference quantities which are virtually identical and therefore insensitive to the details of the constitutive relationship. This approach follows the work of Williams and Leckie.[7]

Consider a structure with volume V and surface S subjected to a cyclic history of loading P_i on S_T, part of S, and a cyclic history of temperature $\theta(x_{ij}t)$ within V. Zero displacements are maintained over the remainder of S.

Consider a cycle which occurs in the time interval $t_0 < t < t_0 + \Delta t$ or, equivalently, $0 < \tau < 1$ where $\tau = (t - t_0)/\Delta t$. If we assume that Δt is very small then the resulting stress history is given by[2]

$$\sigma_{ij}(t_0, \tau) = \hat{\sigma}_{ij}(\tau) + \bar{\rho}_{ij}(t_0) \tag{18.10}$$

where $\hat{\sigma}_{ij}$ denotes the linear elastic solution (equivalent to assuming $v_{ij} = 0$) and $\bar{\rho}_{ij}$ denotes a residual stress field which remains constant within the cycle. The average strain rate over a cycle is given by,

$$\dot{\bar{\varepsilon}}_{ij} = (\varepsilon_{ij}(t_0 + \Delta t) - \varepsilon_{ij}(t_0))/\Delta t = C_{ijkl}\dot{\bar{\rho}}_{kl} + \dot{\bar{v}}_{ij} \tag{18.11}$$

$$\dot{\bar{v}}_{ij} = \int_0^1 \dot{v}_{ij}(\hat{\sigma}_{kl} + \bar{\rho}_{kl}, v_{kl})\, \mathrm{d}\tau \tag{18.12}$$

where $\dot{\bar{v}}_{ij}$, the average creep rate, is computed in (18.12) by assuming that the stress history (18.10) holds during the cycle and that $v_{kl} = v_{kl}(t_0)$. The rate of change of $\bar{\rho}_{ij}(t_0)$ becomes determinate from the condition that $\dot{\bar{\varepsilon}}_{ij}$ shall be compatible. This form of solution is strictly only correct in the limit as $\Delta t \to 0$, but should provide a good approximation for Δt small. In simple experiments on a two-bar structure[8] it was found that a cyclic stress history was achieved within two cycles with a cycle time of 24 hours. The residual stresses remained constant during most of the cycle but suffered a rapid change at each temperature change owing to the presence of anelastic creep, with a very short time scale of the order of 1–2 hours. The rapid cycle solution was shown to be conservative, and the effect of anelastic creep on the average creep rate appeared to be small, effectively producing a delayed thermo-elastic response of the order of approximately 25 per cent of the immediate thermo-elastic response.

The cyclic solution with $\bar{\rho}_{ij}$ constant may be extracted directly from equations (18.11) and (18.12) provided we assume that the initial stress $\sigma_{ij}(0) = \bar{\rho}_{ij}$ and subsequently maintains this value. Under these conditions, equations (18.10), (18.11), and (18.12) become,[2]

$$\sigma_{ij} = \hat{\sigma}_{ij}(\tau) + \bar{\rho}_{ij} \tag{18.13}$$

$$\dot{\bar{\varepsilon}}_{ij} = \dot{\bar{v}}_{ij} = v_{ij}^{*} m t^{m-1} \tag{18.14}$$

where

$$v_{ij}^{*} = (H_{kl}H_{kl})^{(m-1)/2} H_{ij} \tag{18.15}$$

and

$$H_{ij} = \int_0^1 B^{1/m} \phi^{n/m} n_{ij}/(n_{rs}n_{rs})^{(m-1)/2m}\, \mathrm{d}\tau \tag{18.16}$$

Here v_{ij}^{*} is a time constant strain field. The condition that $\dot{\bar{\varepsilon}}_{ij}$ shall be compatible now requires that $v_{ij}^{*}(x_k)$ shall be compatible. Hence we are posed with the problem of finding a constant residual stress field $\bar{\rho}_{ij}$ such that the strain field given by (18.15) and (18.16) shall be compatible, where the stress history in the integral of (18.16) is given by (18.13).

Such a solution would be the same as the solution for the original problem for a specific initial residual stress field. However, for any other initial stress field (and in all practical situations this stress field is unknown), the stresses redistribute to a cyclic state which is provided by our solution. The additional strains induced in the process are generally small, and the deformation in the cyclic state generally dominates the total deformation.

It is of interest to consider the uniaxial form of equations (18.6), (18.14), (18.15), and (18.16). Noting that $\phi=\sigma$ and $n_{11}=1$ and $\sqrt{n_{ij}n_{ij}}=\sqrt{\tfrac{3}{2}}$, these equations become,

$$\dot{v}^s(\sigma)=B\sigma^n m t^{m-1} \quad \text{for } \sigma \text{ constant} \tag{18.6a}$$

$$\dot{\bar{\varepsilon}}=\dot{\bar{v}}=v^* m t^{m-1} \tag{18.14a}$$

$$v^*=\left\{\int_0^1 B^{1/m}(\hat{\sigma}+\bar{\rho})^{n/m}\,\mathrm{d}\tau\right\}^m \tag{18.15a}$$

Further, (18.14a) and (18.15a) may now be written in the form

$$\dot{\bar{\varepsilon}}=\dot{\bar{v}}=\left\{\int_0^1 (\dot{v}^s(\hat{\sigma}+\bar{\rho}))^{1/m}\,\mathrm{d}\tau\right\}^m \tag{18.17}$$

where $\dot{v}^s$ denotes the creep rate assuming the current stress had remained constant from $t=0$. Hence, the average creep rate $\dot{\bar{\varepsilon}}$ becomes a weighted mean of the constant stress creep rates corresponding to the stress history $\hat{\sigma}+\bar{\rho}$. We will show that the form (18.6a) and (18.17) allow a considerable insight into the behaviour of structures. Unfortunately a similar reduction of the complete triaxial equation does not seem possible except for proportional stress histories.

Equation (18.17) is in fact, correct for any length of cycle provided that the history of stress is cyclic. Commencing with the uniaxial equation (18.2) and assuming $v(0)=0$, then

$$v^{1/m}(t)=\int_0^t B^{1/m}\sigma^{n/m}\,\mathrm{d}t$$

Assuming that σ is cyclic with cycle time Δt and that t occurs at the end of a discrete number of cycles then

$$v^{1/m}(t)=t\int_0^1 (B\sigma^n)^{1/m}\,\mathrm{d}\tau$$

from which may be derived equation (18.17) for the special case of $\sigma=\hat{\sigma}+\bar{\rho}$. Hence the rapid cycle solution effectively only assumes this particular form of stress history and that it persists from time $t=0$, as under these conditions equation (18.17) gives the exact creep rate at the end of a discrete number of cycles.

The quantity $B(\theta)$ describes the dependence of the creep rate upon temperature. The form which has received widest acceptance,[6] is given by,

$$B = B'e^{-\Delta G/R\theta} \tag{18.18}$$

where B' denotes a constant, and R the universal gas constant. The activation energy ΔH varies among the pure metals but appears to remain close to the pure metal value for alloys. The magnitude of B' is highly dependent on the alloy structure. For temperatures in excess of a half of the melting temperature (Kelvin), ΔH correlates well with the activation energy of self-diffusion but usually has a smaller value for lower temperatures.

We will be concerned with temperature distributions which are characterized by a single parameter, $\Delta\theta$, the temperature difference between two points where the temperature is assigned. The largest temperature variation will be given by,

$$\theta_0 \leqslant \theta \leqslant \theta_0 + \Delta\theta$$

and, for sufficiently small values of $\Delta\theta$ we may make the approximation,

$$B = (B'e^{-\Delta H/R\theta_0})e^{\gamma(\theta-\theta_0)}, \qquad \gamma = \frac{\Delta H}{R\theta_0^2} \tag{18.18a}$$

By making this approximation the resulting solutions may be made independent of θ_0 and dependent only upon $\gamma\Delta\theta$. We find that the most significant non-dimensional parameter is given by

$$\beta = \frac{\gamma\Delta\theta}{n} \tag{18.19}$$

The quantity β measures the relative sensitivity of creep rate to changes in stress and temperature. Its physical meaning becomes clear when we compare the creep rates (at a prescribed strain) corresponding to a stress $\lambda\sigma$ at a temperature θ_0, and a stress σ at a temperature $\theta_0 + \Delta\theta$. If these two creep rates are to be equal then $\lambda = e^{\beta}$. Hence β may be measured directly from constant stress and temperature creep curves or, alternatively, from measured values of ΔH and n. Although the model assumes that β is constant, inspection of data indicates that β decreases with increasing stress. The values of β which occur in practice appear to lie within the range $0 < \beta < 5$, although the higher values are exceptional.

In the next section the finite element method is described.

18.3 THE FINITE ELEMENT METHOD

The finite element method consists, essentially, of the initial strain method in elastic analysis and this part of the method will not be described in detail.[9] Within the ith element, a displacement matrix N^i is defined which relates the displacements $\mathbf{u}^i(x)$ to a global nodal displacement vector $\mathbf{U}$,

$$\mathbf{u}^i = N^i\mathbf{U} \tag{18.20}$$

From N^i the internal strains are derived by differentiating,

$$\boldsymbol{\varepsilon}^i = B^i\mathbf{U} \tag{18.21}$$

The isotropic stress–strain relationship is written as

$$\boldsymbol{\sigma}^i = D^i\boldsymbol{\varepsilon}^i = D^iB^i\mathbf{U} \tag{18.22}$$

where D^i is a matrix of elastic constants.

Hence the symmetric global stiffness matrix K is defined, following the usual arguments[9] by,

$$K = \sum_i \int_{V_i} B^{iT}D^iB^i \,\mathrm{d}V \tag{18.23}$$

where V_i is the element volume.

For the elastic problem with thermo-elastic strains Δ_{ij} and applied loads $\mathbf{P}$, the total strain is given by,

$$\boldsymbol{\varepsilon}^i = D^i\boldsymbol{\sigma}^i + \boldsymbol{\Delta}^i \tag{18.24}$$

Assuming that $\boldsymbol{\Delta}^i$ is a constant value within each element, the nodal displacements are given by[9]

$$K\mathbf{U} = \mathbf{f} + \sum_i C^i\boldsymbol{\Delta}^i \tag{18.25}$$

where

$$C^i = \int_{V_i} B^{iT}D^i \,\mathrm{d}V \tag{18.26}$$

and

$$\mathbf{f} = \sum_k \int_{S_k} \mathbf{P}^{\mathrm{T}}N^k \,\mathrm{d}S \tag{18.27}$$

Here S_k denotes an element of surface corresponding to the kth element. The stresses are given by

$$\boldsymbol{\sigma}^i = D^iB^i\mathbf{U} - D^i\boldsymbol{\Delta}^i \tag{18.28}$$

Initially the thermo-elastic solution was solved for the problem under consideration and the stress history $\hat{\boldsymbol{\sigma}}(\tau)$ computed for each element. The load and thermal parts were each characterized in terms of a single load and temperature parameter P and $\Delta\theta$ and hence two separate solutions $\hat{\boldsymbol{\sigma}}^P$ and $\hat{\boldsymbol{\sigma}}^\theta$ were generated proportional to P and $\Delta\theta$

$$\hat{\boldsymbol{\sigma}}(\tau) = \hat{\boldsymbol{\sigma}}^P + \hat{\boldsymbol{\sigma}}^\theta \tag{18.29}$$

All the relevant variables were non-dimensionalized, and the variables used are listed in the appendix. This discussion here is carried out in terms of dimensional quantities for the sake of clarity.

Consider now the cyclic creep problem of equations (18.11) and (18.12) for the special case of a viscous material where $\dot{v}_{ij}$ in the integrand of equation (18.12), depends only upon the stress history $\hat{\sigma}_{kl} + \bar{\rho}_{kl}$ and is independent of the accumulated creep strain $v_{kl}(t_0)$. Equation (18.11) defines an elastic rate problem analogous to the thermal elastic problem except that displacements, strains, and stresses are replaced by the time derivative of the corresponding quantities. Further, as $\dot{\bar{\rho}}_{kl}$ is a residual stress field, the corresponding boundary load rates are zero. Hence the average rate of change of the displacement vector is given by the analogous expression to (18.25) with $\mathbf{f} = 0$:

$$K\dot{\mathbf{U}} = \sum_i C^i \dot{\bar{\mathbf{v}}}^i \tag{18.30}$$

where $\dot{\bar{\mathbf{v}}}^i$ is known from (18.12) in terms of the current residual stress $\dot{\bar{\boldsymbol{\rho}}}$ and the elastic stress history (18.29). With $\dot{\mathbf{U}}$ known, the average rate of change of stress, i.e. the rate of change of residual stress is given in the ith element by (cf. 18.27),

$$\dot{\bar{\rho}} = D^i B^i \dot{\mathbf{U}} - D^i \dot{\bar{\mathbf{v}}}^i \tag{18.31}$$

The procedure for finding the cyclic solution was as follows. The two constituents of the thermo-elastic solution were obtained by operating on the two right-hand side components of (18.25) separately using the square root method due to Banachiewicz,[11] to generate $\hat{\boldsymbol{\sigma}}^P$ and $\hat{\boldsymbol{\sigma}}^\theta$. Initially, assuming $\bar{\rho}_{ij} = 0$ everywhere, the average creep rate $\dot{\bar{\mathbf{v}}}$ may be computed from the elastic solution although at a subsequent stage in the iterative process, it depends on $\hat{\sigma}_{kl} + \bar{\rho}_{kl}$. We now introduce two quantities which measure the convergence of the solution to a steady state value. R is defined as the ratio of the numerically largest component of $\dot{\bar{\boldsymbol{\rho}}}$ over all the elements of the structure at time t to the same quantity at time $t = 0$ and may be thought of as the relative maximum stress rate. The definition of C comes straight from equation (18.31) and is

$$C = \frac{\max\limits_{\substack{\text{all components} \\ \text{and all } i}} \{|D^i B^i \dot{\mathbf{U}} - D^i \dot{\bar{\mathbf{v}}}^i|\}}{\max\limits_{\substack{\text{all components} \\ \text{and all } i}} \{|D^i B^i \dot{\mathbf{U}}|, |D^i \dot{\bar{\mathbf{v}}}^i|\}} \tag{18.32}$$

and is a measure of how closely the two constituents of $\dot{\bar{\rho}}_{ij}$ cancel one another out. C can vary between 0 and 2, the nearer to zero it is then the nearer we are to a steady state.

The rate of change of nodal displacement $\dot{\mathbf{U}}$ was found from (18.30) and thence the rate of residual stress change from (18.31). The time was then incremented by Δt so that the largest component of $\hat{\sigma}_{kl} + \bar{\rho}_{kl}$ in any element during the first part of the loading cycle did not change by more than a fixed proportion (usually 5 per cent), provided also that R did not increase. If this second condition was not satisfied, the value of Δt was successively halved until either the condition was satisfied or else Δt was so small that no significant change in any of the values occurred. Provided that a significant change did occur, then the displacement and residual stress were incremented by $\dot{\mathbf{U}}\Delta t$ and $\dot{\bar{\boldsymbol{\rho}}}\Delta t$ and the process was repeated. The iterative scheme essentially consists of a re-evaluation of $\dot{\bar{\mathbf{v}}}$ within each element and matrix multiplication by K^{-1}. The solution process was allowed to continue until the steady cyclic state was reached in which, ideally, $\dot{\bar{\boldsymbol{\rho}}} = 0$. In practice, the process was terminated when C reached a value of 10^{-3}, experience showing that for this value, the resulting stresses were accurate to 0.1 per cent, that is any further lowering of C would not have changed the three most significant figures of any element stress.

It was not attempted to lower the value of C at each iteration since it was found that in the early stages of the process, C oscillated while R decreased. On the other hand R was not used as the only convergence indicator as it is highly dependent on n and β, often reaching a value of 10^{-6} for larger n values well before a steady state solution had been reached. In some cases, but only for larger values of n and β, after a certain stage R did not decrease any further, however small the quantity Δt was chosen, the method suffering from the classical difficulty of solving a parabolic system. The numerical values of $\dot{\bar{\rho}}$ tend to find small equilibrium values arising from the numerical errors of the finite element method. For the strain-hardening case the method can obviously be extended by storing the accumulated creep strain at each stage and including the current value in the evaluation of the average creep rate of equation (18.12). However, we were interested only in the steady state response when $\dot{\bar{\boldsymbol{\rho}}} = 0$, and a simplifying device was adopted.

The average creep rate was computed from v^*_{ij} of equations (18.14) and (18.15),

$$\dot{\bar{v}}_{ij} = v^*_{ij}/t \tag{18.33}$$

The transient solution in this case, of course, has no physical meaning, but the steady state response when $\dot{\bar{\rho}}_{ij} = 0$ will be identical to the solution of the system (18.13) to (18.16). As well as the advantage of simplicity, accumulated errors in $v_{ij}(t_0)$ are also avoided.

18.4 A PLATE PROBLEM

In Figure 18.1 we show the geometry and finite element mesh of the plane stress problem of a plate with a central hole subjected to edge tension P and a temperature difference $\Delta\theta$ between the edge of the hole AB and the edge

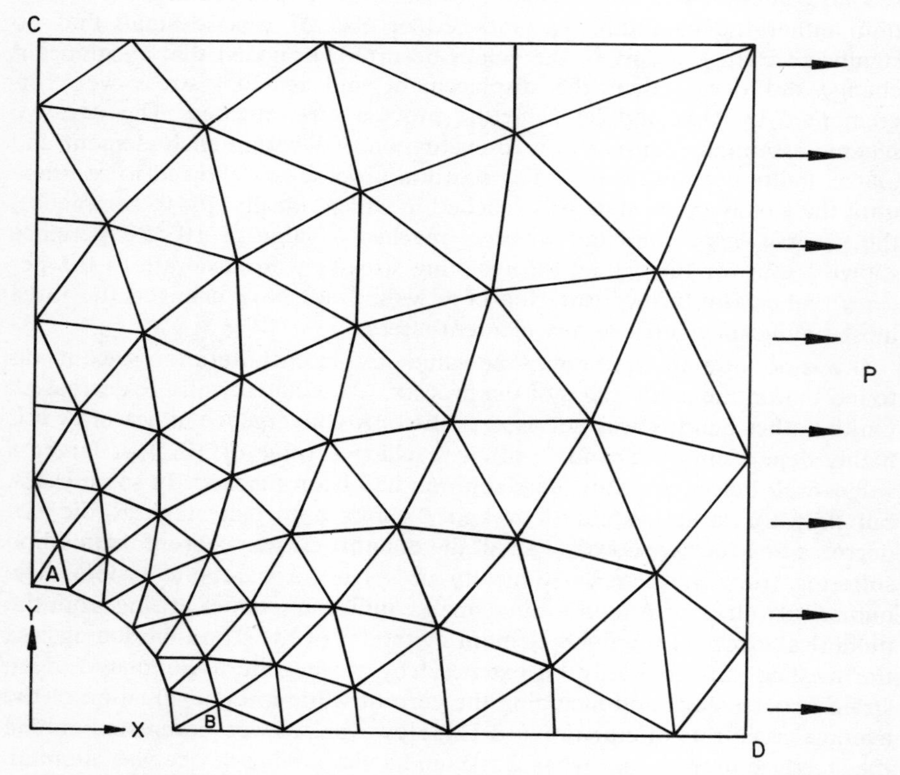

Figure 18.1 Finite element mesh for plate subjected to uniform loading

of the plate. A simple temperature distribution is assumed:

$$\begin{aligned} \theta &= \theta_0 + \Delta\theta \ln(5a/r)/\ln(5), \qquad && a<r<5a \\ &= \theta_0, && r>5a \end{aligned}$$

where r is the radial distance from the centre of the plate and a is the radius of the hole. Generally, the temperature distribution could be found from a finite element solution of the heat conduction problem.

Consider the following problem. The tension P remains constant in time and the temperature difference $\Delta\theta$ fluctuates between zero and some positive value which defines a value of β by equation (18.19).

The thermo-elastic solution was found to yield a maximum tensile stress at the edge of the hole, in element A of the form:

$$\hat{\sigma}^{\theta}_m = 0.246\alpha E \, \Delta\theta$$

where α is the coefficient of thermal expansion and E denotes Young's modulus. The maximum value of the elastic solution was found to be

$$\hat{\sigma}^{P}_m = 2.98P$$

which compares well with the analytic value of $3P$ for an infinite plate. The degree of severity of the thermal loading may be measured by

$$q = \hat{\sigma}^{\theta}_m / \sigma^{P}_m$$

and we shall study the case $q=1$. The proportion of the cycle during which the temperature difference $\Delta\theta$ acts is given by $\mu \, \Delta t$, so that $\mu=1$ corresponds to constant temperature difference and $\mu=0$ corresponds to uniform temperature θ_0.

The stress distribution across the section AC for the case $\mu=0.5$, $n=3$ and $m=\frac{1}{3}$ are shown in Figures 18.2, 18.3, and 18.4 for both the strain-hardening and viscous relationships. The phrase 'heat on' refers to that part of the cycle during which the temperature difference $\Delta\theta$ operates.

In Figure 18.2 the stress distributions are shown, assuming $\beta=0$. Although this assumption is unrealistic, it forms the extreme case when the material is insensitive to temperature and the effect of the temperature difference is manifested only through the thermo-elastic stresses. Two features of these solutions are very noticeable. The strain-hardening and viscous solutions are virtually identical. Further, the maximum values of the stresses are very close to the solution for $\mu=0$, but with a constant load of $1.4P$. The reason for this is fairly obvious. During the cycle the strains produced when the stress is largest dominate the deformation process and the stress distribution assumes maximum values which give rise to a compatible strain rate distribution and we may expect them to be close to a

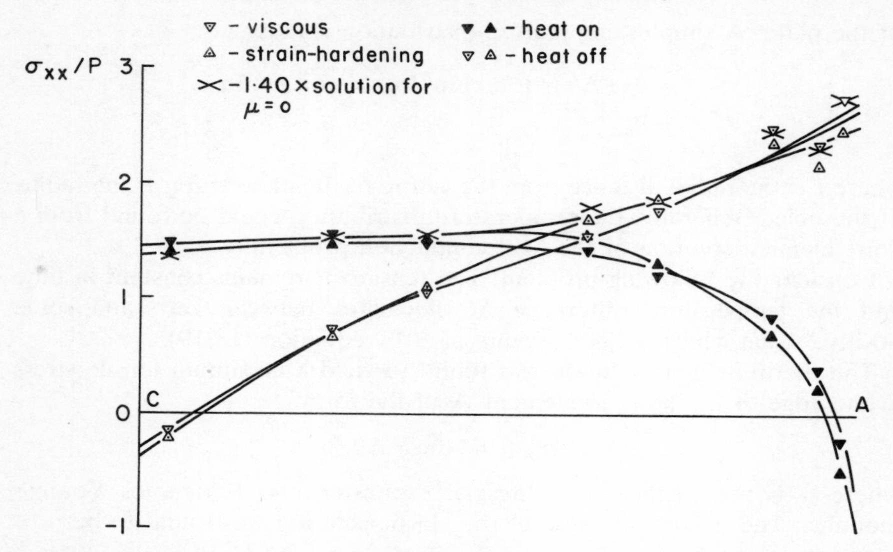

Figure 18.2 Distribution of σ_{xx}/P along AC of Figure 18.1 for $q = 1$, $\mu = 0.5$, $n = 3$, and $\beta = 0$

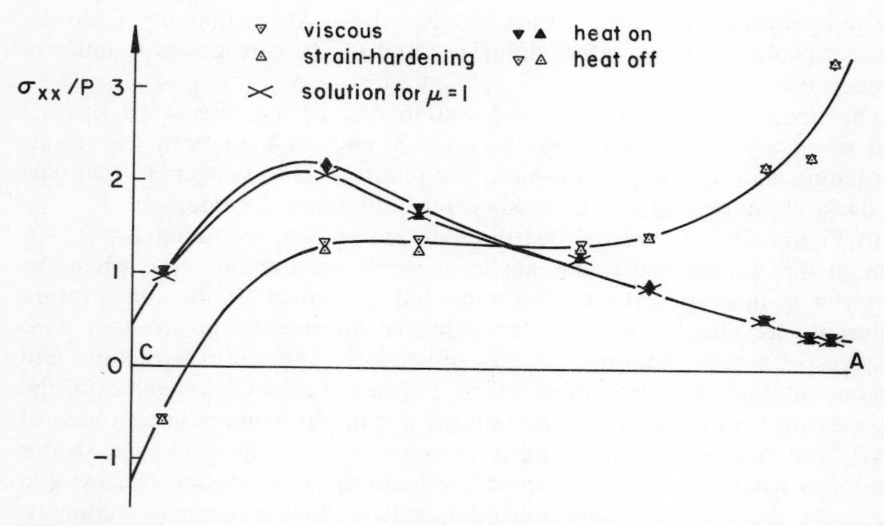

Figure 18.3 Distribution of σ_{xx}/P along AC of Figure 18.1 for $q = 1$, $\mu = 0.5$, $n = 3$, and $\beta = 3$

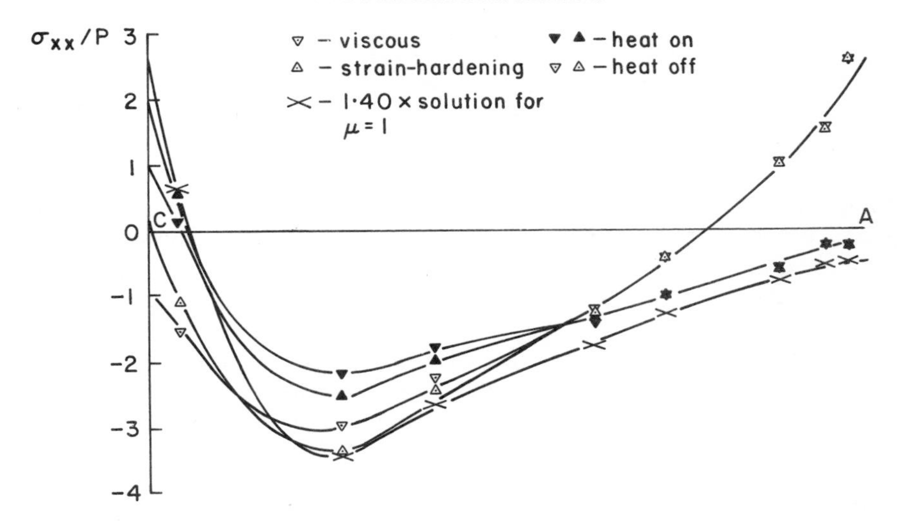

Figure 18.4 Distribution of σ_{xx}/P for negative P for $|q| = 1$, $\mu = 0.5$, $n = 3$, and $\beta = 3$

constant load solution for $\mu = \frac{1}{2}$. Hence the solution has a form which is similar to a plastic shake-down solution except that the maximum stress is governed by the constant load solution and not the yield stress. A perfectly plastic material with yield condition $\phi = \sigma_0$ may be considered as the limit of the viscous material as $n \to \infty$, the viscous solution may be expected to approximate the shake-down solution at large n.

In Figures 18.3 and 18.4, σ_{xx} is shown for $P > 0$, $P < 0$ and $\beta = 3$, a rather large value corresponding to the other extreme of material behaviour where the temperature sensitivity of the material behaviour is important. Again it can be seen that the strain-hardening and viscous solutions are very close to each other. For $P > 0$, however, the values when $\Delta\theta$ operates is nearly identical to the solution with $\mu = 1$ and for load P. The reason is that the deformation is now dominated by that part of the cycle when the higher temperature operates and the strains produced during the remainder of the cycle are negligible. Although the stresses at A increase when $\Delta\theta$ is removed, the drop in temperature effectively 'freezes' the material.

For $P < 0$ the solutions are somewhat more complex as the larger stress at the edge of the plate at C occurs when $\Delta\theta = 0$, whereas the reverse is true at A. Hence the largest negative stress appears to dominate throughout the section producing a correlation with the $\mu = 1$ solution for $1.40P$.

18.5 CONCLUSION

Cyclic creep solutions may be generated with fairly simple finite element methods provided the assumption that the cycle time is small compared with a material time scale that can be considered reasonable (see Reference 8 for further discussion of this point). For a prescribed cyclic history of load and temperature the computation of the cyclic state is comparable in complexity with a constant load 'steady state' viscous solution. Further, the solutions are generated directly from the elastic stiffness matrix and, as a result, the technique may easily be added to a conventional elastic finite element system.

From the discussion of the plate problem it becomes clear that solutions display certain broad trends. For the cases discussed the solutions for both strain-hardening and viscous materials are virtually identical, implying that some details of the material behaviour are relatively unimportant. Further, the stress histories may be correlated with those for constant applied load and constant temperature either for the prescribed load or some augmented value. A more extensive discussion of these solutions may be found in Reference.[10]

APPENDIX: NON-DIMENSIONAL VARIABLES

Variables were non-dimensionalized with respect to a specified uniaxial stress σ_0, temperature θ_0 and temperature difference $\Delta\theta$. In the computations discussed in the paper $\sigma_0 = P$, θ_0 was the smallest temperature in the body and $\Delta\theta$ the largest temperature difference. The non-dimensional variables were identical for the strain-hardening and viscous calculation except for the time variable T.

Strain: $E_{ij} = \varepsilon_{ij}/(\sigma_0/E)$ $\quad E = \text{Young's modulus}$

Stress: $\Sigma_{ij} = \sigma_{ij}/\sigma_0$

Time:

$$T = Eke^{-\Delta H/R\theta_0}\sigma_0^{n-1}t \quad \text{Viscous}$$

$$T = EB'e^{-\Delta G/R\theta_0}\sigma_0^{n-1}t^m \quad \text{Strain-hardening}$$

Temperature: $\theta = (\theta - \theta_0)/\Delta\theta$

Thermal strain: $H = \dfrac{\Delta\theta E\alpha}{\sigma_0}$

Temperature sensitivity:

$$\beta = \frac{\Delta H}{R\theta_0^2}\frac{\Delta\theta}{n} = \frac{\gamma\Delta\theta}{n} \quad \Delta\theta \text{ small}$$

$$= \frac{\Delta H}{Rn}\left(\frac{1}{\theta_0} - \frac{1}{\theta_0 + \Delta\theta}\right) \quad \Delta\theta \text{ large}$$

Spatial: $x_i = x_i/a$ $\quad a = \text{radius of hole}$

Physical interpretation:

$T = 1$: when t equals the time for the accumulated creep strain for constant stress σ_0 to be equal to the elastic strain.
$H = 1$: when the linear thermal expansion equals the elastic strain for stress σ_0.

$\beta = 1$: when the same change in creep rate may be achieved by raising the temperature from θ_0 to $\theta_0 + \Delta\theta$ at constant stress or raising the stress from σ to $e \;\; \sigma$ at constant temperature θ_0, where $e = 2.718$ is the base of the natural logarithms. Hence β measures the relative sensitivity of the material to changes in stress and changes in temperature.

REFERENCES

1. Odqvist, F. K., *Mathematical Theory of Creep and Creep Rupture*, Oxford University Press, London, 1966.
2. Ponter, A. R. S., 'Analysis of cyclically loaded creeping structures', *Int. J. of Solids and Structures*, **12,** 809–825 (1976).
3. Ponter, A. R. S., 'On the stress analysis of creeping structures subjected to variable loading', *Trans. ASME, J. Appl. Mech.*, **40,** Series E, 589–595 (1973).
4. Greenstreet, W. L., 'Structural analysis technology for high temperature design', *Paper L3-2, 3rd International Conference on Structural Mechanics in Reactor Technology*, Imperial College, London, September 1975.
5. Lemaitre, J., 'Creep deformation of a cylinder under complex varying loading', *IUTAM Symposium*, Gothenburg, August 1970. *Creep in Structures, 1970*, Springer-Verlag, Berlin, 1972, pp. 203–219.
6. Garofalo, F., *Fundamentals of Creep and Creep-Rupture in Metals*, Macmillan, New York, 1965.
7. Williams, J. J., and Leckie, F. A., 'A method of estimating creep deformation of structures subjected to cyclic loading', *Trans. ASME, J. Appl. Mech.*, **40,** Series E, 928–934 (1973).
8. Ponter, A. R. S., and Walter, M. H., 'A theoretical and experimental investigation of creep problems with variable temperatures', *Trans. ASME, J. of Appl. Mech.*, **43,** 639–644 (1976).
9. Zienkiewicz, O. C., *The Finite Element Method in Engineering Science*, McGraw-Hill, London, 1971.
10. Ponter, A. R. S., and Brown, P., 'The finite element solution of rapid cycling creep problems', *Int. Jn. for Numerical Methods in Engineering Sciences* (to appear).
11. Banachiewicz, T., 'Méthodes de resolution numérique des équations linéaire, du calcul des determinants et des inverses et de réduction des formes quadratiques', *Bul. Intern. Acad. Polon, Sci. A.*, 393–401 (1938).

Chapter 19

Application of the Biorthogonal Lanczos Algorithm

M. Geradin

19.1 THE GENERALIZED EIGENVALUE PROBLEM

In several fields of engineering (for example, structural dynamics and stability, hydrodynamics) efficient algorithms for extracting the higher eigenspectrum of the generalized eigenvalue problem

$$Ax = \lambda Bx \tag{19.1}$$

are required. If the systems (19.1) results from a finite idealization of the physical problem, it can be of relatively large dimension (up to several thousands) with a very small population pattern.

In structural analysis in particular, there is a strong need for eigenvalue extraction methods specially adapted to very large, sparse matrices. B identifies then itself with the stiffness matrix of the structure, and is thus symmetric and, in general, positive definite. A can be either the mass matrix of the structure or its geometric stiffness matrix and is also symmetric, but not necessarily positive definite.

There are cases, however, where the positive definite property of B is also lost. It will be so if the structure possesses kinematical freedoms, or if linear constraints between components of x are assumed using the Lagrangian multiplier method.

Performing the Choleski factorization of $B(B = LL^{\mathrm{T}})$ to recast (19.1) in the simpler form

$$Sy = \lambda y \quad \text{with} \quad y = L = L^{\mathrm{T}}x \quad \text{and} \quad S = L^{-1}AL^{-\mathrm{T}} \text{ symmetric}$$

is thus not necessarily the proper way of solving (19.1), and cannot be recommended as a general procedure.

Let us rather transform (19.1) in the still more classical form

$$Dx = \lambda x \text{ with the iteration matrix } D = B^{-1}A \tag{19.2}$$

According to (19.2) one power iteration consists of two steps

$$\text{(i) performing the matrix product } y_i = Ax_i \tag{19.3}$$

$$\text{(ii) solving the linear system } x_{i+1} = B^{-1}y_i \tag{19.4}$$

The equation solver recommended to solve (19.4) successfully is Gauss's elimination with a maximum pivot strategy. As shown in References 13 and 14, it can be implemented even in positive semi-definite and non-positive definite cases without any artificial transformation of the initial matrix B such as frequency shifting.

Among the eigenvalue solvers based on power iterations (19.3–19.4), subspace iteration is probably the most favoured one because of its ability to extract accurately the lower eigenspectrum in an arbitrary number of iterations.[12]

Recent studies[6–11] have however demonstrated that the same goal could be reached with much greater economy using the Lanczos method of minimal iterations,[1–3] provided that its unstable behaviour can be overcome.

Our purpose is to implement a truncated version of the Lanczos algorithm for non-symmetric matrices by constructing biorthogonal sequences of right- and left-hand iteration vectors according the scheme (19.3–19.4) and to bring a remedy to the rapid loss of orthogonality observed when the method is applied in its crude form. The reorthogonalization procedure described in Sections 19.4 and 19.5 generalizes to the unsymmetrical case the sequence of Hermitian transformations proposed by Golub *et al.*[6] and applied with success to large-scale eigenproblems in Reference 9.

The numerical behaviour of the method has been tested on a structural eigenvalue problem of moderate size, the results of which are reported in Section 19.7.

19.2 THE BIORTHOGONAL LANCZOS ALGORITHM

19.2.1 The iteration scheme

Let D be a non-symmetric square matrix, and let x_0 and y_0 be two arbitrary intial vectors. We set up two vector sequences

$$\begin{aligned} x_1 &= Dx_0 - \alpha_0 x_0 & y_1 &= D^T y_0 - \alpha_0^* y_0 \\ x_{r+1} &= Dx_r - \alpha_r x_r - \beta_{r-1} x_{r-1} & y_{r+1} &= D^T y_r - \alpha_r^* y_r - \beta_{r-1}^* y_{r-1} \end{aligned} \tag{19.5}$$

where the constants (α_r, β_{r-1}) and (a_r^*, β_{r-1}^*) are evaluated so as to maintain biorthogonality:[3]

$$x_{r+1}^T y_j = 0 \quad \text{and} \quad y_{r+1}^T x_j = 0 \qquad j < r \tag{19.6}$$

we obtain

$$\alpha_r = \alpha_r^* = \frac{y_r^T D x_r}{y_r^T x_r}$$

and

$$\beta_{r-1} = \beta^*_{r-1} = \frac{y^T_{r-1} D x_r}{y^T_{r-1} x_{r-1}} = \frac{y^T_r D x_{r-1}}{y^T_{r-1} x_{r-1}} = \frac{y^T_r x_r}{y^T_{r-1} x_{r-1}}$$

The sequences (19.5) are continued until $y_r^T x_r = 0$ (breakdown) or $i = n-1$ (normal termination).

The recurrence relations (19.5) can be rewritten in matrix form

$$\begin{aligned} D[x_1 \ldots x_r] &= [x_1 \ldots x_r]T_r + [0 \ldots 0 x_{r+1}] \\ D^T[y_1 \ldots y_r] &= [y_1 \ldots y_r]T_r + [0 \ldots 0 y_{r+1}] \end{aligned} \tag{19.7}$$

with the tridiagonal symmetric matrix

$$T_r = \begin{pmatrix} \alpha_1 & \beta_1 & & & 0 \\ \beta_1 & \alpha_2 & \beta_2 & & \\ & \ddots & & \ddots & \\ & \beta_2 & & & \beta_{r-1} \\ & & \ddots & & \\ 0 & & & \beta_{r-1} & \alpha_r \end{pmatrix}$$

19.2.2 Interaction eigenvalue problem

The matrix T_r is an interaction matrix containing the approximations to the eigenvalues of D obtained by projecting the left- and right-hand eigenvectors of D onto the subspaces $[y_1 \ldots y_r]$ and $[x_1 \ldots x_r]$. The proof holds by observing that if either one of equations (19.7) is multiplied at any step r by the other sequence of vectors and if use is made of (19.6), we obtain the matrix equation

$$[y_1 \ldots y_r]^T D[x_1 \ldots x_r] = [y_1 \ldots y_r]^T [x_1 \ldots x_r] T_r \tag{19.8}$$

To interpret the equation above we denote by X_r and Y_r the (nxr) matrices collecting the successive directions x_i and y_i, and by a and b, the right- and left-hand eigenvectors of T_r associated to an eigenvalue λ. Equation (19.8) yields

$$Y_r^T D X_r a = Y_r^T X_r T_r a = \lambda Y_r^T X_r a \tag{19.9}$$

and, premultiplying (19.9) by b^T we obtain

$$\lambda = \frac{b^T Y_r^T D X_r a}{b^T Y_r^T X_r a} = \frac{v^T D u}{v^T u} \tag{19.10}$$

with the vectors $u = X_r a$ and $v = Y_r b$.

The eigenvalues of T_r appear to be the stationary points of the generalized Rayleigh quotient[3] (19.10) in the orthogonal subspaces spanned by the orthogonal matrices X and Y, and are thus expected to converge rapidly to those of the initial matrix D.

19.2.3 Breakdown of the iteration process

The iteration process can break down in several ways. Bilateral breakdown occurs if $x_{r+1}=y_{r+1}=0$ for $r<n$, and means that a subset of r eigenvectors has been separated from the complete eigenspace. If it is encountered, reinitiating the iteration process is always possible by choosing, for the current vectors x_{r+1} and y_{r+1}, two vectors orthogonal to the preceding ones X_r and Y_r.

According to Faddeev and Faddeeva,[2] dead-end breakdown ($y_{r+1}^T x_{r+1}=0$, $x_{r+1}\neq 0$ and $y_{r+1}\neq 0$ for $r<n$) and unilateral breakdown ($x_{r+1}=0$ or $y_{r+1}=0$ for $r<n$) are very unlikely to occur and can always be prevented through a suitable choice of the starting vectors x_0 and y_0.

19.2.4 Application to the symmetric generalized eigenvalue problem $\boldsymbol{Ax=\lambda Bx}$[4]

If both matrices A and B are symmetric, the left- and right-hand eigenvectors of $D=B^{-1}A$ are related by $y_{(r)}=Ax_{(r)}$. The process of setting up biorthogonal sequences (19.6) can then be simplified by taking $y_0=Ax_0$. The successive iterates result from the simpler iteration rule

$$x_{r+1}=B^{-1}y_r-\alpha_r x_r-\beta_{r-1}x_{r-1} \tag{19.11}$$

$$y_{r+1}=Ax_{r+1} \tag{19.12}$$

with the coefficients

$$\alpha_r=\frac{y_r^T B^{-1} y_r}{y_r^T x_r} \quad \text{and} \quad \beta_{r-1}=\frac{y_{r-1}^T B^{-1} y_r}{y_{r-1}^T x_{r-1}}=\frac{y_r^T x_r}{y_{r-1}^T x_{r-1}} \tag{19.13}$$

The linear relation (19.12) between left- and right-hand vectors implies that the only form of breakdown that can occur in this case is bilateral breakdown.

19.3 NUMERICAL SCALING

Repeated multiplication of the x_r by $B^{-1}A$ leads to an unnecessary growth or decay of the vectors and can lead to either overflow or underflow in the computation scheme. It can be avoided through a proper scaling of the successive iterates X_r and y_r, but has to be applied in different ways according to the signature properties of A.

19.3.1 The matrix A is positive definite

Then $y_r^T x_r = x_r^T A x_r \geq 0$, and the iteration process can be transformed into

(i) $x_{r+1}^* = B^{-1} y_r - \alpha_r x_r - \beta_{r-1} x_{r-1}$ (19.14)

(ii) $y_{r+1}^* = A x_{r+1}$ (19.15)

(iii) $\gamma_{r+1}^2 = x_{r+1}^{*T} y_{r+1}^*$ (19.16)

(iv) $x_{r+1} = \dfrac{1}{\gamma_{r+1}} x_{r+1}^*$ and $y_{r+1} = \dfrac{1}{\gamma_{r+1}} y_{r+1}^*$ (19.17)

with the coefficients

$$\alpha_r = y_r^T B^{-1} y_r \quad \text{and} \quad \gamma_r = \beta_{r-1} = y_{r-1}^T B^{-1} y_r = (y_r^{*T} x_r^*)^{1/2} \qquad (19.18)$$

The resulting tridiagonal matrix T_r remains symmetric and is positive definite if B is so.

19.3.2 The matrix A is not positive definite

If scaling is performed, the iteration matrix is no longer transformable into symmetric tridiagonal form. The normalized iteration scheme can be written as above (19.14–19.17) with the definitions

$$\gamma_{r+1} = |y_{r+1}^{*T} x_{r+1}^*|^{1/2} \qquad \varepsilon_{r+1} = \frac{y_{r+1}^{*T} x_{r+1}^*}{\gamma_{r+1}^2} = \pm 1 \qquad (19.19)$$

and the resulting coefficients

$$\alpha_r = \varepsilon_r y_r^T B^{-1} y_r$$

$$\varepsilon_{r-1} \beta_{r-1} = \varepsilon_r \gamma_r = y_{r-1}^T B^{-1} y_r = \frac{y_r^{*T} x_r^*}{|y_r^{*T} x_r^*|^{1/2}} \qquad (19.20)$$

Changes of sign appear thus between upper and lower diagonals of T_r according to those in the successive scalar products $y_r^T x_r$.

19.4 REORTHONORMALIZATION

As observed by Golub *et al.*,[6] departure from orthogonality is the result of cancellation when computing x_{r+1} and y_{r+1} from (19.5) and not, as is generally believed, the result of accumulation of rounding errors.

In order to be certain of obtaining the full set of eigensolutions it is necessary to ensure that the computed x_r are orthogonal to working accuracy. The conventional way of restoring the orthogonality with all previously computed vectors is the well-known Schmidt process. According to Ojalvo and Newman,[7] it has to be applied in an iterative way and leads thus to an amount of extra computation difficult to estimate.

In the symmetrical case Golub *et al.*[6] achieved the same end with greater economy and elegance using elementary Hermitian matrices. Their procedure can be generalized to the unsymmetrical case as follows.

Elementary transformations are applied to the orthogonal sequences $[x_1, \ldots, x_r]$ and $[y_1, \ldots, y_r]$ in order to transform them into two sequences of base vectors $[\alpha_1 e_1, \ldots, \alpha_r e_r]$ and $[\beta_1 e_1, \ldots, \beta_r e_r]$.

To this purpose we consider elementary operators $P_1, \ldots, P_r$ of the form

$$P_r = I - 2r_r q_r^{\mathrm{T}} \qquad r_r^{\mathrm{T}} q_r = 1 \tag{19.21}$$

where the $r-1$ first components of r_r and q_r are zero:

$$r_{r,j} = 0 \quad \text{and} \quad q_{r,j} = 0 \qquad j < r$$

and constructed such that

$$\begin{aligned} P_r P_{r-1} \ldots P_1 x_1 &= \alpha_e e_r \\ P_r^{\mathrm{T}} P_{r-1}^{\mathrm{T}} \ldots P_1^{\mathrm{T}} y_1 &= \beta_r e_r \end{aligned} \tag{19.22}$$

α_r and β_r are unknown constants. We will show in the next section how to construct these operators.

From their very definition, we deduce their following main properties

(1) Self-inversion:

$$P_r P_r = I \text{ and, more generally } P_1 \ldots P_r P_r \ldots P_1 = I \tag{19.23}$$

(2) Invariance of base vectors:

$$P_r e_j = P_r^{\mathrm{T}} e_j = e_j \qquad r > j \tag{19.24}$$

and consequently

$$P_r \ldots P_1 x_j = \alpha_j e_j \quad \text{and} \quad P_r^{\mathrm{T}} \ldots P_1^{\mathrm{T}} y_j = \beta_j e_j \qquad r \geq j \tag{19.25}$$

(3) With exact computation the r first components of the vectors

$$v_{r+1} = P_r \ldots P_1 x_{+1} \quad \text{and} \quad w_{r+1} = P_r^{\mathrm{T}} \ldots P_1^{\mathrm{T}} y_{r+1} \tag{19.26}$$

would be zero. A lack of orthogonality can thus be removed by cancelling them out.

(4) Restoration of the vectors x_r and y_r: making use of (19.23) and (19.25) we obtain

$$x_r = \alpha_r P_1 \ldots P_r e_r \quad \text{and} \quad y_r = \beta_r P_1^{\mathrm{T}} \ldots P_r^{\mathrm{T}} e_r \tag{19.27}$$

and, as a consequence of (19.24)

$$\begin{aligned} X &= [x_1 \ldots x_r] = P_1 \ldots P_r E_\alpha \\ Y &= [y_1 \ldots y_r] = P_1^{\mathrm{T}} \ldots P_r^{\mathrm{T}} E_\beta \end{aligned} \tag{19.28}$$

with the matrices of base vectors

$$E_\alpha = [\alpha_1 e_1 \ldots \alpha_r e_r] \quad \text{and} \quad E_\beta = [\beta_1 e_1 \ldots \beta_r e_r]$$

The reorthogonalizing scheme that can be deduced from (19.23) to (19.28) proceeds in three steps:

(i) transformation phase

$$v_{r+1} = P_r \ldots P_1 x_{r+1} \quad \text{and} \quad w_{r+1} = P_r^{\mathrm{T}} \ldots P_1^{\mathrm{T}} y_{r+1} \tag{19.29}$$

(ii) purification

$$v_{r+1,j} = 0 \quad \text{and} \quad w_{r+1,j} = 0 \qquad j = 1, \ldots, r \tag{19.30}$$

(iii) construction of the Hermitian matrix P_{r+1} such that

$$P_{r+1} v_{r+1} = \alpha_{r+1} e_{r+1} \quad \text{and} \quad P_{r+1}^{\mathrm{T}} w_{r+1} = \beta_{r+1} e_{r+1} \tag{19.31}$$

(iv) restoration phase

$$x_{r+1} = \alpha_{r+1} P_1 \ldots P_{r+1} e_{r+1} \quad \text{and} \quad y_{r+1} = \beta_{r+1} P_1^{\mathrm{T}} \ldots P_{r+1}^{\mathrm{T}} e_{r+1} \tag{19.32}$$

by performing one step (19.29–32) after each Lanczos iteration.

The numerical experiments described in Section 19.7 show that biorthogonality can be preserved up to a very high level of accuracy even after a relatively large number of steps.

19.5 CONSTRUCTION OF THE HERMITIAN OPERATORS P_r

The construction of the successive projection operators P_r is based on the following lemma:

Lemma 19.1 *Let* (x, y) *and* (a, b) *be pairs of vectors such that*

$$x^{\mathrm{T}} y = a^{\mathrm{T}} b \quad \text{and} \quad x^{\mathrm{T}} b = y^{\mathrm{T}} a \tag{19.33}$$

Then, a projection operator of the form

$$P = I - 2qr^{\mathrm{T}} \tag{19.34}$$

exists such that

$$Px = a \quad \text{and} \quad P^{\mathrm{T}} y = b$$

Proof: Let us try the particular projection directions

$$r=\frac{y-b}{[(y-b)^{\mathrm{T}}(x-a)]^{1/2}} \quad \text{and} \quad q=\frac{x-a}{[(y-b)^{\mathrm{T}}(x-a)]^{1/2}} \tag{19.35}$$

we obtain

$$Px=\left[I-2\frac{(x-a)(y-b)^{\mathrm{T}}}{(x-a)^{\mathrm{T}}(y-b)}\right]x$$

$$=x-2\frac{y^{\mathrm{T}}x-b^{\mathrm{T}}x}{x^{\mathrm{T}}y-a^{\mathrm{T}}y-b^{\mathrm{T}}x+a^{\mathrm{T}}b}(x-a)=a$$

and similarly

$$P^{\mathrm{T}}y=\left(I-2\frac{(y-b)(x-a)^{\mathrm{T}}}{(x-a)^{\mathrm{T}}(y-b)}\right)y=b$$

The uniqueness of the solution results from the observation that there exist other vectors (u, v) such that

$$(I-2qr^{\mathrm{T}})x=(I-2uv^{\mathrm{T}})x=a$$

and

$$(I-2rq^{\mathrm{T}})y=(I-2vu^{\mathrm{T}})y=b$$

we must have

$$q(r^{\mathrm{T}}x)=u(v^{\mathrm{T}}x) \qquad r(q^{\mathrm{T}}y)=v(u^{\mathrm{T}}y)$$

and thus

$$u=\delta q \qquad v=\frac{1}{\delta}q$$

One uses the lemma above to construct the operator (19.34) such that

$$Px=\alpha e_k \quad \text{and} \quad P^{\mathrm{T}}y=\beta e_k$$

with, as a consequence of (19.33), the constraints

$$\alpha\beta=x^{\mathrm{T}}y \quad \text{and} \quad \alpha y_{,k}=\beta x_{,k} \tag{19.36}$$

The equalities (19.36) involve that the choice of the base vector e_k into which x and y can be transormed is not fully arbitrary: the products $\alpha\beta$ and $y_{,k}x_{,k}$ must be of same sign, but there is obviously at least one component k for which this condition is met.

The best choice of k corresponds to the maximum ratio $x_{,k}y_{,k}/\alpha\beta$. According to (19.36) and (19.35) the matrix P is next obtained from the following sequence of operations:

(i) calculate the normalizing constants

$$\beta=\left(\frac{y_{,k}}{x_{,k}}\cdot x^{\mathrm{T}}y\right)^{1/2} \qquad \alpha=\frac{x^{\mathrm{T}}y}{\beta} \tag{19.37}$$

and

$$\phi = |\gamma|^{-1/2} \qquad \psi = \gamma\phi \quad \text{with} \quad \gamma = 2x^{T}y - \alpha y_{,k} - \beta x_{,k} \tag{19.38}$$

(ii) from the projection vectors q and r

$$\begin{aligned} q_{,k} &= \phi(x_{,k} - \alpha) \qquad & r_{,k} &= \psi(y_{,k} - \beta) \\ q_{,i} &= \phi x_{,i} & r_{,i} &= \psi y_{,i} \qquad i \neq k \end{aligned} \tag{19.39}$$

19.6 CALCULATION OF EIGENSOLUTIONS

Let $\lambda_1^{(r)} \geqslant \lambda_2^{(r)} \geqslant \ldots \geqslant \lambda_r^{(r)}$ and $[z_{(1)}, \ldots, z_{(r)}]$ be the eigensolutions of the tridiagonal matrix T_r at step r

$$T_r z_{(k)} = \lambda_k^{(r)} z_{(k)} \qquad k = 1, \ldots, r \tag{19.40}$$

To extract the eigenvalues of T_r, algorithms specialized to tridiagonal matrices such as implicit QL iteration with shifts (procedure TQL1, Reference 5) or bisection (procedures TRISTURM and BISECT, same reference) are recommended.

Convergence of the Lanczos sequence is reached when

$$|\lambda_k^{(r+1)} - \lambda_k^{(r)}| < \varepsilon\, |\lambda_k^{(r)}| \qquad k = 1, \ldots, m$$

for the number $m < r$ of eigenvalues to be extracted. ε denotes the precision required on the eigenvalues.

Once the eigenvalues $(\lambda_1, \ldots, \lambda_m)$ of T have been determined, inverse iteration provides an efficient algorithm for computing the corresponding eigenvectors $(z_{(1)}, \ldots, z_{(m)})$. A proved version of it can also be found in Reference 5.

To restore the eigenvectors of the initial problem (19.1) we return to equation (19.9) from which we deduce that the approximations to $x_{(k)}$ contained in the subspaces X and Y are

$$x_{(k)} = X z_{(k)}, \text{ and similarly } y_{(k)} = Y z_{(k)} \tag{19.41}$$

As we have seen in Section 19.4, full information about the x_i is contained in the successive operators P_i. The products (19.41) can thus be performed recursively according to equations (19.28).

If error bounds to the initial eigenproblem (19.1) are wished they can be obtained from the bracketing algorithm[13,14]

$$\rho_1 - \sqrt{\rho_1(\rho_0 - \rho_1)} \leqslant \rho_k \leqslant \rho_1 + \sqrt{\rho_1(\rho_0 - \rho_1)} \tag{19.42}$$

where ρ_0 is the Rayleigh quotient obtained from the approximation $x_{(k)}$ to the eigenmode k:

$$\rho_0 = \frac{x_{(k)}^T A x_{(k)}}{x_{(k)}^T B x_{(k)}} \tag{19.43}$$

and ρ_1, the next Schwarz quotient

$$\rho_1 = \frac{x_{(k)}^T A B^{-1} A x_{(k)}}{x_{(k)}^T A x_{(k)}} \tag{19.44}$$

Computing the bounds (19.41) requires only one more power step (19.4–19.5) for each eigenmode.

19.7 NUMERICAL EXPERIMENTS

A fairly simple problem has been adopted in order to exhibit the convergence and accuracy properties of the Lanczos method applied to the generalized eigenvalue problem $Ax = \lambda Bx$. It consists of a uniform fixed–free bar in longitudinal vibration, and idealized using finite elements of degree 1. The stiffness and mass matrices for one element of length l are

$$K = \frac{EA}{l}\begin{bmatrix} 1 & -1 \\ -1 & 1 \end{bmatrix} \qquad M = ml\begin{bmatrix} \frac{1}{3} & \frac{1}{6} \\ \frac{1}{6} & \frac{1}{3} \end{bmatrix}$$

and the structural eigenvalue problem takes thus general form (19.1) with the matrices

$$B = \begin{bmatrix} 1 & -1 & & & \\ -1 & 2 & -1 & & \\ & -1 & 2 & \ddots & \\ & & \ddots & \ddots & -1 \\ & & & -1 & 2 \end{bmatrix} \qquad A = \frac{1}{6}\begin{bmatrix} 2 & 1 & & & \\ 1 & 4 & 1 & & \\ & 1 & 4 & \ddots & \\ & & \ddots & \ddots & 1 \\ & & & 1 & 4 \end{bmatrix}$$

and the frequency parameter

$$\lambda = \frac{EA}{\omega^2 m l^2}$$

It can be solved analytically for an arbitrary number N of elements by assuming an eigensolution of the form $x_{,k} = \sin k\mu$. The successive eigenvalues are found to be

$$\lambda_r = 6\,\frac{1 - \cos\,(2r-1)\dfrac{\pi}{2N}}{2 + \cos\,(2r-1)\dfrac{\pi}{2N}} \qquad r = 1, \cdots N$$

Owing to the availability of an exact solution the example chosen is thus ideally suited to an evaluation of the accuracy properties of the algorithm.

The calculation of eigenvalues with the biorthogonal algorithm implemented as described above has been repeated for $N=10$, $N=50$, and $N=100$ on an IBM 370/156 computer using a double precision arithmetic (15 significant digits). The results obtained are the following.

19.7.1 Convergence

A convenient way of representing the convergence properties of the algorithm consists to plot the number of iterations needed to stabilize the successive eigenvalues λ_r of the system.

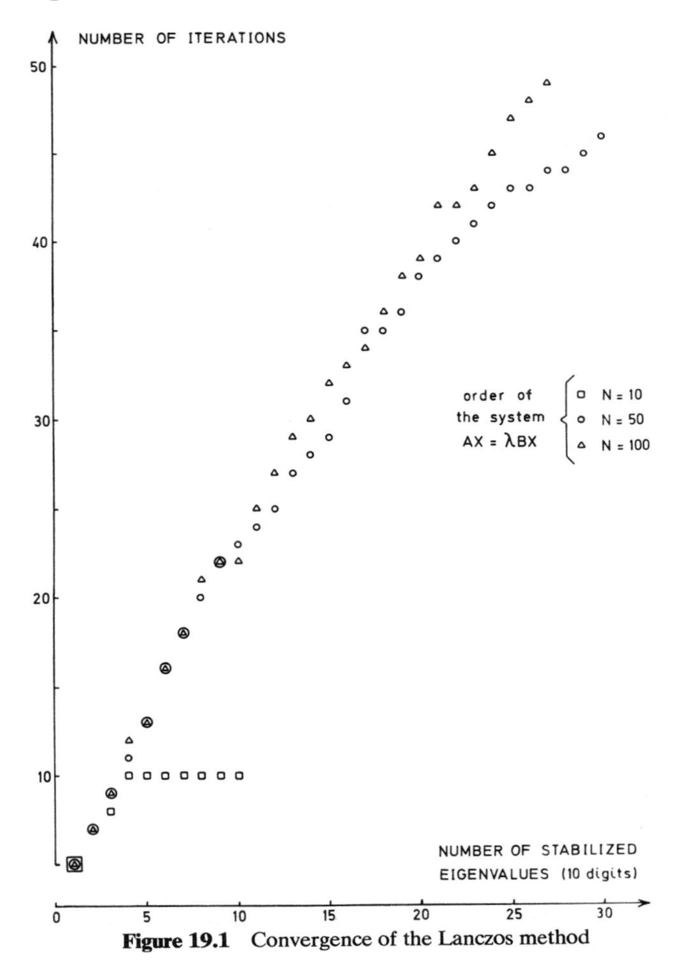

Figure 19.1 Convergence of the Lanczos method

To obtain the curves of Figure 19.1, convergence is supposed to be effective when $(\lambda_r^{(k+1)} - \lambda_r^{(k)}) < \lambda_r^{(k)} \cdot 10^{-10}$. They show that the rate of convergence of the algorithm to a given a eigenvalue is remarkably constant over a large range of the spectrum, whatever be the size of the eigenproblem. About $2(r+2)$ iterations required to extract r eigenvalues, the rate of convergence tending to speed up when reaching the upper eigenspectrum.

Some experiments of the algorithm have been performed without orthogonalization procedure. For example, in the case $N = 50$, full cancellation of the starting vector occurs after ten iterations, giving again convergence to λ_1. From then spurious roots in the lower frequency range appear repeatedly, confirming thus the fact that no confidence can be accorded to the algorithm without reorthogonalization procedure.

19.7.2 Accuracy

The accuracy properties of the algorithm are illustrated in Figure 19.2. The diagram gives, for all three cases, the maximum accuracy (measured in significant digits) that has been obtained in evaluating the successive eigenvalues λ_r, as of function of the order r of the eigenvalues.

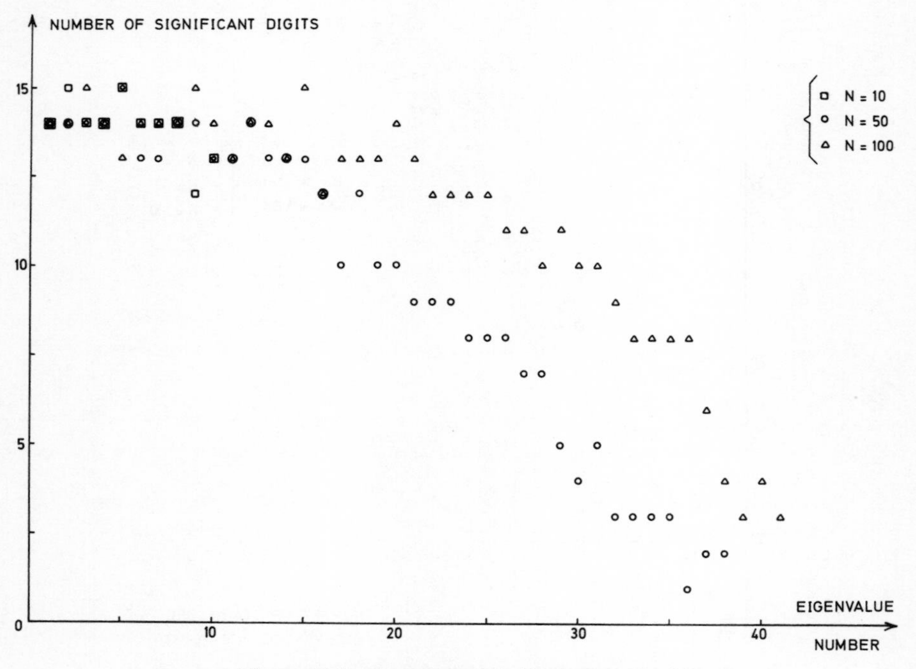

Figure 19.2 Accuracy of the Lanczos method

From it, one can assert that the Lanczos iteration scheme can be relied to extract a limited number of eigenvalues (say, up to 40) with a sufficient accuracy for engineering purposes. The effectiveness of the algorithm to extract a larger number of eigensolutions seems to increase with the order of the system, but such a favourable behaviour should still be confirmed by other numerical experiments on larger-scale problems.

19.8 CONCLUSION

Owing to its high and remarkably constant rate of convergence the Lanczos algorithm seems ideally suited to the efficient solution of very large eigenvalue problems $Ax = Bx$ such as encountered in engineering analysis of structural vibration and stability, where only the few dominant eigenvalues of the eigenspectrum are generally needed. If associated with an efficient equation solver, there is almost no practical limit on the size of the systems that can be solved, the computer requirements being limited to about $(4 \times N + 8 \times NVAL)$ core storage positions, N being the size of the system and $NVAL$, the number of required eigenvalues.

REFERENCES

1. C. Lanczos, 'An iteration method for the solution of the eigenvalue problem of linear differential and integral operators' *J. Res. Nat. Bureau Standards*, **45,** 255–282 (1950).
2. D. K. Faddeev and V. N. Faddeeva, *Computational Methods of Linear Algebra*, W. H. Freeman and Co., San Francisco, 1963.
3. J. H. Wilkinson, The Algebraic Eigenvalue Problem, Clarendon Press, Oxford, 1965.
4. S. H. Crandall, *Engineering Analysis*, McGraw-Hill, New York, 1956.
5. J. H. Wilkinson and C. Reinsch, *Handbook for Automatic Computation*, Vol. 2., *Linear Algebra*, Springer-Verlag, Berlin, 1971.
6. G. H. Golub, R. Underwood, and J. H. Wilkinson, 'The Lanczos algorithm for the symmetric $Ax = \lambda Bx$ problem', Stanford University, Computer Science Dpt, STAN–CS-72-270, 1972.
7. I. U. Ojalvo and M. Newman, 'Vibration modes of large structures by an automatic reduction method', *AIAA Jnl*, **8,** 7, 1234–1239 (1970).
8. I. U. Ojalvo, 'Alarm—a highly efficient eigenvalue extraction routine for very large matrices', *Shock and Vibration Digest*, **72,** 12 (1975).
9. P. C. Chowdhury 'The Truncated Lanczos Algorithm for partial solution of the symmetric eigenproblem', *Computers and Structures*, **6,** 439–446 (1976).
10. W. Kahan and B. N. Parlett, 'How far should we go with the Lanczos process' in *Sparse Matrix Computations* (edited by J. R. Bunch and B. J. Rose, Academic Press, New York, 1976.
11. A. K. Cline, G. H. Golub, and G. W. Platzman, 'Calculation of normal modes of oceans using a Lanczos method' in *Sparse Matrix Computations* (edited by J. R. Bunch and B. J. Rose), Academic Press, New York, 1976

12. K. J. Bathe and E. L. Wilson, 'Solution methods for eigenvalue problems in structural mechanics', *Intl. Jnl. Num. Meth. Engg,* **6,** 213–216 (1973).
13. M. Geradin, 'Eigenvalue analysis by matrix iteration in the presence of kinematical modes', *Shock and Vibration Digest,* **6,** 3 (1974).
14. M. Geradin, B. M. Fraeijs de Veubeke, and A. Huck, *Structural Dynamics,* CISM lecture notes no. 126, 1972, Udine, edited by Springer-Verlag.

Appendix: The Main Publications of Professor Fraeijs de Veubeke

0. 'L'Aérodynamique des Planeurs', *Revue des Elèves des Ecoles Spatiales de l'Université de Louvain,* II et IV Trimestre 1939 et II Trimestre 1940.
1. 'Effet gyroscopique des rotors sur les vitesses critiques de flexion', *Bulletin Technique des Ingénieurs de Louvain,* 1942, p. 63–76.
2. 'Déphasages caractéristiques et vibrations forcées d'un système amorti', *Académie royale de Belgique, bulletin de la Classe des Sciences,* XXXIV-5, 1948, p. 626–641.
3. 'Application des équations de LAGRANGE au calcul des forces de liaison dans les moteurs en étoile', *Bulletin* 22 *du Service Technique de l'Aéronautique,* Bruxelles, 1949.
4. 'Crankshaft-propeller vibration modes as influenced by the torsional flexibility of the engine suspension', *Journal of the Aeronautical Sciences,* May 1950, p. 288–296.
5. 'Quelques applications de la méthode des déphasages caractéristiques', Communication at the second Congrès National des Sciences, Bruxelles, June 1950.
6. 'Calcul des cadres bordant une ouverture circulaire dans un champ plan de tensions', *Mécanique,* September 1950, p. 240–248.
7. 'Diffusion des inconnues hyperstatiques dans les voitures à longerons couplés', *Bulletin* 24 *du Service Technique de l'Aéronautique,* Bruxelles, 1951.
8. 'Dispositifs stabilisateurs entre les couples moteurs variables', *Bulletin Annales de l'union Belge des Ingénieurs Navals* 1951, p. 123–142.
9. 'Aérodynamique instationnaire des profils minces déformables', *Bulletin* 25 *du Service Technique de l'Aéronautique,* Bruxelles, 1953, 1–108.
10. 'Iteration in semi-definite eigenvalue problems', *Journal of the Aeronautical Sciences,* October 1955, p. 710–720.
11. 'Quelques problèmes de grandes vitesses en aéronautique', *Revue des Questions Scientifiques,* October 1955, p. 488–507.
12. Matrices de projection et techniques d'itération', *Annales de la Société Scientifique de Bruxelles,* 70-1, 1956, p. 37–61.

13. 'A variational approach to pure mode excitation based on characteristic phase lag theory', Report 39 of the AGARD Working Group on Materials and Structures,Washington, 1956.
14. 'Méthodes variationnelles et performances optimales en aéronautique', *Bulletin de la Société Mathématique de Belgique,* VIII-2, 1956, p. 136–157.
'Trajectoires optimales des fusées', *Revue Générale des Sciences Appliquées,* 3, No. 2, U.L.B., p. 13–26.
15. 'La balistique externe. Principes de guidage. Accélérations transversales', Ch. XI, p. 334–354.
'La balistique externe. Trajectoires optimales', Ch. XII, p. 355–383, in *La propulsion par fusées,* Berieu, Jaumotte, Fraeijs de Veubeke, Vandenkerckove, Sciences et Lettres, Liège, 1956.
15*bis*. *La propulsion par fusées,* 2nd edition, M. Barrere, A. Jaumotte, B. Fraeijs de Veubeke, J. Vandenkerckove, Dunod, Paris, 1957.
15*ter*. 'La propulsion par fusées' Traductions. *Rocket Propulsion* Ch. XI: 'Elementary problems of several rocket performance' (p. 712–771); Ch. XII: 'Variational methods in optimizations rocket performances' (p. 772–812); and *Raketen Antriebe,* Elsevier, 1961.
16. 'Aspects cinématique et énergétique de la flexion sans torsion', *Académie Royale de Belgique, mémoire 8, no.* 1657, 1955.
17. 'Le problème du maximum de rayon d'action dans un champ de gravitation uniforme', *Astronautica Acta,* IV-1, 1958, p. 1–14.
18. 'Performances balistiques des fusées', *Revue Universelle des Mines,* Series 9, XIV, June 1958, p. 174–193.
19. Creep buckling', Chapter 13 of *High Temperature Effects in Aircraft Structures,* AGARDograph, *no.* 28, N. J. Hoff, ed. Pergamon Press, 1958, p. 267–287.
19*bis*. 'Quelques propriétés des structures ayant deux ou plusieurs fréquences propres confondues', *Compte-Rendus du 3ème Congrès Aéronautique Européen,* Bruxelles, 1958, p. 424–441.
20. 'Flexion et extension des plaques d'épaisseur modérée', *Académie Royale de Belgique, mémoire 8, no.* 1698, 1959, p. 1–38.
21. 'Influence de l'amortissement interne sur la résonance d'un avion' p. 1–40, *Manuel d'aéroélasticité de l'AGARD* (*OTAN*), vol. 1, 1st part, Ch. III, (> 1958).
22. 'L'énergie potentielle complémentaire dans les problèmes dynamiques. Un principe de variation des accélérations', *Annales de la Société Scientifique de Bruxelles,* 73-III, 1960, p. 327–344.
23. 'La méthode des fonctions propres dans les problèmes de transmission de chaleur en régime transitoire', *Bulletin de la Société Royale des Sciences de Liège,* 7–8, 1960, p. 173–195.

24. ‘Principes variationnels en mécanique des fluides compressibles’, *Annales de la Société Scientifique de Bruxelles,* 74-III, 1960, p. 157–174.
25. ‘Sur certaines inégalités fondamentales et leur généralisation dans la théorie des bornes supérieures et inférieures en élasticité’, *Revue Universelle des Mines,* May 1961, p. 305–314.
26. ‘Théorie des coques prismatiques minces renforcées’, *Académie Royale de Belgique, mémoire 8, no.* 1729, 1961.
27. ’L’étagement optimum des groupes de fusées en fonctionnement parallèle’, *Astronautica Acta,* 5–6, 1961, p. 359–375.
28. ‘Les grandes méthodes de calcul des modes et fréquences propres dans les structures complexes’, *Revue Universelle des Mines,* 18, *no.* 2, February 1962, p. 1–13.
29. ‘Approximation par le calcul des variations’, in *Quelques formes modernes des mathématiques, utiles à l’ingénieur,* OCDE, Colloque 1962, SRB 11, p. 581–615.
30. ‘Optimization des systèmes’, in *Quelques formes modernes des mathématiques, utiles à l’ingénieur,* OCDE, Colloque 1962, SRB 11, p. 615–633.
31. ‘The inertia tensor of a liquid bounded by walls in rigid body motion’, *International Journal of Engineering Science, no.* 1, 1963, p. 23–32.
32. ‘Upper and lower bounds in matrix structural analysis’, 14th meeting of the AGARD Working Groups on Materials and Structures, July 1972, AGARDograph *no.* 72, Pergamon Press, 1964, p. 165–201.
33. ‘Displacement and equilibrium models in the finite element method’, Symposium ‘Numerical Methods in Elasticity’, University College of Swansea, January 1964, Chapter 9 of *Stress Analysis,* J. Wiley & Holister, 1965, p. 145–197.
34. ‘Optimization of multiple-impulse orbital transfers by the maximum principles’, B. Fraeijs de Veubeke and J. Geerts, Proceedings XVth IAF Congress in Warsaw, 1964, vol. 1, p. 589–616.
35. ‘Régularisation des réticences et réduction du principe du maximum de Pontriagin au calcul des variations’, *Report OF-3,* May 1964.
36. ‘Considérations sur la recherche spatiale technique’, *Revue des questions scientifiques,* January 1965, XXVI, p. 5–34.
37. ‘Les déphasages caractéristiques en présence de modes rigides et de modes non amortis’, *Académie Royale de Belgique, Bulletin de la Classe des Sciences,* 5th series, LI, 1965-5, p. 525–540.
38. ‘Variational principles in fluid mechanics’, 7th Scientific Symposium on Fluid Mechanics, Polish Academy of Sciences Jurata, September 1–8 1965, p. 111–126.
39. ‘Analyse de la réponse forcée des systèmes amortis par la méthode des

déphasages caractéristiques', *Revue Française de Mécanique*, 1965, *no.* 13, p. 49–58.

39*bis*. 'Second variation test with algebraic and differential constraints', Symposium on advanced problems and methods for space flight optimization, Liège, 1967, Pergamon Press, 1969, p. 189–217.

40. 'Canonical transformations and the thrust-cost-thrust optimal transfer problem', *Astronautica Acta*, 11, *no.* 4, 1965, p. 271–282.
41. 'Optimal steering and cutoff-relight programs for orbital transfers', *Astronautica Acta*, 12, *no.* 4, July–August 1966, p. 323–328.
42. 'Le problème de Newton du solide de révolution présentant une traínée minimum', *Académie Royale de Belgique, Bull. Cl. Sc.*, 1966-2, p. 171–182.
43. 'Une condition nécessaire et suffisante de minimum relatif dans le problème des extrêmes liés', *Académie Royale de Belgique, Bull. Cl. Sc.*, 1966-4, p. 512–517.
44. 'Bending and stretching of plates, special models for upper and lower bounds', AFFDL-TR-66-80, p. 863–886, 1966.
45. 'Upper and lower bonds to structural deformation by dual analysis in finite elements', AFFDL-TR-66-199, January 1967.
46. The second variation test with algebraic and differential constraints', Colloquium on Optimization in Space Flight Mechanics, Liège, June 19–23, 1967, Pergamon Press.
47. 'Basis of a well conditioned force program for equilibrium models via the Southwell slab analogies', AFFDL-TR-67-10, p. 1–20.
48. 'Strain energy bounds in finite element analysis by slab analogy', B. Fraeijs de Veubeke and O. C. Zienkiewicz, *Journal of Strain Analysis*, 2, *no.* 4, 1967, p. 265–271.
49. 'Applications of the slab analogy to the finite element method with particular reference to strain energy bounds', by B. Fraeijs de Veubeke and O. C. Zienkiewicz, Rep. SA11, 1967.
50. 'A conforming finite element for plate bending, *Int. Journal of Solids and Structures*, 1968, 4, p. 95–108.
51. 'An equilibrium, model for plate bending', B. Fraeijs de Veubeke and G. Sander, *Int. Journal of Solids and Structures*, 1968, 4, p. 447–468.
52. 'Dual analysis by finite elements linear and nonlinear applications', (4 chap.) B. Fraeijs de Veubeke, G. Sander et P. Beckers, AFDL-TR-72-93, Wright-Patterson AFB, 1968.
53. 'A maximum-minimum principle for bang-bang systems', *Lecture Notes in Mathematics*, vol. 112, Colloquium on Method of Optimization at Akademgorodok, 1968, Springer-Verlag Berlin, 1969, p. 225–247.
54. Natural strains and stresses for trapezoidal structure analysis', by B. Fraeijs de Veubeke and Nguyen Dang Hung, Rep SA12, November 26, 1968.

55. 'The theoretical design laws of warping-free multicellulus box beams', *Problems of Hydrodynamics and Continuum Mechanics*, SEDOV Anniversary Volume, SIAM, 1969, p. 287–305.
56. 'Natural stresses and strains for trapezoidal structures analysis', B. Fraeijs de Veubeke (in collaboration with H. D. Nguyen) Air Force Flights Dynamics Laboratory, Technical Report, AFFDL-TR-69-18, 1969.
57. 'Une généralisation du principe du maximum pour les systèmes bang-bang avec limitation du nombre de commutations', Colloque sur la Théorie Mathématique du Contrôle Optimal, Centre Belge de Recherches Mathématiques, 1970, p. 56–57.
58. 'Stability analysis by finite elements', A. P. Kabaila and B. Fraeijs de Veubeke, AFFDL-TR-70-35, Wright-Patterson AFB, 1970.
59. 'Bifurcation of space frames', A. P. Kabaila and B. Fraeijs de Veuseke, AFFDL-TR-70-36, Wright-Patterson AFB, 1970.
'Matrix structural analysis'.

59*bis*. 'The analysis of equilibrium bifurcation problems by finite element methods' A. P. Kabaila and B. Fraeijs de Veubeke, Rep. SA19, 1969, (–1970?).

60. 'The analysis of equilibrium bifurcation problems by finite element methods'. A. P. Kabaila and B. Fraeijs de Veubeke, Rep. SA49, (> 1969).
61. 'Large displacement formulations for elastic bodies', AFFDL-TR-70-34, Wright-Patterson AFB 1970.

61*bis*. 'Variational principles in fluid mechanics and finite element applications', Progress in Numerical Fluid Dynamics, Rhode Ste Genèse, 1974, Springer, 1975, p. 226–259, *Lecture Notes in Physics*, vol. 41.

62. The numerical analysis of structures', (> 1968).
63. 'The dual principles of elastodynamics finite element applications', p. 357–377, Nato Advanced Studies Institute Lecture Series on Finite Element Methods in Continuum Mechanics, Lisbon, 1971, ed. Oden and Oliveira, UAH Press, 1973.
64. Duality in structural analysis by finite elements', Nato Advanced Studies Institute Lecture Series on Finite Element Methods in Continuum Mechanics Lisbon, 1971, ed. Oden and Oliveira, UAH Press, 1973, p. 323–355.
65. 'Static-geometric analogies and finite element models', Nato Advanced Studies Institute Lecture Series on Finite Element Methods in Continuum Mechanics, Lisbon, 1971, ed. Oden and Oliveira, UAH Press, 1973, p. 299–322.
66. 'On frequency shifting by elementary modifications of inertia and stiffness', *Prof. van der Neut's Anniversary Volume*, 1972, Brockhuis, Rotterdam, p. 413–421.

67. 'A new variational principle for finite elastic displacements', *International Journal of Engineering Sciences*, 1972, 10, p. 745–763.
68. 'Dual analysis for heat conduction problems by finite elements', B. Fraeijs de Veubeke and M. A. Hogge, *Int. Journal for Num. Methods in Eng.*, 5, p. 65–82, 1972.
68*bis*. 'Heat conduction', B. Fraeijs de Veubeke, M. A. Hogge, Rep. SA24, '1. Dual minimum theorems in heat conduction problems and finite element solution based on compatible temperature fields.' '2. Finite element solution of heat conduction problems bounded balanced heatflow fields bounds to the dissipation functional.'
69. 'Structural dynamics—Heat conduction', B. Fraeijs de Veubeke, M. Geradin, A. Huck, M. A. Hogge, Courses held at the Dept. of Mechanics of Solids, CISM, Courses and Lectures no. 126, Udine, Springer-Verlag, Vienna–New York, July 1972.
70. 'The numerical analysis of structures', *Proceedings of Thirteenth International Congress of Theoretical and Applied Mechanics.* Moscow, August 1972, Springer-Verlag, Berlin, p. 20–28.
71. 'Static-geometric analogies and finite element models', (> 1968); in Russian, 1975.
72. 'Nonlinear theory of shells', Rep. SA31.
73. 'The dynamics of flexible bodies', in *Gyrodynamics*, Euromech 38, Louvain-la-Neuve, 1973, Springer 1974, pp. 1 and 2.
74. 'Finite elements method in aerospace engineering problems, in *Lecture Notes on Computer Science*, Vol. 16, 'Computing methods in applied sciences and engineering' part 1, Springer, 1974, p. 224–258.
75. 'Variational principles in fluid mechanics and finite element applications', *Progress in Numerical Fluid Dynamics*, Von Karman Institute 1974, Lecture Notes in Physics, Springer-Verlag, Berlin, 1975, p. 226–259.
76. 'Variational principles and the patch test', *Int. Journal for Num. Methods in Eng.*, 8, no. 4, p. 783–801, 1974.
77. 'Les conversions cinématiques co-déformables des éléments de coque plans', Rep. SF28, 1974.
78. 'Structural Optimization', C. Fleury et B. Fraeijs de Veubeke, *Lecture Notes in Computer Science*, Vol. 27, 'Optimization Techniques', IFIP Technical Conference, Novosibirsk, 1974, Springer-Verlag, 1975, p. 314–326.
79. 'The numerical integration of laminar boundary layer equations', B. Fraeijs de Veubeke and C. Delcourt-Bon *Comp. and Math. with Appl.*, 1, p. 167–173, 1975.
80. 'Un nouveau principe variationnel pour les déplacements élastiques finis', (in Russian), *Volume anniversaire de la naissance de Galerkin,*

Institute des Problèmes de Mécanique, Académie des Sciences d'U.R.S.S., Moscow, 1975, p. 194–210.

81. Convexity properties in structural optimization', in 'Optimization techniques in modeling and optimization in the service of man', Part 1, 1975, *Lecture Notes in Computer Science* Vol. 40, Jean Cea, Ed., Springer-Verlag, 1976, p. 1–12.
82. 'Stress function approach', *Proceedings of the World Congress on Finite Element Methods in Structural Mechanics*, Bournemouth, Dorset, England, 12–17 October 1975, Vol. 1, p. J. 1–J.51.
83. 'Discretization of rational equilibrium in the finite element method', in 'Mathematical Aspects of Finite Element Methods, Rome 1975', I. Galligani and E. Magenes, Eds., *Lecture Notes in Mathematics*, Vol. 606, Springer, 1977, p. 87–112.
84. 'Non linear dynamics of flexible bodies', *Proceedings of the Symposium on Dynamics and Control of non-rigid Spacecraft* held in Frascati, Italy, 24–26 May 1976, ESA SP 117, p. 11–19, July 1976.

Author Index

The page numbers in light face indicate where the author's work is mentioned in the text. The numbers in italic refer to the pages where the journal or book reference is given.